Oil and the American Century

The Johns Hopkins University Studies in Historical and Political Science
104th Series (1986)

Oil and the American Century: The Political Economy of U.S. Foreign Oil Policy, 1941–1954
by David S. Painter

Oil and the American Century

The Political Economy of U.S. Foreign Oil Policy, 1941–1954

David S. Painter

The Johns Hopkins University Press
Baltimore and London

The Johns Hopkins University Press, 701 West 40th Street, Baltimore, Maryland 21211
The Johns Hopkins Press Ltd, London

The paper in this book is acid-free and meets the guidelines for permanence and durability of the Committee on Production Guidelines for Book Longevity of the Council on Library Resources.

Library of Congress Cataloging-in-Publication Data

Painter, David S.
  Oil and the American century.

  Bibliography: p.
  Includes index.
    1. Petroleum industry and trade—Government policy—United States—History—20th century. 2. Petroleum industry and trade—Near East—History—20th century.
3. United States—Foreign economic relations—Near East. 4. Near East—Foreign economic relations—United States. I. Title.
HD9566.P24 1986   382'.42282'0973   85-24216
ISBN 0-8018-2693-4 (alk. paper)

This book is dedicated to my parents,
Eva Ritchie Painter and Joseph Graham Painter
(1899–1965)

Contents

Acknowledgments

In the course of this project I have incurred a great many scholarly and personal debts which I am pleased to be able to acknowledge.

I have been fortunate over the years to have had a number of teachers whose dedication to the pursuit of knowledge set high standards I have tried to follow. In particular, I want to express my deep appreciation to Graham Landrum, the late Paul B. Patterson, and Thomas R. Peake, King College; T. W. Mason, St. Peter's College, Oxford University; and Samuel F. Wells, Jr., Woodrow Wilson International Center for Scholars. My years at the University of North Carolina would not have been as pleasant without the friendship and support of Frank W. Klingberg.

Over the years my understanding of U.S. foreign policy has been greatly enriched by discussions with Melvyn P. Leffler, Michael J. Hogan, William Burr, and Lynn R. Eden. Their willingness to exchange ideas with me, their concern for my work, and their continual encouragement of my efforts have been greatly appreciated. In addition, I have benefited from the writings of Irvine H. Anderson, John A. DeNovo, Robert Engler, Ellis W. Hawley, Burton I. Kaufman, Clayton R. Koppes, Aaron D. Miller, Gerald D. Nash, Edith T. Penrose, Stephen G. Rabe, Michael B. Stoff, and William Appleman Williams. All scholars working on U.S. foreign oil policy owe a special debt to the Senate Subcommittee on Multinational Corporations for the enormous amount of material it made available to scholars and the public.

I would also like to express my gratitude to Warren Lenhart, Congressional Research Service; Henry G. Bartholomew and Jan Marfyak, Department of Energy; and Paul Claussen, Department of State; for being good friends as well as colleagues. In addition, I wish to thank my colleagues at the Office of the Historian, Department of State for their support; in particular, Nina J. Noring, James E. Miller, and David S. Patterson who read and commented on portions of the manuscript; and Carol Becker and Kay Herring who provided advice on its preparation. I have been employed by the Department of State since October 1982. This book is based entirely on declassified and unclassified documents and secondary literature and does not draw on material unavailable to other scholars. The views expressed in this book are solely my own and do not necessarily reflect the views or policies of the U.S. government or its agencies.

Although I have received courteous attention and helpful advice wherever I have carried on my research, I particularly want to express my gratitude to Dennis Bilger of the Harry S. Truman Library, and John E. Taylor, Shaun Shaunessey, Katherine NiCastro, and John Pontius of the National Archives. I am grateful to the International Security Studies Program of the Woodrow Wilson International Center for Scholars, the Harry S. Truman Library Institute, the Eleanor Roosevelt Institute, and the Graduate School of the University of North Carolina for financial assistance that helped make possible the research necessary for this work. A grant from the Mellon Fund helped defer the cost of publishing this book.

No book is complete without a publisher, and I have been fortunate to be associated in this endeavor with the Johns Hopkins University Press. I especially want to thank Henry Y. K. Tom, George F. Thompson, Jean B. Toll, and Carol Ehrlich for their efforts.

Most of all, I am indebted to my wife, Flora Montealegre, for her support, encouragement, assistance, scholarly advice, patience, and love.

Abbreviations

AD	Acción Democrática
AIOC	Anglo-Iranian Oil Company
ANPB	Army-Navy Petroleum Board
APOC	Anglo-Persian Oil Company
ARAMCO	Arabian American Oil Company
bpd	barrels per day
Caltex	California-Texas Oil Company
CASOC	California Arabian Standard Oil Company
CFP	Compagnie Française des Pétroles
CIA	Central Intelligence Agency
ECA	Economic Cooperation Administration
ECEFP	Executive Committee on Economic Foreign Policy
ERP	European Recovery Program
f.o.b.	free on board
IPC	Iraq Petroleum Company
IPPA	Independent Petroleum Association of America
JCS	Joint Chiefs of Staff
Jersey	Standard Oil Company (New Jersey)
MER	maximum efficient rate
NIOC	National Iranian Oil Company
NSC	National Security Council
NSRB	National Security Resources Board
OPC	Office of the Petroleum Coordinator
OSS	Office of Strategic Services
PAW	Petroleum Administration for War
PED	Petroleum Division
PEMEX	Petróleos Mexicanos
PIWC	Petroleum Industry War Council

PRC   Petroleum Reserves Corporation
SOCAL   Standard Oil Company of California
Socony   Socony-Vacuum Oil Company
SWNCC   State-War-Navy Coordinating Committee
TAPLINE   Trans-Arabian Pipeline
Texas   The Texas Company
USMC   United States Maritime Commission

Oil and the American Century

Corporatism and U.S. Foreign Oil Policy

Three interrelated, long-term developments have decisively shaped the political economy of the United States in the twentieth century: the rise of the large corporation to dominance in the U.S. economy, the assumption by the federal government of an increasingly important role in the nation's economic life, and the outward thrust of U.S. political and economic power.[1] The history of U.S. foreign oil policy during World War II and the early cold war contributes important insights into these developments and their impact on U.S. democracy. Faced with the vital role of petroleum products in modern warfare and economic life and a growing recognition that the nation's historic self-sufficiency in petroleum could soon come to an end, U.S. policymakers chose to utilize private oil corporations to protect and promote the national interest in foreign sources of petroleum.[2] As a result, a symbiosis developed between public and private interests that safeguarded and advanced the private interests of the oil companies while furthering U.S. efforts to control world oil reserves, combat economic nationalism, and contain the Soviet Union.

This outcome evolved out of prolonged struggle among government bureaucracies and between domestic and multinational oil companies, and was further shaped by conflicts with Congress, disagreements with Great Britain, challenges from the producing countries, and the nascent cold war with the Soviet Union. The result was a public-private partnership in oil that achieved U.S. political, strategic, and economic goals, accommodated the desires of the various private interests, conformed to U.S. ideological precepts, and palliated congressional critics.[3] Even though private interests rather than government agencies were given primary responsibility for implementing U.S. foreign oil policy, the U.S. government was nonetheless deeply involved in maintaining an international environment in which private companies could operate with security and profit, assuring the security and stability of the Middle East, containing economic nationalism, and sanctioning and supporting private control of the world's oil.

One of the most important developments in recent writing on U.S. history and politics has been the "discovery and study of a corporate liberalism" or corporatism.[4] For many years the term *corporatism* was largely restricted to discussion of authoritarian regimes and thus had pejorative connotations. In the last decade, however, an increasing number of scholars have expanded its meaning to include the cooperative use of public and private power in pluralistic, democratic societies.[5] Focusing on the process through which U.S. values and institutions have adapted to the requirements of a corporate-controlled economy, historians and other social scientists have traced the development of public-private cooperative arrangements and structures in various versions of Progressive Era reformism, the "planned economy" of World War I, Herbert Hoover's "associationalism," and certain aspects of the New Deal. Common to these efforts has been the goal of resolving the tension between increasing corporate domination of the U.S. economy and the nation's liberal, democratic heritage by finding alternatives to the destructive disorder of unregulated free enterprise on the one hand and the dangers of statist regimentation on the other.

As yet only a few scholars have extended this reinterpretation beyond 1940.[6] And, with the exception of impressive work on the 1920s, scholars have neglected the international dimensions of the emerging corporatist order.[7] This neglect is unfortunate because World War II reinvigorated corporate liberalism and "created, at least temporarily, the kind of partnership between business and government that corporate liberals had long extolled."[8] Moreover, it is in the realm of foreign policy that the corporatist vision of public-private partnership has had its greatest success as the political and military strength of the state has supported, and, in turn, has been supported by, the economic power of U.S. corporations.

As Philippe Schmitter has noted, study of the origins and nature of corporatism "leads one very quickly to the constraints, opportunities and contradictions placed on political actors by the operation of the economic system."[9] Thus a corporatist framework provides a way for diplomatic historians to come to grips with one of the key developments in the history of the twentieth century and to reintegrate their work with that of scholars in other fields. Moreover, a corporatist focus provides a framework for the analysis of specific issues and events that avoids an "overemphasis on personalities and events as compared to structures and social forces." A "corporatist synthesis" can also help diplomatic historians go beyond the controversies that have thrown more heat than light on the history of U.S. foreign policy in the era of the cold war.[10]

This study employs a corporatist framework to help illuminate questions of private power and public policy that are crucial to understanding U.S. foreign oil policy during World War II and the early cold war. Viewed from the perspective of the nature and development of corporatist ideas and institutions in the United States, the period 1941–1954 emerges as a time when the strategy for managing the world oil economy which took shape in the 1920s broke down. The result was an intense but confused search for a new strategy, one that necessarily involved a larger role for government. This search was tied to the past in the sense of trying to cope with continuing problems and in the sense that most of the proffered solutions were derived from past experience.[11] For these reasons it is worth reviewing earlier efforts at business-government cooperation in the oil industry. This history not only serves to put post-1941 events into perspective, but also helps clarify the nature and development of corporatist ideology and institutions in the United States.

During the first two decades of the twentieth century, the structure of the U.S. oil industry was transformed by the discovery of new fields in Texas and California, the emergence of vast new markets for petroleum products created by low-priced automobiles, and the breakup of the Standard Oil Trust in 1911. In place of an industry dominated by a single company, around twenty large, vertically integrated firms carried on the bulk of the industry's activities. Surrounding this core of major firms was a fringe of hundreds of smaller companies engaged in only a single aspect of the oil business. These companies were known as independents.[12] These structural changes were accompanied by changes in public policy toward the oil industry as adversary relationships gave way to cooperation because of concern over oil depletion, the growing importance of oil to the nation's economy, and the conversion of the navy from coal to oil power. The new emphasis on public-private partnership, which paralleled changes in the larger political economy, was strengthened during World War I as government and the oil industry cooperated to meet the nation's wartime needs.[13]

The 1920s witnessed a search for "secondary organizations" capable of coordinating the activities of an economy of large corporations which had, to varying degrees, freed themselves from traditional market restraints. The oil industry provides a classic example of the kind of industrial self-regulation or "associationalism" favored by Secretary of Commerce, and later President, Herbert Hoover. Impressed by the success of intra-industry cooperation during the First World War, oil industry leaders created the American Petroleum Institute (API) in 1919. Through such activities as gathering and publishing statistics on the industry's operations, developing a uniform system of accounting, promoting the standardization of oil industry equipment, and representing the industry before the public and the government, the API sought to "create coordination" in an industry noted for its individualism and thus avert the need for government regulation. The API was successful in most of its endeavors, including securing favorable tax treatment for the oil industry, and helped the industry weather the storm created by the Teapot Dome scandal.[14]

Foreign oil policy in the 1920s also illustrates the larger development of corporatist ideas and institutions in the United States. Oil emerged as a factor in U.S. foreign policy after World War I because of the growing importance of oil to modern warfare, fears of the pending exhaustion of U.S. reserves, and the needs of U.S. oil companies with foreign markets for additional sources of supply. A clash with the British over access to Middle East oil was averted when the U.S. government rejected the alternatives of government intervention through state ownership or of division of the world into exclusive spheres of influence, and sought an "informal entente" with the British based on public support for private arrangements between U.S. and British corporations.[15] In Middle East oil cooperation took the shape of a multinational consortium established in July 1928. The British-owned Anglo-Persian Oil Company (APOC), Royal Dutch/Shell, the French Compagnie Française des Pétroles, and an "American Group" (composed of five companies) each received a 23.75 percent share in the consortium, and Calouste Sarkis Gulbenkian, an Armenian entrepreneur who had negotiated the original concession before the war, received the remaining 5 percent (see table 1). Anglo-Persian received an overriding royalty of 10 percent of all production in the original concession

Table 1. Ownership of the Iraq Petroleum Company

Company	Nationality	Share(%)
D'Arcy Exploration Co., Ltd. (subsidiary of Anglo-Persian Oil Company)[a]	British	23.75
Anglo-Saxon Petroleum Co., Ltd. (subsidiary of the Royal Dutch / Shell group)[b]	British/Dutch	23.75
Compagnie Française des Pétroles[c]	French	23.75
Near East Development Corporation[d] (Standard Oil Company [New Jersey], 50% Socony-Vacuum Oil Company, Inc., 50%)	U.S.	23.75
Participations and Investments Company (wholly owned by C. S. Gulbenkian)	—	5.00

Source: U.S. Congress, Senate Select Committee on Small Business, *The International Petroleum Cartel: Staff Report to the Federal Trade Commission* (Washington, D.C.: U.S. Government Printing Office, 1952). pp. 54, 65. At the time of the group agreement, 31 July 1928, the consortium was still called the Turkish Petroleum Company.

[a] The British government held a 56 percent interest in Anglo-Persian. In 1935, Anglo-Persian changed its name to Anglo-Iranian.

[b] Ownership of the Royal Dutch / Shell group was divided between Dutch interests (60 percent) and British interests (40 percent). Operating and commercial headquarters were in London.

[c] The French government acquired a 25 percent interest in CFP in 1929; in 1931, it increased its holdings to 35 percent.

[d] At the time of the 1928 agreement, five U.S. companies shared in the ownership of NEDC: Jersey, 25 percent; Standard Oil Company of New York, 25 percent; Gulf Oil Corporation, Atlantic Refining Company, and Pan American Petroleum and Transport Company (a subsidiary of Standard Oil of Indiana), each with 16.66 percent. In 1930, Atlantic and Pan American sold their shares to Jersey and Socony, which in 1931 merged with the Vacuum Oil Company and became Socony-Vacuum. In 1934, Gulf sold its interest to Jersey and Socony-Vacuum.

area as compensation for giving up half of its original share to make room for the American Group.[16]

Between 1919 and 1928, oil production increased sharply in the United States, Venezuela, the Soviet Union, Romania, Iran, and the Netherlands East Indies. In this context, the prospect of uncontrolled production in the Middle East was unwelcome. Therefore, the consortium agreement contained a self-denying ordinance, which prohibited the members from engaging except as part of the consortium in oil development within the area of the old Ottoman Empire, which was marked on a map with a red line. The purpose of the so-called Red Line Agreement was to ensure that the development of the region's oil took place in a cooperative, rather than a competitive manner. In 1929, the consortium changed its name to the Iraq Petroleum Company (IPC). IPC, like its predecessor, was a British corporation, and its legal provisions, including the Red Line Agreement, were enforceable in British courts.[17]

The Red Line Agreement sought to control production through joint ownership of reserves. To avoid undue competition for markets the major oil companies tried to work out their problems cooperatively through the Achnacarry or "As-Is" Agreement. In the summer of 1928, Walter Teagle of Jersey, Sir Henri Deterding of Shell, and Sir John Cadman of Anglo-Persian met at Achnacarry Castle in the highlands of Scotland, ostensibly to hunt grouse. As Teagle later admitted, however, "the problems of the world's petroleum industry naturally came in for a great deal of discussion." The resulting agreement pledged the three companies to accept their present volume of business and their proportion of any future increase in production, to share existing facilities and supplies, and to develop new production and facilities only as required by increases in overall demand. In the crucial area of prices, the companies agreed to maintain the system of "Gulf-plus" pricing which already existed. Under this system, prices were calculated on the basis of the prices at the Gulf Coast of the United States so that the delivered price of oil shipped to any country in the world was equal to the delivered price of oil shipped from the U.S. Gulf, regardless of differing production or transportation costs. This system prevented price cutting to gain markets, which could be disastrous given the tremendous excess of potential productive capacity over potential demand. It also protected higher-cost U.S. producers and allowed companies with low-cost production to make higher profits. Because of the difficulties in bringing production under control, the "As-Is" partners, joined by Standard Oil of New York, entered into three supplementary agreements to control markets—the Memorandum for European Markets (1930), the Heads of Agreement for Distribution (1932), and the Draft Memorandum of Principles (1934)—which spelled out in progressively greater detail the means to allocate markets, fix prices, eliminate competition, and avoid duplication of facilities. The agreements enabled the partners to stabilize markets and maintain prices in countries where they were the only companies selling petroleum products.[18]

Although the United States had been explicitly excluded from the As-Is and subsequent agreements to avoid running afoul of U.S. antitrust laws, bringing U.S. production—70 percent of world production in 1925—under control was crucial to dealing with the problem of overproduction. Otherwise, there was always the danger of "outsiders" using U.S. oil to break into world markets. The U.S. oil "shortage" had ended in 1921 with the discovery of several rich new fields in California. With additional discoveries in Oklahoma in 1926, the growth in U.S. reserves outstripped demand for the rest of the decade. The discovery of the great East Texas oil field in 1930, coming at a time of depression and overproduction, gave added urgency to the search for means to bring U.S. production under control.[19]

President Herbert Hoover, however, rejected industry recommendations that the federal government enforce compulsory production control to deal with the situation as incompatible with the proper role of the government in a free society.[20] Hoover's scruples forced the industry to look to the governments of the producing states for relief. Oklahoma and Texas had been involved in regulating their oil industries for many years, and in September 1928 Oklahoma issued the first statewide proration order. Prorationing became caught up in the chronic conflict between majors and independents, however, with the majors supporting efforts to bring production under control and the independents seeing it as another conspiracy by the majors to ruin them. In any event, the efforts of both states were overcome by the magnitude of new discoveries. In 1931, the governors of Texas and Oklahoma declared martial law in the oil fields and closed wells by military order in an effort to bring order to an increasingly anarchic situation. Nevertheless, illegally produced, or "hot," oil managed to find its way out, leading to demands for some kind of federal role in restoring order to the oil industry.[21]

Federal control of oil came with the New Deal and the National Industrial Recovery Act, which authorized state-sanctioned cooperative arrangements to control the oil industry. Reflecting the depth of the crisis, the National Recovery Administration (NRA) oil code provided for production quotas, import limitations, and minimum prices, and thus contained the potential for extensive federal control. Moreover, President Franklin D. Roosevelt named Secretary of the Interior Harold L. Ickes, an advocate of federal control, to administer the petroleum code, rather than lodging responsibility with the newly created National Recovery Administration. In practice, however, the NRA relied on production control to restore oil prices. The Bureau of Mines furnished forecasts of petroleum demand, which the NRA oil administrator used in setting quotas for each of the oil-producing states. The states, in turn, allocated the quotas among individual fields and wells, while the federal government cooperated by prohibiting the shipment of hot oil in interstate commerce. The oil industry, with the help of sympathetic congressmen, beat back attempts by Ickes to extend federal control. Moreover, a minimal response was sufficient; by October 1933, the price of crude oil had recovered dramatically. In January 1935, however, the Supreme Court ruled that presidential jurisdiction over hot oil, a key feature of the control system, was an unconstitutional delegation of legislative power. A few months later, the Court invalidated all the NRA codes.[22]

After the rejection of the New Deal solution, the federal government, state governments, and the oil industry worked out a cooperative control system which retained the key elements of the NRA system, but eliminated the potential for federal control. On the state level, proration laws, now more uniform thanks to the efforts of the newly formed Interstate Oil Compact Commission, regulated production. Of crucial importance in this regard, the Texas Railroad Commission, through a historic compromise, satisfied the majors by controlling production, while gaining the confidence of the independents by shielding their relatively expensive production from the rigors of the free market. At the federal level, the Bureau of Mines continued to issue forecasts of demand, while the "Connally Hot Oil Act," passed by Congress in February 1935, prohibited hot oil shipments. Through the tariff law and a voluntary agreement with the main importing companies, the secretary of the interior limited and controlled oil imports. In addition, the major companies cooperated to achieve market stability. The API helped out by maintaining an extensive statistical service and providing advice on how to balance supply and demand. This system, with the Texas Railroad Commission as the balance wheel, regulated the domestic oil industry until the early 1970s.[23]

The New Deal oil settlement, with its emphasis on public-private partnership, foreshadowed the pattern of business-government relations that would become dominant in the postwar years.[24] The solution of the problems of the domestic oil economy in this manner would have a powerful impact on later attempts to deal with the problems facing the United States in the world oil economy.

With the situation in the United States finally under control, attention turned once more to control of foreign production. Increasingly, the major companies turned to joint ownership of reserves as a means of regulating the world oil economy. In 1934, Gulf and Anglo-Persian joined forces to develop Kuwait's oil. The agreement establishing the Kuwait Oil Company contained a provision binding the partners not to use the oil produced in Kuwait to injure each other's trade or marketing position. In addition, Anglo-Persian was given the option to supply Gulf's requirements for oil from Iran or Iraq in lieu of producing additional oil in Kuwait. In effect, this provision allowed Anglo-Persian to control the amount of oil produced in Kuwait, and thus prevent overproduction.[25]

Standard Oil of California (SOCAL) had not been a party to any of the cooperative agreements since it had little foreign production and no established marketing position in Europe. SOCAL had obtained a concession on the island of Bahrain, off the coast of Saudi Arabia in 1930, and in 1932 brought in its first commercial well. Lacking markets and capital, SOCAL made a deal with the Texas Company in July 1936 whereby SOCAL received a half interest in Texas's marketing positions east of Suez in return for a half interest in the Bahrain concession and facilities. A new marketing company, the California Texas Oil Company Ltd. (Caltex), was set up to handle marketing. In December, the SOCAL-Texas partnership was extended to Saudi Arabia, where SOCAL had obtained rights in May 1933 to the region of al-Hasa along Saudi Arabia's eastern coast for sixty years. Soon after the discovery of oil in March 1938, it became apparent that Saudi Arabia's reserves were much larger than Caltex's marketing position could handle. Moreover, the king of Saudi Arabia, Abd al-Aziz ibn Abd al-Rahman al Faisal al Saud (known in the West as Ibn Saud) had only a decade earlier consolidated his authority over the nation that bore his name and needed money to maintain his rule. Therefore, he wanted production expanded rapidly. Concerned about the impact of uncontrolled production in Saudi Arabia on world markets, the chief As-Is partners sought to work out arrangements with SOCAL and Texas to integrate Saudi production into their system. In this case, the Red Line Agreement backfired, as the French and Gulbenkian, concerned about the effects of such arrangements on their interests in Iraq, prevented their more powerful partners from making a deal. Although IPC eventually reached an internal compromise which would have allowed the company to form a partnership with SOCAL and Texas for the development of Saudi Arabia, the outbreak of the Second World War ended negotiations, leaving open the question of where SOCAL and Texas would find the capital and markets needed to develop their potentially rich holding.[26]

How to integrate Saudi Arabian oil into world markets without disruption was a key problem facing the major companies on the eve of World War II. A more direct threat to the international oil companies, however, was the growing desire of producing countries to gain control over their resources. One of the chief goals of the Mexican Revolution was to reassert national control over Mexico's economic life. On March 18, 1938, over twenty years of struggle between Mexico and the international oil companies culminated in the nationalization of the Mexican oil industry. As Clayton Koppes has pointed out, nationalization was more than expropriation. Not only did the companies lose their property but henceforth foreign capital was denied access to a basic sector of the Mexican economy. Moreover, the resource in question was an exportable commodity in great demand by the developed countries. Thus the Mexican action challenged not only the position of the international oil companies but also the role of multinational corporations in the economic development of what would become known as the Third World.[27] The Mexican action had been preceded by a short-lived expropriation attempt by Iran in 1932, and by the successful nationalization of the small Bolivian oil industry in 1937.[28]

In addition to these changes in the international arena, the transformation of the domestic oil economy raised questions about the nation's traditional self-sufficiency in oil. On the one hand, U.S. proved reserves of oil increased from 6.7 billion barrels in 1919 to over 19.5 billion barrels in 1941, some 46 percent of total world reserves. Production increased from 3.7 million barrels in 1919 to over 1.4 billion barrels in 1941, around 63 percent of total world production for that year.[29] On the other hand, the tremendous increase in reserves and production was paralleled by increases in consumption. Between 1919 and 1941, oil increased its share of total U.S. energy supply from 12.1 percent to 31.4 percent. A dramatic increase in the demand for motor fuels led the way, as the number of registered motor vehicles in the United States increased from just under 7.6 million in 1919 to 34.47 million in 1941. Moreover, the average annual consumption of gasoline per vehicle increased from about 400 gallons to 648 gallons. On the farm, the number of oil-powered tractors rose from about 158,000 in 1919 to nearly 1.68 million in 1941. In addition, there was a large increase in the consumption of fuel oil by railroad locomotives, ships, and gas and electric plants, and in the use of oil for commercial and domestic heating. As the U.S. population increased from 106 million in 1920 to 133 million in 1941, annual per capita consumption of oil and oil products increased from 4.3 barrels to 11.2 barrels.[30]

By 1941, the first "Anglo-American petroleum order" had broken down under the impact of depression, world war, and the growing ability and desire of producing countries to control their economic destiny.[31] Moreover, the dominant position of the United States in world oil production and reserves, though still not surpassed, was already beginning to be undermined by the vast discoveries in the Persian Gulf and the relentless rise of domestic oil consumption. These developments would be intensified by U.S. entry into World War II, forcing U.S. policymakers to consider the potential importance of foreign oil to the nation's security and economic well-being. U.S. foreign oil policy would be strongly influenced by the previous history of business-government relations in the oil industry, concerned that the United States might one day be unable to supply its oil needs from domestic sources, and significantly shaped by the structure and power of the U.S. oil industry.

Chapter 1

Origins of Foreign Oil Policy, 1941–1943

The spectacular success of German forces in the first two years of World War II highlighted the vital role of oil in modern warfare. Airplanes, tanks, and motorized troop carriers provided Hitler's armies with the tactical power and speed central to the strategy of Blitzkrieg—lightning war. The German successes pointed up the fact that "if the internal combustion engine was the heart of the modern military machine, its life blood was oil."[1] The military importance of oil forced the U.S. government to create new organizations to deal with oil matters and to take an active interest in foreign oil. Early efforts to develop and implement a foreign oil policy focused on Venezuela and Mexico, and were marked by conflict between the State Department and Secretary of the Interior Harold L. Ickes.

Oilmen in Government

On May 28, 1941, President Roosevelt named Secretary of the Interior Harold L. Ickes petroleum coordinator for national defense. Though Washington had been giving increasing attention to oil matters for some time, the immediate cause of the president's action was the emergency diversion of fifty U.S. flag oil tankers to Great Britain, one-fifth of the tonnage available to serve the East Coast of the United States. Over 95 percent of the East Coast's oil requirements was brought in by tanker, and Roosevelt's action highlighted the potential vulnerability of U.S. oil supplies.[2] Roosevelt appointed Ickes by letter, and assigned him "to obtain . . . information as to (a) the military and civilian needs for petroleum and petroleum products and (b) any action proposed which will affect such availability of petroleum and petroleum products," and "to make specific recommendations . . . as to action which is necessary or desirable . . . to insure the maintenance of a ready and adequate supply of petroleum and petroleum products." The president's letter also authorized Ickes to appoint a deputy coordinator and to employ the necessary staff to carry out his duties. Ickes's authority was limited to gathering information and making recommendations. He was given no power to implement his directives. The emphasis on voluntary cooperation was, in effect, a "peace offering" to an industry that had often clashed with the New Deal. On the other hand, the appointment of the pugnacious Ickes, long an advocate of federal regulation, to head the oil agency was a warning of what could happen if voluntarism proved ineffective.[3]

Ickes had not endeared himself to the oil industry in the 1930s, and, as the president of the American Petroleum Institute later remarked, his appointment was "viewed with alarm as the possible beginning of another drive for Federal control." Though nominally a Republican, Ickes was more accurately "a Bull Moose Progressive who, like Theodore Roosevelt, saw the federal government as the steward of the public welfare." Schooled in the rough and tumble world of Chicago politics, Ickes was a formidable bureaucratic infighter. As secretary of the interior and as National Recovery Administration oil administrator, he fought to expand federal powers over the oil industry. Later, Ickes served as chairman of the Natural Resources Committee, and endorsed its 1939 report *Energy Resources and National Policy*, which recommended the creation of a federal oil conservation board to regulate the oil industry. On the other hand, by 1940, Ickes was beginning to make his peace with the oil industry. Concerned over the worsening international situation, Ickes and many other New Dealers moved to smooth over old divisions. These concerns and fear of the oil industry's impact on the 1940 presidential election led Ickes to question antitrust prosecution of the oil industry.[4]

Aware of the oil industry's mistrust and the need for full cooperation, Ickes moved quickly to gain its confidence. On the recommendation of California oilman and Democratic party treasurer Edwin W. Pauley, Ickes appointed Ralph K. Davies, senior vice-president of Standard Oil of California (SOCAL), deputy coordinator, and vested him with full power to manage the operations of the Office of Petroleum Coordinator (OPC). A native of Virginia, Davies had worked his way up from a junior clerk at a SOCAL branch office to a directorship of the company before he was forty. He quickly gained Ickes's confidence and became a close friend of the self-styled curmudgeon, who was twenty years his senior.[5] Davies took charge of assembling OPC's staff and, carefully balancing independents and majors, recruited over three-quarters of his personnel from the oil industry. To ensure that no one suffered a reduction in pay or a loss in retirement benefits, Ickes paid his staff the maximum allowed by law and worked out arrangements whereby the employee's former employer made up the difference between the employee's government salary and his or her former salary. Davies, for example, received $47,500 annually from SOCAL in addition to his government salary of $10,000. Although all OPC staff were full-time government employees, the salary arrangements obviously created the potential for conflict of interest. Ickes and Davies were sensitive to this problem and tried to ensure that no one,

Davies included, worked on matters that directly affected his or her former employer. Ickes stoutly defended his agency against criticism, and maintained that the use of so many oilmen was necessary since only the industry could supply the trained technicians and experienced managers necessary for swift and effective mobilization.[6]

To allay the industry's apprehensions, Ickes summoned over a thousand of its leaders to a conference in Washington on June 19, 1941, to discuss OPC's plans. Ickes made it plain to the assembled oilmen that he envisaged a partnership between the government and the industry and that he had no intention of imposing plans not worked out in cooperation with industry members. Davies explained to his former colleagues that even though government direction was necessary to effect a quick and successful mobilization of the oil industry's resources industry participation in decision making would ensure that government coordination would not become government control. To underline the new spirit of cooperation, Ickes presented Acting Attorney General Francis Biddle, who repeated pledges made earlier in a letter to Ickes that the Justice Department would take no action in oil matters without consulting OPC.[7]

The OPC and its successor agency the Petroleum Administration for War (PAW, created in December 1942), consisted of a central office, five district offices, and various suboffices. Both the central office and the field offices were organized along the lines of an integrated oil company, with functional divisions for production, refining, transportation, and distribution. Paralleling the OPC's structure was a private industry advisory system headed by a national Petroleum Industry Council for National Defense (later Petroleum Industry War Council or PIWC) and five district committees. The council's membership was drawn largely from the executives of the major oil companies and from production and marketing associations and served without pay, as did the members of the district committees. Meeting at least once a month throughout the war, the council played a key role in the operations of OPC and its successor agency. Although the committees were supposed to be purely advisory in nature, the Justice Department later charged that "during the operations of the committee system fundamental questions of basic policy were initially resolved by these committees and . . . resulting government action amounted to no more than giving effect to decisions already made by such committees."[8]

OPC's initial efforts focused on the transportation problems that had brought it into existence. Ickes's interest soon turned to foreign oil matters, however, and he moved to carve out a leading role for OPC. He established a Foreign Division in August 1941 to handle transportation and supply problems with U.S. allies, and directed U.S. oil companies with operations in Latin America to form a Petroleum Supply Committee for Latin America. In mid-December, Ickes decided to form a Foreign Operations

Committee composed of the heads of the major U.S. oil companies with overseas interests to advise OPC on foreign oil matters. This decision brought OPC directly into conflict with the State Department, which considered foreign oil matters part of its domain. The ensuing conflict poisoned relations between OPC and the State Department for the rest of the war.[9]

Foreign oil had emerged as a factor in U.S. foreign policy after World War I as the growing importance of oil to modern warfare, fears of the pending exhaustion of U.S. reserves, and the needs of U.S. oil companies with foreign markets for additional sources of supply led to a clash with the British over access to Middle East oil. Rejecting government intervention through state ownership of oil reserves or dividing the world into exclusive spheres of influence, the State Department threw its support behind cooperative arrangements worked out by the major international oil companies. Throughout the rest of the 1920s and 1930s, the State Department left foreign oil matters largely to private initiative. By the eve of World War II, U.S. oil companies had acquired concessions in all the major oil-producing countries. Standard Oil of New Jersey held a dominant place in Venezuelan oil (through a subsidiary, Creole Petroleum), and shared with the Socony-Vacuum Oil Company a 23.75 percent interest in Iraq and a 25 percent interest in the oil of the Netherlands East Indies (NEI). Standard Oil of California (SOCAL) and the Texas Company shared in the ownership of the Bahrain Petroleum Company and the California Arabian Standard Oil Company (CASOC), which controlled the oil resources of Bahrain and Saudi Arabia respectively. SOCAL also had interests in the NEI. Gulf Oil Company (through a subsidiary, Mene Grande) was the third largest producer in Venezuela (behind Jersey and Shell), and shared the potentially rich reserves of Kuwait with the British-owned Anglo-Iranian Oil Company (AIOC). Altogether, U.S.-owned oil companies accounted for nearly 40 percent of oil production abroad in 1939, and their foreign holdings represented around half of total reserves outside the United States.[10]

In July 1941, with war clouds on the horizon, Max Weston Thornburg, a vice-president of the Bahrain Petroleum Company, was brought into the State Department to serve as an adviser on oil matters. Under an arrangement similar to that employed in OPC, Thornburg received $29,000 annually from his former company in addition to his government salary of $8,000. Thornburg had been involved in international oil since 1919, and was deeply concerned about the "pathology" of U.S. oil operations abroad, which seemed to reveal a pattern whereby U.S. companies would obtain rights to reserves only to lose them a few years later. Mexico's nationalization of its oil industry had challenged not only the position of the international oil companies but also the role of multinational corporations in the economic development of the Third World. Moreover, the

Mexican action had been preceded by a short-lived expropriation attempt by Iran, and by the successful nationalization of the small Bolivian oil industry. With this history in mind, Thornburg concluded that the main problem facing the United States in foreign oil policy was not how to acquire rights to more foreign reserves but rather how to secure the rights already acquired by U.S. oil companies.[11]

Another of Thornburg's concerns was the stiff competition U.S. oil companies faced from the British, though in 1941 he was also worried about the Germans and their "apt pupils," the Japanese. The British government owned a majority interest in the Anglo-Iranian Oil Company, and both AIOC and Shell had traditionally received strong backing from Whitehall. Thornburg expected British oil policy after the war to be even more aggressive and state-sponsored than in the past because of Britain's wartime experiences and its need to regain its position in world trade. To meet the prospective British challenge, U.S. oil companies would have to work together and in close cooperation with the U.S. government. An ambitious man, Thornburg did not keep his concerns to himself. He wrote publicist Edward Bernays in early 1941 that, though he would not argue for giving business a "free hand for combination within this country," the reasons that could be advanced against combination within the country did not apply to combinations between U.S. companies operating solely outside the United States and competing with strong groups of foreign interests. Thornburg warned that, unless U.S. business found a way to work together in the foreign field, "we shall lose this war, in a commercial sense, just as we did the other one."[12]

The trend toward nationalization forced oil problems into the intergovernmental, as opposed to the purely commercial, field. Moreover, competition from foreign companies strongly backed by their own governments had convinced U.S. oilmen of the "necessity for similar collaboration with our own Government." Thornburg explained that large corporations took a long-range view of profits and placed a high value on security of basic assets and diversification of risks. Only over a long period could they develop their extensive projects and offset losses in one area with profits elsewhere. Recognizing the prospective decline in U.S. oil reserves, the major oil companies had acquired rights to oil all over the world. Foreign operations, however, entailed great risks, political as well as economic, and even though the large companies had developed certain methods to minimize these risks, they remained vulnerable to retaliatory actions by host governments. Sovereign governments and private corporations, however large, were not "homologous" from the standpoint of agreements, and in most disputes between them, the governments would ultimately prevail.[13]

Thornburg believed that the answer to these problems was for the government, through such means as intergovernmental agreements and strong diplomatic support, to ameliorate the risks faced by the companies, particularly "those which threaten the right to remain in business over a long period." In return, the companies could afford to deal more generously with host countries since their long term position would be assured. Thornburg believed that the major U.S. companies were ready to cooperate with the government: "Despite the companies' reluctance to see control of private business pass into the hands of Government, there is a very strong present tendency to collaborate in shaping an increased governmental control which they recognize as inevitable."[14]

Although hired as a staff member of Economic Adviser Herbert Feis, Thornburg soon broke away and, in Feis's words, began "operating as a semi-independent branch of the department." At his urging, the State Department began systematically collecting information on all aspects of the foreign oil business and began assigning petroleum attachés to U.S. diplomatic missions in key countries to gather information and report on oil matters.[15] In October 1941, Thornburg acquired the assistance of Walton C. Ferris, a career foreign service officer, who began a study of U.S. foreign oil policy under Thornburg's tutelage. Ferris's initial report, in late November, argued that in view of the prospective decline in U.S. reserves predicted by experts the government should consider a policy of using oil imports to conserve domestic resources. U.S. oil companies operating abroad were the "supply organizations" upon which the nation would have to depend for access to foreign oil, and should be strongly supported by the government. Therefore, Ferris recommended the negotiation of commercial agreements granting U.S. nationals the right to explore and develop oil resources in other countries and strong diplomatic backing of U.S. companies operating abroad. As the nation's domestic reserves declined, its foreign oil policy would have to become "more and more aggressive" in order to assure access to the oil the United States would need for its security and economic well-being.[16]

Ferris's paper, which Thornburg circulated among the department's various divisions, elicited a mixed response. On the one hand, Leroy Stinebower, an economist who worked on international economic affairs, proposed that in view of the wartime drain on U.S. reserves any petroleum products that the United States provided Britain through lend-lease be repaid in kind after the war. Alternatively, Stinebower suggested that the British could turn over some of their foreign oil properties to the United States. In contrast, Harry Hawkins, chief of the Division of Commercial Policy and Trade Agreements and like Secretary of State Cordell Hull a firm believer in free trade, felt that Thornburg and Ferris did not have enough confidence in the ability of the Atlantic Charter's principles to protect U.S. interests. Rather than make plans on the assumption that the postwar world would be characterized by ruthless competition for concessions and markets, the United States should "formulate and advocate policies now which are designed to bring about and maintain a better world order than the one which set the stage for the present war." Specifically, Hawkins proposed an agreement with the British on the principles that should govern the international oil trade and reciprocal most-favored-nation agreements with the governments of producing countries to provide U.S. companies the freedom to do business. Arrangements that guaranteed "wide space" to private enterprise in the oil business would be in the interest of the United States since the U.S. oil industry, with its tremendous financial resources and strong domestic base, would prevail in open commercial competition.[17]

Despite these criticisms, Thornburg continued his work. In November 1942, he took his case to influential Assistant Secretary of State for Economic Affairs Dean Acheson. The United States, he informed Acheson, was running short of oil and would in the future have to obtain an increasing share of its requirements from abroad. At the same time the United States was faced with important choices in Venezuela, Mexico, and the Middle East that could affect the nation's position in world oil for years to come. If the United States were to maintain its dominant position in world oil, it would need a "positive" foreign oil policy that protected its interests and anticipated problems between U.S. companies and foreign governments before they developed into crises. Close collaboration between the U.S. government and U.S. oil companies would be necessary to develop and implement such a policy. It was time to begin: "Nothing essential to such a program is imposssible nor incompatible with our historical ideas as to the functions of government or the nature of private enterprise."[18]

Preserving Access to Venezuelan Oil

Close collaboration between the U.S. government and U.S. oil companies proved possible in Venezuela, where a conservative government was seeking to increase its revenues from the nation's oil industry. U.S. oil companies had achieved a strong position in Venezuela in the 1920s by purchasing existing concessions from capital-short British companies and by obtaining new concessions. With Standard Oil of Indiana and Gulf leading the way, U.S. companies increased their share of Venezuelan production from 31.5 percent in 1925 to 55 percent in 1929. Standard Oil of New Jersey, unsuccessful in discovering oil on its own, bought out Standard of Indiana's holdings in 1932, and by the end of the decade Jersey dominated Venezuelan oil, with Shell second and Gulf a poor third. Further concentrating control was a series of agreements, concluded just before the war, that effectively merged the Venezuelan operations of the three major companies. On the eve of World War II Venezuela had become the third leading oil producer in the world and the leading exporter.[19]

The oil companies had enjoyed good relations with the Venezuelan government during the twenty-seven-year dictatorship of Juan Vicente Gómez (1908–1935). After Gómez's death, a potential clash between the Venezuelan government and the oil companies over revision of the laws governing relations between the companies and the Venezuelan government was averted because the increased demand for Venezuelan oil brought about by the onset of World War II increased Venezuela's revenues and reduced tensions. The entry of the United States into the war, however, brought German submarines to the Caribbean sea lanes, where they menaced the tankers that carried Venezuela's oil to its overseas markets. Venezuelan production fell by one-third in 1942 as tanker losses and the convoy system cut into the amount of oil that could be exported. This reduction caused severe unemployment, reduced Venezuelan government revenues by 22 percent, and cut deeply into the country's foreign exchange reserves.[20]

The decline in oil revenue presented Venezuelan President General Isaías Medina Angarita with a serious dilemma. Venezuela desperately needed more money, and politically Medina, who had taken office in April 1941 after rigged elections, was under pressure to take a strongly nationalistic stance on the oil issue. Both organized labor and the developing middle class viewed the oil companies as linked with the old oligarchy and felt that a larger portion of the companies' profits should go to Venezuela. A member of the military clique from the Andean state of Táchira, which had ruled Venezuela since 1899, Medina was concerned not to damage the oil industry or drive the oil companies out of the country. The Venezuelan economy was completely dependent on oil in 1942. Oil exports accounted for 90 percent of total exports, oil revenues for 40 percent of total government revenues, and the oil industry for some 25 to 30 percent of gross domestic product. Venezuela possessed neither the people nor the resources to replace the foreign companies. As Stephen Rabe has pointed out, "Expropriation would have turned a troubled economy into a bankrupt one."[21]

Medina turned to the United States for help. In March 1942, he sent his attorney general, Gustavo Manrique Pacanins, to Washington to present Venezuela's case. Manrique Pacanins, a former oil company lawyer, met with the head of the Division of American Republic Affairs (ARA), Philip W. Bonsal, and complained that the oil companies' insistence on retaining the rights they had acquired during the Gómez era was preventing Venezuela from establishing a new framework for the oil industry that would be beneficial to both sides. Unless the companies cooperated in this effort, Venezuela would be forced to take unilateral action, and many of the companies' concessions would not stand up in court.[22]

Manrique Pacanins returned to Washington in August with a letter from President Medina to President Roosevelt that pointed out that he was firmly resolved on revising petroleum policy to "rectify illegitimate, illicit or distressing conditions in order that Venezuela may receive a truly just share in the exploitation of the riches of its subsoil." While complaining of the "obstinacy" of the oil companies, Medina assured his North American counterpart that Venezuela had "no intention in any way of nationalizing the industry or likewise of despoiling from their concessions those who have acquired them legally." Rather, the purpose of revising the terms under which the companies operated was to guarantee long-lasting industrial peace and the future of the industry. When he delivered the letter, Manrique Pacanins assured Under Secretary of State Sumner Welles that Venezuela did not have the "remotest intention of expropriating foreign-owned oil properties" and that any action Venezuela took would be in strict accordance with international law as well as with the laws of Venezuela. A career diplomat and an old friend of President Roosevelt, Welles was the president's principal adviser on Latin America and one of the main architects of the "Good Neighbor policy." With the outbreak of World War II, Welles had turned his energies toward obtaining Latin American support for U.S. policies. He therefore promised that if Venezuela abided by these assurances, the U.S. government "would do whatever it appropriately could to facilitate a friendly adjustment." In a September 14 letter to the Venezuelan chief executive, President Roosevelt also welcomed Medina's "categoric and satisfactory assurances" regarding Venezuela's intentions and expressed his hope that the present problems could be resolved in a manner "which will benefit Venezuelan development and at the same time preserve the rights of the private parties concerned."[23]

Surveying the situation in mid-September, Thornburg concluded that the "primary interest" of the United States was that Venezuelan oil remain available for the war effort. Venezuela had supplied around 80 percent of Britain's oil needs early in the war, and this contribution had helped take pressure off U.S. production. It was important that Venezuelan oil be available not only at the reduced 1942 level of 200,000 to 300,000 barrels per day (bpd) but at much higher levels in the event that the Allies lost all access to Middle East oil. For this reason, Thornburg felt that it was of the greatest importance that arrangements made between Venezuela and the oil companies "not only insure uninterrupted supplies from Venezuela at their present level, but also provide for sufficient development work and adequate personnel to enable practically our entire European and African war effort to be supplied from Venezuela on short notice."[24]

To accomplish these objectives, Welles invited several oil executives to Washington where they were informed of the seriousness of the situation and urged to replace their representatives in Venezuela with new men who better understood the Venezuelan position. In addition, Thornburg began meeting informally with Manrique Pacanins and with "personal friends who hold positions of high responsibility in the companies chiefly concerned." As a result of these discussions, Thornburg drew up an outline of the terms he felt the oil companies would accept, which he sent to Manrique Pacanins on October 30. The oil companies were ready "to sit down around the table as partners with Venezuela, and as partners work out plans for the long range development of Venezuela's oil industry."[25]

On the oil company side, Thornburg met mainly with his "trusted friend" Wallace E. Pratt, a vice president and director of Standard Oil of New Jersey. Pratt, in turn, argued the case for revision before the Jersey Board. Many Jersey officials, including the senior executives of the company in Venezuela, opposed revision of the terms of Jersey's concessions, arguing that the principle of the inviolability of contracts was an essential safeguard of company interests and should not be abandoned. Accepting changes in the company's contracts would weaken the contracts, with serious consequences to Jersey's interests in Venezuela and elsewhere. Pratt, on the other hand, argued that the issue was one of equity because Jersey was making higher profits than had been anticipated. Mutuality of interests between the company and the host country, not contracts, would provide the essential safeguard for company interests in the future. In addition, Pratt pointed out that some of Jersey's concessions, which had been negotiated with the Gómez regime by third parties, might have trouble standing up in court, and that many of Jersey's concessions would expire before it would be physically possible to withdraw from them the oil and gas reserves already developed. With the Venezuelans offering new long-term concessions in exchange for the old ones, it was to Jersey's advantage to convert the old concessions into new ones.[26]

At the end of November, Jersey's Executive Committee decided in favor of negotiations with the Venezuelan government, and sent Pratt to Caracas with full authority to act for the company. Shell, most of whose concessions were due to expire in December, preferred a harder-line approach, but Jersey's decision, coupled with State Department pressure for a quick and amicable settlement, left the British little choice but to follow the U.S. lead. Venezuela hired the U.S. oil consulting firm of Hoover and Curtice to represent it in the negotiations, and invited Thornburg to attend as an unofficial observer. Working from the outline that Thornburg had prepared in October, the two sides quickly reached agreement on general principles. Disagreement over the pace of refinery construction in Venezuela (most Venezuelan oil was refined offshore on the Dutch-owned islands of Aruba and Curaçao) and public outcry over the secrecy of the negotiations delayed the final settlement until March 1943, when the Venezuelan Congress passed a new petroleum law.[27]

The 1943 oil law completely reformed the Venezuelan oil industry. The new law was applicable to old as well as new concessions and permitted the companies to convert their existing titles to its terms. Royalties were raised to a uniform 16.66 percent of total production (the 1922 oil law had set royalties of 7.5 to 11 percent), and exploration and exploitation taxes were increased and fixed on a progressive scale to encourage production. These provisions, in conjunction with a new income tax law passed in January, were designed to give Venezuela approximately half of the companies' earnings from oil production. In return, the companies received confirmation of their old concessions and their renewal for forty years, and the opportunity to acquire new concessions with forty-year leases. The Venezuelan government also agreed to terminate pending and proposed legal challenges to the validity of certain concessions and to halt proceedings concerning back taxes. Secure, long-term leases enabled the companies to justify the heavy investment required for continued development of existing concessions and the acquisition and development of new concessions. Jersey was especially concerned about safeguarding its position, as it had discovered a new producing formation in the huge fields of the Maracaibo Basin and had extensive holdings in the proved but largely undeveloped fields of eastern Venezuela.[28]

The companies quickly took advantage of the new law. Jersey converted 3.3 million acres of its old concessions to the new law, and during 1943 and 1944 obtained 900,000 acres of new land. Because of the wartime demand for Venezuelan oil and the law's limit on exploration titles, Jersey stepped up its exploration and development activities. By 1945, the net production from Jersey's Venezuelan interests was larger than the combined net of its affiliates in the United States. In all, total acreage held by the oil companies in Venezuela rose from 13 million acres at the end of 1943 to 26.5 million acres at the end of 1944. Production increased from 177 million barrels in 1943 to 323 million barrels in 1945. Venezuela's oil revenue also dramatically increased, as the government received a larger share of company profits as well as payments for new concessions. Thus it is no wonder that a Jersey executive later described the 1943 oil law as "a business deal profitable to both sides."[29]

Public-private partnership, though of a less formal nature than envisaged by Thornburg, preserved the U.S. stake in Venezuelan oil. Thornburg and the State Department convinced the oil companies of the need to accept reform if they were to receive backing from the government. State Department policy strengthened the position of oil executives like Jersey's Wallace Pratt, who, pointing to the Mexican nationalization and Venezuela's oil potential, argued for accommodating reform in order to secure a firm foundation for expanded operations. On the other side, the State Department restrained Venezuelan desires for control over their oil and channeled them into demands for an increased share of the companies' profits. In this regard, U.S. policy strengthened the hand of Venezuelan conservatives against the nationalists in the Acción Democrática party, whose chief concern was control over Venezuela's resources.[30]

In his influential study *The Making of the Good Neighbor Policy*, Bryce Wood argued that the 1943 oil law and the negotiations leading to it supported his thesis that the United States, in carrying out the Good Neighbor policy, subordinated the interests of U.S. corporations to larger foreign policy goals. It is clear that in Venezuela, at least, U.S. foreign policy goals and the interests of the U.S. oil companies involved did not conflict but rather were complementary. Both aimed at the continuation of the companies' control of Venezuelan oil. As Stephen Rabe has noted, the 1943 oil law, far from representing a defeat for the oil companies, "reinforced Venezuela's dependence on the United States, the foreign oil companies, and a capricious world economy."[31]

Meanwhile, in January 1943, prompted by suggestions from Wallace Pratt of Jersey and Socony-Vacuum President John Brown that the State Department should take the lead in foreign oil policy, Secretary of State Cordell Hull had created a Committee on International Petroleum Policy to coordinate State Department policy on foreign oil matters. Economic Adviser Herbert Feis chaired the group, which included representatives from the department's geographical divisions as well as Thornburg, who had assumed the title "Petroleum Adviser." Thornburg planned to involve the leaders of the major oil companies in the committee's efforts, but, before this could be arranged, he was forced to leave the government because of allegations that he put oil company interests ahead of the national interest.[32] These allegations grew out of a conflict between Thornburg and the State Department on the one hand and Secretary of the Interior Harold L. Ickes on the other for control of U.S. foreign oil policy. The focus of conflict was Mexico.

Frustration in Mexico

After the nationalization of the Mexican oil industry on March 18, 1938, the state-owned oil company Petróleos Mexicanos (PEMEX) took over the nationalized properties, and henceforth private capital was denied access to the petroleum sector of the Mexican economy. The oil companies reacted vigorously to the nationalization. Claiming that the Mexican action constituted a major threat to the entire system of foreign investment throughout the world, they instituted a boycott of Mexican oil and pressured oil-equipment manufacturers not to sell equipment to Mexico. Also concerned about the impact of nationalization on U.S. interests abroad, the State Department supported the key points of the companies' position. Although officially acknowledging the right of a sovereign nation to nationalize foreign-owned property, the State Department insisted on fair and immediate compensation. Although President Roosevelt softened "immediate" to "prompt," it was clear that Mexico could raise the funds for compensation only through long-term operation of the industry. Since the companies, with the active support of the State Department, were working to prevent Mexico from selling its oil, the demand for full and prompt compensation was tantamount to preventing Mexico from carrying out nationalization.[33]

The boycott was initially effective in hindering the export of Mexican oil, and PEMEX turned to Germany, Italy, and Japan for sales and equipment. In addition, the effects of the boycott were softened because exports, though reduced, no longer had to support interest and dividend payments to foreign investors. By the end of 1938, it was clear that Mexico was not going to give in. Nevertheless, the State Department continued to back the companies until the summer of 1941, when the worsening international situation forced the United States to reconsider its policy toward Mexico.[34]

On November 19, 1941, the United States and Mexico signed a "global" settlement which established a commission to evaluate the expropriated properties and to decide on compensation. Morris L. Cooke, a prominent engineer and Democrat who had headed the Rural Electrification Administration from 1935 to 1937, was selected as the United States expert, and General Manuel J. Zevada represented Mexico. Their report, released in April 1942, set the total compensation figure at $23,995,991, plus slightly more than $5 million in interest since March 1938. Jersey received the lion's share, $18,391,641, and Standard Oil of California was awarded $3,589,158. The remainder was divided among several companies. The companies, however, continued to oppose a settlement until the fall of 1943, when they accepted the Cooke-Zevada awards on the condition that all claims against them by Mexico be dropped. A major factor leading Jersey to accept the settlement may have been a desire to retain State Department support for its position in Venezuela.[35]

In recent years, a growing number of scholars have pointed out that the November 1941 settlement of the Mexican oil controversy did not signify U.S. acceptance of the nationalization of the Mexican oil industry or abandonment of the U.S. oil companies. Rather the agreements represented, in one historian's words, "a strategic withdrawal from an untenable position to one which permitted a subsequent regrouping and redefinition of the problem." Thus the November agreements and the subsequent Cooke-Zevada agreement on the value of the expropriated properties were limited to the compensation issue, and were significantly silent on the question of the future status of foreign participation in the Mexican oil industry. Indeed, the State Department hoped that settlement of the compensation issue would clear the way for U.S. oil companies to resume operations in Mexico. As Under Secretary of State Sumner Welles explained to President Roosevelt: "Although the award brought to an end the intergovernmental controversy created by the expropriation, the more basic question of the terms and conditions under which United States interests would be permitted to participate in the Mexican petroleum industry remained to be solved." Veteran Ambassador George S. Messersmith, who replaced Josephus Daniels in early 1942, was instructed that "above all other duties he should give first place to endeavoring . . . to work out a plan . . . under which United States interests could again participate in the Mexican oil industry."[36]

The State Department was beginning the process of reaching a settlement of the oil controversy when Max Thornburg was appointed petroleum adviser in the summer of 1941. On learning of the department's plans, Thornburg tried to convince the proponents of a settlement to reconsider, arguing that what they proposed would harm U.S. oil interests all over the world. As an alternative, Thornburg proposed the establishment of a Mexican corporation to own the properties, but with management by a U.S.-owned and -controlled management corporation under a long-term contract. Compensation would be paid out of gross revenues, as would the management corporation's fees, and the Mexican government and the management corporation would split net profits. According to Thornburg, this plan would meet the political goal of achieving a settlement of the issue, while "preserving certain principles upon which the rights of lawful American enterprise abroad depend." Thornburg's management contract plan was part of an effort by the U.S. oil companies, led by Jersey, to head off the State Department's plans for a settlement. The companies were willing to give up ownership of the oil as long as they retained effective control over its development.[37]

As part of this effort to derail the proposed settlement, Thornburg convinced the oilmen in OPC's Production Division to produce a study on the U.S. reserve position which would demonstrate that Latin America's oil, including Mexico's, was crucial to U.S. security and prosperity. According to the resulting report, the U.S. production-reserves ratio— proved reserves divided by annual production—had been steadily diminishing since 1933, and since 1938, discoveries of new reserves had not equaled the quantity of oil produced. Deputy Petroleum Coordinator Davies passed these conclusions along to Ickes in mid-October, and stressed that the United States "must have extra-territorial petroleum reserves to guard against the day when our steadily increasing demand can no longer be met by our domestic supply." Looking south, Davies suggested that the "petroleum resources of Mexico, Colombia, Venezuela, and other Caribbean countries must be considered to be reserves for the United States." This oil was more important to the United States than to the countries in which it was located because it was "more vital to the life of the consumer than to the producer." Davies stressed that what was done in regard to the Mexican situation would "very largely determine the course that will be pursued by the other Latin American countries having large oil reserves of equal or even greater importance to us." Ickes repeated these warnings in a note to Secretary of State Hull which cautioned that great care should be exercised in the settlement with Mexico lest it encourage other Latin American nations to nationalize their oil industries. It was essential that Latin America's oil remain in the hands of U.S. companies since through them the U.S. government could control and coordinate oil development throughout the hemisphere.[38]

This report had a major impact on thinking within the U.S. government on oil matters. Ickes sent a copy of it to President Roosevelt with a note saying that it was "one of the most important documents I have ever sent you." Therefore, it is important to note that OPC's statistics were derived by an unusual method of calculating the discovery of reserves. In contrast to the standard method which attributed reserves to the year in which they were developed, OPC credited reserves backward to the year in which a field was first discovered. This procedure probably gave a more accurate picture of future possibilities of discovering *new* fields than the conventional method, but it tended to inflate the figures for earlier years and to understate them for subsequent years. According to the standard method of calculating reserves, additions to reserves between 1935 and 1939 averaged about 2.4 billion barrels annually and exceeded current output by such a margin that there was a net increase in proved reserves averaging 1.3 billion barrels per year.[39]

Thornburg soon found that involving Harold Ickes in the Mexican oil situation was at best a mixed blessing. Ickes soon concluded that the United States had a national interest in Mexican oil separate from the interests of the oil companies. In a pair of letters to President Roosevelt in early December, Ickes complained that the United States had "no adequate national policy with respect to petroleum, and no international policy that I know of except to protect the interests of our nationals." Ickes believed that the interests of the oil companies would not always run parallel to those of the nation and argued that the United States should approach the problem of foreign oil "from a nationalistic point of view," and begin to formulate a "national" foreign oil policy, "our own reserves being what they are, and the importance of oil to civilization being what it is." Turning to Mexico, Ickes noted that "the time may come when our government will wish that it owned these properties itself in order to exploit them in the national interest."[40] Two months later, Ickes recommended that the government buy the expropriated properties from Mexico and hold them as a petroleum reserve. Roosevelt, in a letter drafted by the State Department, rejected the idea, pointing out to his enthusiastic cabinet officer that it was inconceivable that the Mexicans, who had gone to so much trouble to gain control of their oil, would agree to such a plan.[41]

In the meantime, Ickes had turned to a scheme devised by California oilman and Democratic party fund-raiser Edwin W. Pauley whereby a group of U.S. independent oil companies would build a high octane gasoline refinery in Mexico in return for a percentage of its output. Ickes presented the plan at a meeting of the cabinet on February 6, 1942, and President Roosevelt immediately seized on it as a means of improving relations with Mexico. A promise of U.S. assistance for the construction of a high octane gasoline plant in Mexico was included in the agreements surrounding the conclusion of the Cooke-Zevada negotiations in April 1942.[42]

As might be expected of an independent oilman, Pauley was apparently looking for a cheap source of crude oil. Ickes's deputy, Ralph Davies, had worked closely with Pauley when Davies was vice-president of SOCAL. Evidence presented during the 1946 hearings over Pauley's nomination for appointment as under secretary of the navy indicates that that any oil that Pauley obtained might have been shared with SOCAL. Thus Davies may have seen the project as a means of obtaining access to oil for his former company. In addition, Davies supported the plan as a means for getting U.S. oil companies back into Mexico. Pauley and Davies gained the support of Ickes, who apparently felt that the project was an effective means of promoting the development of oil the United States might need in the future.[43] The State Department, on the other hand, opposed any aid to Mexico, including Pauley's project, because it would undercut emerging plans to use Mexico's need for assistance as a lever to reverse nationalization. In addition, Thornburg argued that Pauley's proposal was unfair to Mexico and could lead to a repeat of earlier problems with Mexico.[44]

The link between assistance and reversing nationalization was made clear in instructions sent to Ambassador Messersmith in August 1942, which noted that the United States, while recognizing Mexico's need for help in rehabilitating its oil industry and desiring to help develop "the full export possibilities of the industry," nevertheless felt that when it was "called upon to consider questions involving financial or other substantial assistance, or involving hemispheric security both military and economic, it must be in a position to judge the long-range consequences of its own actions." As Thornburg, who drafted the instructions later explained, what they meant was that "until we know whether or not Mexico would permit American participation in some way, we were not ready to say how much help we would be to Mexico."[45]

Ickes stood in the way of this strategy with his new-found interest in Mexican oil. His suspicions fueled by Pauley's insinuations that Thornburg's opposition to the project was part of a plan to get Jersey and the other big companies back into Mexico, Ickes pressed for immediate action on the refinery. On April 17, after listening to Pauley complain about State Department indifference to his project, Ickes snapped off a note to the president complaining that although the president had approved the idea efforts to move ahead were being blocked. Then, on August 4, he reminded Hull that the president was interested in the project, and asked if there were any reasons of "high political importance" having to do with U.S. foreign policy that were precluding action on Pauley's proposal.[46]

To spur matters along, Ickes sent an OPC technical mission to study the Mexican oil industry's needs. The Technical Mission's report, submitted in late October, pointed out that the Mexican oil industry was "sadly in need of materials for ordinary maintenance and repairs." Assistance was needed to continue current operations and to meet possible wartime requirements. Mexico's refining sector especially needed help, though the mission felt that the high octane gasoline refinery should be postponed until existing refinery capacity was repaired and additional equipment installed in Mexico City. Most importantly, unless Mexico wanted to "court disaster," an expanded program of exploration had to be undertaken immediately to strengthen Mexico's fast declining reserve position.[47]

The report rekindled the president's interest in the project, and on November 23, Roosevelt wrote Hull, Ickes, and Secretary of Commerce Jesse Jones that in view of its "imperative character" for U.S.-Mexican relations, he wanted the 100 octane project "disposed of, on its merits, with the least possible delay. . . . It should not be allowed to become involved in the expropriation or other extraneous matters." The State Department tried to hold the president off by pointing to the shortage of materials for such a project and to the Technical Mission's recommendation to delay the 100 octane refinery until other matters were disposed of. Moreover, the department argued that Pauley's proposal suggested "an exploitation which might easily invite a repetition of the difficulties which have caused so many problems for our government in its relations with Mexico." Roosevelt was not satisfied with these explanations, and on December 7 instructed Hull to get together with Ickes and Jones and "put this through . . . without further delay."[48]

The president's instructions threatened to undermine the State Department's strategy. Construction of a 100 octane refinery involved much more than building a single facility. Constituents had to be made in other specialized plants, and an entire program of pipelines, preliminary processing plants, and other facilities would have to precede the 100 octane facility. Thus the high octane project was "inseparably bound up" with the rehabilitation of the Mexican petroleum industry. With such a program completed, Thornburg feared that Mexico "would be in a position to show far greater independence of our ideas than she is now." Rather than committing itself to allowing the reentry of U.S. oil companies in order to get the materials it urgently needed to rehabilitate its declining oil industry, Mexico could then afford to wait until after the war, when other sources of outside assistance might be available.[49]

To compound the State Department's problems, the Mexican government was reluctant to change its policy and let foreign oil companies back into Mexico. Elected in 1940, Mexican President Manuel Ávila Camacho saw his task as consolidating the gains of the revolution. To this end he wanted to improve relations with the United States and was inclined to favor the participation of foreign capital in all areas of the Mexican economy. He did not have a free hand in oil policy, however. Nationalization of the oil industry had been tremendously popular in Mexico. With ex-President Lázaro Cárdenas in the wings to mobilize opposition to any attempt to undo his greatest accomplishment, the mere hint of permitting foreign oil companies back into Mexico was "political dynamite." Therefore, Ambassador Messersmith, a former high school teacher whose long rambling dispatches earned him the nickname "Forty-page George," continually advocated patience to give the "inescapable facts" of the international oil economy time to bring the Mexican government around to the U.S. position. Messersmith also believed that a "formula" that would allow U.S. companies to participate in the Mexican oil industry without violating, at least openly, the principle that Mexico had to be in control of its petroleum resources would be necessary. In early November 1942, he wrote Welles that "the question is to determine what the Mexican government can do and what it can get away with at this time and what is really in our interests."[50]

Thornburg began meeting with Ickes in early 1943 in an attempt to win over his chief opponent. Ickes, however, kept up the pressure on the State Department. On February 16, 1943, he met with President Roosevelt and complained about the department's continued inaction on the refinery project. Ickes claimed to agree with the State Department's objective of getting U.S. oil companies back into Mexico, but felt that it was moving too slowly. In response, Roosevelt snapped off a sharp note to the State Department demanding an explanation for the delay. Welles tried to placate the angry chief executive by carefully explaining the relationship of the project to the department's overall strategy on Mexican oil. If the United States forced the issue the result could be "the exclusion of American interests from participation in the Mexican oil industry for many years to come when such participation will be more than ever necessary for our own national interest." President Roosevelt replied on February 19 that, while he also agreed that Mexico would have to determine its fundamental petroleum policy before the United States would provide assistance for the general development of the Mexican oil industry, he did not consider the 100 octane project relevant to the question of

Mexico's future oil policy. Therefore work on it could and should begin immediately.[51]

Before carrying out the president's instructions on the refinery, the State Department made one last attempt to develop a "formula" that would allow U.S. oil companies to take part in Mexican oil development. Working from an outline that Thornburg had prepared while in Venezuela for the conclusion of negotiations there, the department proposed that U.S. companies could participate through contracts with Petróleos Mexicanos (PEMEX), the national oil company of Mexico. Mexico would retain all subsoil rights as well as overall responsibility for oil development. It would have this control "only for a moment," however, as the actual operation of its properties would be contracted out to an operating company composed of several U.S. companies. Thornburg argued that the companies that had been nationalized had a "preferential right" to profit from their former investments. The creation of a new company, jointly owned by several oil companies, would enable them to avoid the unfavorable popular reaction in Mexico that their return in a "popularly recognizable form" would arouse. The operating company would be paid through a share in the oil produced, and in general would be run as an independent company. The State Department hoped that such an arrangement would provide U.S. oil companies with the necessary incentives, while not offending Mexican pride or violating Mexican law.[52]

As part of his attempt to make peace with Ickes, Thornburg sent Ickes copies of this and other memorandums. Rather than winning Ickes over, however, these papers fueled Ickes's suspicions of Thornburg's plans and motives. Ickes promptly passed the "Caracas memorandum" along to President Roosevelt, noting that it might well be entitled "an indiscreet memorandum from the State Department" since it "scarcely even attempts to disguise the fact that it is . . . the State Department's policy not to permit any settlement of the Mexican oil question unless the expropriated companies, either openly or in disguise, get back into that field."[53]

Negotiations with Mexican government representatives on the 100 octane project finally began in Washington on March 20. Although the Mexicans wanted to expand the talks to include the large program of rehabilitation outlined in the Technical Mission's report, the State Department made it clear from the outset that the talks would be strictly limited to arrangements for the 100 octane refinery. In addition, to make sure that Pauley would not be involved, the State Department proposed that the U.S. government finance the project through the Export-Import Bank, with repayment through the sale of products to the U.S. government. The actual construction and operation of the facility would be handled by private contractors hired by PEMEX. The Petroleum Administration for War (PAW), which had replaced OPC, would supervise the technical aspects of the project, and the State Department would have the right to review all contracts. The Mexicans readily agreed to these terms, and on March 22 selected Universal Oil Products (UOP) of Chicago to serve as consultants on the project.[54]

Ickes, disgruntled at the State Department's control of the negotiations and at the exclusion of Pauley's proposal from consideration, seized on the selection of UOP as evidence of State Department favoritism toward the big companies. UOP had been formed by the major oil companies in the 1920s as a joint pool company to manage their patent rights and to conduct research. Informed by Davies of UOP's ties to the majors, Ickes charged that Thornburg was attempting to use UOP as a means of slipping the expropriated oil companies back into Mexico in a preferential position. Ickes wrote the president that the State Department's promotion of UOP as consultants to PEMEX appeared to be the "not popularly recognized means by which the initial step is to be taken to return the expropriated companies to Mexico with preference." In contrast, the State Department had been blocking the plans of a "group of independent oilmen" to build a high octane refinery in Mexico.[55]

Ickes held up completion of the negotiations with the Mexicans until mid-May, while PAW and the State Department traded charges of favoritism. The impasse was broken on May 13, when Dean Acheson, who was a personal friend of Ickes, took Thornburg to see Ickes and Davies. Thornburg explained that he opposed Pauley's project because of its "one-sided and exorbitant terms," and could not understand why Ickes kept pushing it. Ickes admitted that he had never seen a copy of Pauley's proposal, and when Thornburg reviewed the project's terms for them, he agreed that they sounded "pretty bad." Next, Acheson brought the Mexican representatives in to see Ickes. The Mexicans explained that they were aware of UOP's big-company connections and had no objection. They had selected UOP because they wanted to use UOP patents in the refinery. Nevertheless, they agreed to select another firm in order to please Ickes. With these matters out of the way, negotiations on the project proceeded to a successful conclusion.[56]

Talks with the Mexican government on general oil policy had been put on hold by the impasse in the refinery negotiations. Although the Mexicans accepted the "formula" presented by Messersmith on April 1 as a basis for discussions, they cautioned that discussions on general oil policy could not be completed until the 100 octane matter was successfully concluded. After the completion of the 100 octane negotiations, however, the Mexicans seemed to lose interest in working out a means for U.S. oil companies to return to Mexico. Whether the Mexican government had ever seriously considered changing its oil policy is not clear. It is clear, however, that the disarray on the U.S. side allowed Mexico to get assistance for the rehabilitation of its oil industry without having to make concessions affecting the industry's nationalized status. To underscore this point, the Mexicans named the 100 octane plant "18 de Marzo," commemorating the date of the nationalization of their oil industry.[57]

Close cooperation between the government and the oil companies worked in the case of Venezuela because the government and the companies shared the same goal—maintaining U.S. control of Venezuelan oil. Another factor working for success was that, in this case, the policy required no government action to implement beyond the informal efforts of Thornburg and limited diplomatic support from the State Department. Thus, neither Congress nor other executive agencies were involved. These favorable factors were absent in Mexico. There were no U.S. oil companies to work through since they had been expropriated in 1938. In addition, Ickes's interest in Mexican oil upset State Department strategy for reopening Mexico to U.S. companies. Most important, President Roosevelt was determined not to let the oil issue disturb mutually beneficial wartime cooperation between the United States and Mexico. On April 20, 1943, Roosevelt made the first official visit by a U.S. president to Mexico when he met with Ávila Camacho in Monterrey. Stressing the two nations' common cause in the war effort, Roosevelt pointed to the "mutual interdependence of our joint resources," and noted that "the day of the exploitation of the resources and the people of one country for the benefit of any group in another country is definitely over."[58]

By the middle of 1943, Harold Ickes emerged as a force to be reckoned with in foreign, as well as domestic, oil policy. Ickes's influence had been significantly enhanced in December 1942 with the establishment of the Petroleum Administration for War (PAW). With far wider powers than OPC, PAW quickly developed into one of the most important independent wartime agencies, having authority over all aspects of the oil business.[59] In addition, Ickes was able to eliminate his chief rival in foreign oil matters. On June 30, apparently convinced by Ickes's repeated allegations, President Roosevelt wrote Under Secretary of State Sumner Welles about his concern that Thornburg was "representing the oil people as well as the State Department." This action sealed Thornburg's fate, and he left the State Department on July 17, 1943.[60]

Chapter 2

Mr. Ickes's Arabian Nights

The emergence of oil as a security issue during 1943 forced the U.S. government to take an active interest in oil matters and to consider the potential significance of Middle East oil to U.S. security. Efforts to secure the U.S. stake in Middle East oil were marked by conflict between contrasting conceptions of the proper relationship between the government and an industry whose operations were increasingly vital to the nation's security and economic well-being. Drawing on the statist strand in New Deal thought, Secretary of the Interior Harold L. Ickes and elements of the armed forces argued that U.S. government ownership of overseas oil reserves was necessary to maintain U.S. access and control. The State Department, in contrast, feared that government ownership of foreign oil reserves would encourage similar measures by other countries, and sought a solution that would accomplish the same ends without involving the government directly in the oil business.

Oil, War, and Security

Although the California Arabian Standard Oil Company (CASOC) had held a huge oil concession in Saudi Arabia since 1933, few people in government outside the Near Eastern Division of the State Department had given much thought to the desert kingdom or its oil. This state of affairs changed dramatically during World War II as efforts by representatives of the Standard Oil Company of California (SOCAL) and the Texas Company, joint owners of the concession, to obtain U.S. government support for their operations brought the new nation and its oil to the attention of top policymakers in Washington. In the spring of 1941, King Ibn Saud, his finances decimated by the war-induced curtailment of CASOC's operations, the near stoppage of the pilgrimage to the holy cities, and greatly increased expenditures because of the war, demanded $6 million from CASOC. The company had already advanced the king some $6.8 million, but fearing the loss of its concession in which it had invested approximately $27.5 million, turned to the U.S. government for assistance.[1]

Working through James A. Moffett, chairman of the board of two of CASOC's affiliates and a personal friend of President Franklin D. Roosevelt, CASOC was able to reach the highest levels of the government with its message that without financial assistance Saudi Arabia "and perhaps with it the entire Arab world will be thrown into chaos." CASOC's owners proposed that the U.S. government advance Saudi Arabia $6 million annually for the next five years against the security of oil products, which the company would deliver to the U.S. government for the king's account. In addition, the oilmen suggested that the State Department be instructed to urge the British to increase their subsidies to Ibn Saud. They cautioned, however, that "any British advances should be on a political and military basis and should not involve their getting any oil from this concession."[2]

Although President Roosevelt favored the proposal and company officials lobbied extensively, the plan collapsed when Secretary of the Navy Frank Knox reported to the president on May 20, 1941, that the quality of the products refined from Saudi Arabian crude oil was not suitable for navy use. As an alternative, the State Department suggested limited purchases of Saudi Arabian oil coupled with lend-lease assistance, which could be extended in return for political assurances. After the demise of the purchase scheme, President Roosevelt instructed his special assistant, Harry Hopkins, to find a means to provide aid to Saudi Arabia. On June 14, Hopkins wrote Federal Loan Administrator Jesse Jones that some of the aid to Saudi Arabia might be channeled through lend-lease, "although just how we could call that outfit a 'democracy' I don't know." CASOC had also thought of lend-lease and had broached the idea to Ibn Saud, who requested lend-lease assistance on June 26. U.S. Minister to Saudi Arabia Alexander Kirk strongly supported the Saudi request, warning that aid to Saudi Arabia was "not merely a question of buying support but chiefly one of preventing a recourse to other sources of support."[3]

This plan also failed. President Roosevelt and his advisers, already under fire for their increasingly activist foreign policy, became concerned about the possible political consequences of aiding distant Saudi Arabia. Rather than risk a setback to his larger foreign policy goals, the president decided that the matter could be handled through the British. On July 18, he wrote Jesse Jones, who was handling a $425 million loan to the British: "Will you tell the British I hope they can take care of the King of Saudi Arabia. This is a little far afield for us." Jones showed the president's note to his British counterparts and asked them to continue taking care of Saudi Arabia. The British, who had already provided Saudi Arabia with grants of £200,000 in January and May, provided an additional £250,000 in November 1941. Whether Jones merely suggested that the British take care of Ibn Saud or stipulated that they use part of the funds that the United States was providing for this purpose was not clear. According to Aaron Miller, who has closely studied the matter, Jones probably tried to arrange a "gentleman's agreement" in which Great Britain would fund Ibn Saud with some of the aid it would receive from the United States.[4]

British aid to Saudi Arabia helped CASOC in the short run by maintaining stability and relieving the financial pressure on the company, but CASOC officials were concerned that the British might decide to use their growing influence to advance the commercial interests of British nationals. Although these concerns were based more on past British actions and the potential for harm than on actual British policies, company officials were nervous about the British being the main source of income for the king, fearing that they might "attempt to extract certain *quid pro quos*," as CASOC head Fred A. Davies told the State Department in November 1941.[5]

Although unsuccessful in securing direct U.S. assistance for Ibn Saud, CASOC's lobbying efforts played an important role in educating U.S. policymakers to the political and strategic significance of Saudi Arabia, an importance that increased with U.S. entry into the war. Though oil was not a matter of strategic concern to the United States before Pearl Harbor, it became an increasingly important element in State Department thinking about Saudi Arabia during 1942.[6] Washington's increased interest in Saudi Arabia was due to growing concern over the adequacy of U.S. oil reserves, growing awareness of the extent of Saudi Arabia's oil potential, and growing fear that the British might use their influence in Saudi Arabia to the detriment of U.S. oil interests.

When the United States entered the Second World War, hardly anyone in the oil industry or the government anticipated any problems in achieving sufficient oil production to meet future demand. Robert E. Wilson, the Office of Production Management's oil expert, testified to Congress in late March 1941 that because of proration and other conservation activities the U.S. oil industry was able to meet any increase in demand that could conceivably be made upon it. Proved reserves were the highest in history, and the maximum efficient rate of production (MER) stood at almost 1 million barrels per day (bpd) in excess of actual production. Wilson believed that production could be increased by 30 percent without any major new discoveries or any unusual amount of drilling. A "tremendous underground reserve" had been built up, and was available "merely by the opening of valves."[7]

The war, with its tremendous demand for oil, changed this assessment. Between 1939 and 1943 U.S. oil consumption increased 28 percent while reserves increased only 15.7 percent. By the end of 1942, some studies were predicting a decline in the ability of U.S. oil fields to produce "within a few years" and shortages of crude oil "within the relatively near future." In the longer run, there was concern that the United States, with around 45 percent of the world's known oil reserves, accounted for over 60 percent of total world production, and was depleting its reserves at a much faster rate than the rest of the world.[8]

Concern over the adequacy of U.S. oil reserves highlighted the importance of foreign sources of production. As State Department Economic Adviser Herbert Feis later noted, "In all surveys of the situation, the pencil came to an awed pause at one point and place—the Middle East." The Middle East not only contained over one-third of the world's known oil reserves; it also offered better prospects for the discovery of additional reserves than any other area. Saudi Arabia's production had risen from around 500,000 barrels per day (bpd) in 1939 to over 5 million bpd in 1940 before the effects of the war forced a curtailment of CASOC's operations. This production had been achieved with a relatively small amount of drilling, and it was clear that Saudi Arabia's reserves were at least as large as those of Iran and Iraq, and thus measured in billions of barrels.[9]

Thus when the Near East Division of the State Department recommended extending lend-lease aid to Saudia Arabia in December 1942, oil was a key concern. The memorandum recommending lend-lease pointed out that the Dhahran oilfield promised to be one of the world's great fields. In contrast, estimates as to the time when the United States would have to look abroad for oil supplies because of the exhaustion of domestic reserves could "only be described as startling." The potential importance of an oil-rich Saudi Arabia to a United States that seemed to be running out of oil heightened fears of British encroachment. Echoing CASOC's concerns, the memorandum pointed out that the British had been supplying Saudi Arabia with £3-4 million in assistance annually, and warned that there was a "definite possibility that the British will desire a *quid pro quo* at the end of the war." Therefore the Near East Division recommended that the development of Saudi Arabia's petroleum resources "be viewed in the light of broad national interest," and that the United States safeguard its interest in every way possible.[10]

Fears of British encroachment seemed to be materializing in late 1942 as the British government advanced plans to set up a British-controlled bank of issue in Saudi Arabia. CASOC officials feared that the plan would draw Saudi Arabia into the sterling bloc and that they might not be able to maintain the "American character" of the company. U.S. Minister to Saudi Arabia Alexander Kirk, who was always suspicious of British motives, saw the plan as an example of an "increasingly discernible tendency toward British economic intrenchment in this area under [the] impact [of] war to a degree which might materially negate [the] best intentioned postwar agreements for equality of opportunity." To protect U.S. interests, Kirk recommended that the United States insist on participating in the plan.[11]

British activity also spurred CASOC officials into action. In early

February 1943, W.S.S. Rodgers, president of the Texas Company, and SOCAL President Henry D. Collier traveled to Washington to convince the government of the vital importance to the United States that the large Saudi Arabian oil reserves remain exclusively in U.S. hands. Collier and Rodgers met with Secretary of the Interior and Petroleum Administrator Harold L. Ickes on February 5, and told him that recent British actions were a threat to their concession in Saudi Arabia. Ickes was convinced that action should be taken to protect the concession and urged the oilmen to supply him with the facts so he could take the matter up with President Roosevelt. Accordingly, on February 8, Rodgers sent Ickes a memorandum that proposed direct aid from the U.S. government to Saudi Arabia to check rapidly increasing British influence and to give some assurance that the oil reserves of Saudi Arabia would remain under U.S. control. Rodgers suggested that such aid could be extended by crediting the amount owed by Saudi Arabia to the British to Britain's lend-lease debt to the United States. If these suggestions were acted upon CASOC would set aside as a U.S. government reserve an amount of oil of equivalent value and make this oil available to the U.S. government at prices well below world market levels. In addition, CASOC would be willing to set aside more oil on similar terms and to give the U.S. government an option on the oil at an agreed-on percentage of world prices. Ickes's deputy, Ralph K. Davies, a former vice-president of SOCAL, told his chief that the proposal was not feasible and Ickes failed to pursue the matter.[12]

Rodgers, Collier, and CASOC President Fred A. Davies met on February 10 with Under Secretary of the Navy William C. Bullitt, Rear Admiral H. A. Stuart, director of the Naval Petroleum Reserve, and Captain Andrew F. Carter, executive officer of the Army-Navy Petroleum Board. After the oilmen had made their presentation, the group agreed that the problem should be approached as a possible establishment of a large naval petroleum reserve. Bullitt suggested that the simplest approach might be for the U.S. government to acquire a majority interest in CASOC, noting that the British government owned a majority interest in the Anglo-Iranian Oil Company. Bullitt later told Ickes that the oilmen did not seem opposed to the idea though they felt that a 25 percent interest for the government would be more reasonable. This idea spurred Ickes's interest in the matter, and he agreed to cooperate with Bullitt on the problem. Davies was enthusiastic about this new idea, and pointed out that they should act during the war so they could plead the necessity of getting supplies from Saudi Arabia for the war effort.[13]

The oilmen presented their case to State Department Economic Adviser Herbert Feis on February 11. They explained that it would be sound policy for the United States to look to Saudi Arabia as a source of supply for Europe, and thereby lessen the drain on U.S. and other nearby reserves.

To maintain U.S. control of Saudi Arabian oil, the oilmen suggested that the U.S. government could acquire a share in the ownership of CASOC. Alternatively, the government could contract with CASOC for the maintenance of an underground oil reserve which it could draw on at reduced prices. Feis, an international economist who had left academics for government service in 1931, felt that the oilmen's fears of British encroachment were connected with vague possibilities in the future, rather than with any immediate problems, and suggested working out Middle East oil problems through an agreement with the British. Nevertheless, he assured them that the State Department would continue to monitor the situation in Saudi Arabia closely and would give their suggestions careful consideration.[14]

While the State Department deliberated, Ickes brought the matter to President Roosevelt's attention. During a lunch meeting on February 16, Ickes told the president that the British were trying to "edge their way into" Saudi Arabia at the expense of CASOC. Saudi Arabia was "probably the greatest and richest oil field in all the world," and the British "never overlooked any opportunity to get in where there was oil." Ickes argued that the U.S. government should have a financial interest in CASOC's concession in order to safeguard it, and "there would probably never be a better time to do it than now."[15]

Meanwhile, the State Department's recommendation that Saudi Arabia be made eligible for lend-lease had made its way to the president's desk. The State Department had approved the Near East Division's recommendation and sent it to Lend-lease Administrator Edward R. Stettinius on January 9, 1943. Stettinius had forwarded the recommendation to the president three days later. No action was taken on the recommendation, however, until mid-February when the oil companies' lobbying efforts focused high-level attention on Saudi Arabia and its oil. After making sure that the British had been consulted and had no objection, President Roosevelt, on February 18, found "the defense of Saudi Arabia . . . vital to the defense of the United States," thereby clearing the way for lend-lease aid.[16]

Lend-lease for Saudi Arabia enhanced CASOC's position since it was a concrete expression of U.S. interest in Saudi Arabia's welfare. It also indirectly benefited CASOC by providing the king with another source of support, thus reducing British influence and relieving some of the pressure on the company to provide aid. What the companies had hoped to achieve with their February lobbying efforts was a definite commitment by the United States to maintain the security and well-being of their concession in Saudi Arabia. Lend-lease helped meet the king's needs, but the companies also needed money and markets to make CASOC a success. In any event, CASOC's owners soon learned that they had interested some people in the government in Saudi Arabian oil more than they had intended, for plans were soon afoot in Washington to buy the company. As Herbert Feis later wrote, the oilmen "had gone fishing for a cod and had caught a whale."[17]

The Petroleum Reserves Corporation

The State Department's newly created Committee on International Petroleum Policy had been closely following the situation in Saudi Arabia since mid-January, and had already concluded that CASOC's position should be preserved as a means of assuring U.S. access to foreign oil reserves in view of the prospective decline of domestic reserves.[18] In late March, the committee recommended to Secretary of State Cordell Hull that the government create a "petroleum reserves corporation," directed by representatives of the State, War, Navy, and Interior departments, to oversee and coordinate foreign oil policy. According to the plan, which had been developed by Oil Adviser Max Thornburg, the corporation would negotiate option contracts with U.S. oil companies holding concessions abroad giving the government the right to purchase large amounts of oil. In this way the government would obtain rights to billions of barrels of oil and the companies would gain increased security for their holdings. The government would have to pay for any oil it took, but the rights to buy the oil could be acquired without cost since the oil the companies would allocate for ultimate disposition by the government would be oil they probably would not be able to produce commercially during the life of their concessions. The committee recommended that the first of these contracts should be with CASOC, which was eager to negotiate an agreement.[19]

Despite concern that such a plan might stimulate other nations to enact similiar nationalistic measures, Hull approved the committee's report. On March 31, he sent copies of the report to the War, Navy, and Interior departments. Ickes, who was in the midst of a nasty battle with the State Department over Mexican oil policy, found the report a "wishy-washy document that did not seem to take any kind of position one way or other," and set it aside.[20] The Navy, however, reacted strongly. Captain Carter of the Army-Navy Petroleum Board, who had worked for Shell for many years before striking out on his own as an independent, criticized the report's option contract proposal pointing out that "accepting a few million, or even a hundred million barrels of petroleum products as a *quid pro quo* for guaranteeing what now appears to be billions to the oil companies would be dynamite for some public forum." Instead, Carter argued that "U.S. Government participation should be on a *quid pro quo* basis whereby the Government would assume a role similar to the one held by the British Government with respect to the Anglo-Iranian Oil

Company." Any lesser degree of participation by the government, he felt, would not adequately protect the public interest.[21]

Carter's concerns were echoed by Under Secretary of the Navy William C. Bullitt, scion of an old Philadelphia family who had served as Roosevelt's ambassador to the Soviet Union (1933–1936) and France (1936–1940). Bullitt enjoyed personal access to the president and drafted a letter to Roosevelt warning that the United States was running out of oil for the war effort and that at existing consumption levels would use up its known reserves in fourteen years. Therefore, *"to acquire petroleum reserves outside our boundaries has become a vital national interest"* (emphasis in the original). Bullitt attacked the State Department proposal that the government limit its participation to reserving an interest in oil in the ground as "contrary to the interests of the people of the United States" and as a "relapse into the dollar diplomacy of a dead era," since the plan would oblige the government to protect the company without corresponding benefit to the government. On the other hand, if the people of the United States were the owners of a majority interest in the reserves, the government would be justified in using the nation's resources for the defense of a vital interest of all the people of the United States. In addition, government participation in ownership and management was necessary to preserve the U.S. position in Saudi Arabia. Without strong government backing, the concession was in danger of being lost either in possible disorders after Ibn Saud's death, or to the British, who might "be able to lead either Ibn Saud or his successors to diddle them [CASOC] out of the concession."[22]

These concerns led Secretary of the Navy Frank Knox to write Secretary of State Hull on May 24 that the United States could "no longer safely leave its foreign oil security solely to the initiatives and resources of private enterprise." What Knox had in mind became clear a week later when Vice-chief of Naval Operations F. J. Horne submitted a plan to the Joint Chiefs of Staff recommending the establishment of a government corporation to acquire petroleum reserves outside the United States. Drawing on studies by the Army-Navy Petroleum Board, Horne warned that "before the end of 1944 the supply of crude oil will be insufficient to meet U.S. requirements for the Armed Services and the civilian economy." The nation also faced a long term oil shortage as consumption was outpacing the discovery and development of new reserves. U.S. proved reserves had decreased by 964 million barrels in 1942, and unless greater additions to reserves were obtained in 1943, the decrease would exceed a billion barrels. These statistics made it clear that corrective measures had to be taken immediately in the interests of national security. Horne therefore recommended that the proposed corporation, as its first act, acquire a controlling interest in Saudi Arabia's petroleum reserves. The

Joint Chiefs of Staff approved Horne's plan on June 8, and sent President Roosevelt a memorandum recommending the creation of a government corporation "specifically for the purpose of acquiring proven foreign oil reserves." This corporation should immediately take steps to obtain a controlling interest in Saudi Arabian oil.[23]

Meanwhile, Admiral William D. Leahy, Roosevelt's chief of staff, erroneously informed the president that CASOC had only a small part of Saudi Arabia's potential oil lands under concession, leading Roosevelt to toy with the idea of the United States obtaining an oil reserve directly from the king. Roosevelt explained his idea to Assistant Secretary of State Adolph A. Berle on June 10, pointing out that in order to make the idea appealing to the king, the United States could pay a flat rental on the land in addition to a flat royalty per barrel on any oil produced. Berle agreed that "this might save our lives" since it would provide the United States with the means to provide aid to Saudi Arabia and thus counter increasing British influence, but he pointed out that he understood that CASOC had a blanket concession to all the oil lands in Saudi Arabia. Secretary of State Hull confirmed Berle's understanding four days later, and warned that any negotiations that disturbed CASOC's concession could have adverse results. Faced with these objections, the president quietly dropped his plan.[24]

The Navy's actions alerted Ickes to the potential importance of the proposed petroleum reserves corporation as a means for asserting the national interest in foreign oil. Ickes began "stirring about to gang up on Hull," but because of the coal strike he was not able to give the matter his full attention until early June. On June 10, Ickes wrote the president that an impending shortage of oil supplies made it "imperative" for the United States to acquire a "proprietary and managerial interest" in foreign oil reserves. He urged the president to order the immediate creation of a petroleum reserves corporation for this purpose, to be headed by the secretaries of interior, war, and the navy. The corporation's first order of business should be the acquisition of a "participating and managerial interest" in CASOC's concession in Saudi Arabia. Acquisition was necessary not only to assure needed oil for the armed forces, but also "to counteract certain known activities of a foreign power which presently are jeopardizing American interests in Arabian oil reserves."[25]

Alerted to Ickes's and the Navy's plans by Secretary of War Henry L. Stimson, the State Department tried to regain the initiative. At a June 11 White House meeting, Feis explained that the department wanted to limit government involvement to the *quid pro quo* offered by CASOC— government rights to some underground oil and an option to buy it at a low price—and to seeking an international agreement on the exploitation of oil reserves and the free movement of oil in international commerce.

Feis stressed the delicate nature of oil questions in international relations and warned of the possibility of diplomatic repercussions and "unfortunate complications in Latin America which might result from direct government involvement in the oil business." In addition, Secretary Hull, an old Wilsonian and firm believer in free trade, had grave doubts about government ownership and, along with others in the State Department, felt that direct government involvement was contrary to the underlying intent of the Atlantic Charter and the idea of equality of access to raw materials. Hull reiterated these points in a June 14 memorandum to President Roosevelt which reminded the chief executive that "in many conferences after the last war the atmosphere and smell of oil was almost stifling." Therefore, it was essential that U.S. efforts in regard to oil "be so directed as to achieve our ends without stimulating new restrictive moves on the part of other countries and creating intense new disputes."[26]

Ickes and Knox, however, felt that U.S. control of adequate reserves of foreign oil was so vital to national security that they were willing to go against the U.S. tradition of private enterprise and to involve the government directly in the oil business. Therefore, they remained adamant that the situation required government participation in the ownership of U.S. companies possessing oil reserves abroad. Since the authority to set up a petroleum reserves corporation expired July 1, both sides agreed to the establishment of the corporation, leaving the question of its precise course of action for later decision.[27]

At a series of meetings representatives of the State, War, Navy, and Interior departments debated the divergent approaches, with victory going to the proponents of direct government involvement. On June 26, the committee submitted a report to President Roosevelt urging him to instruct the Reconstruction Finance Corporation to organize a Petroleum Reserves Corporation. The board of directors of the corporation, the committee advised, should consist of the secretaries of state, war, navy, and interior, and the corporation's first order of business should be to undertake negotiations for the purchase of 100 percent of the stock of the company holding the concession to Saudi Arabia's oil. The State Department's plan of merely acquiring an option on reserves made its way into the report as an alternative method of obtaining an interest in Saudi Arabian oil. The State Department was able to salvage some of its position as the committee recommended that all major projects had to receive the approval of the secretary of state and that negotiations with foreign governments were to be conducted by the Department of State.[28]

The committee's decision was basically pragmatic—the United States needed the oil and direct government participation was simple, straightforward, and understandable, and would be difficult to criticize as involving the utilization of public funds for private benefit. It gave the government full control of the reserves; the only cash payment involved would be the reimbursement of amounts actually expended by the companies; and the property should produce substantial revenues for the government. Moreover, the oil produced would be available to all companies on equal terms, and the government could even grant rights to operate the properties on the basis of competitive bids. The contract method was less likely to be criticized as government entry into the oil business and to stimulate restrictive measures by other governments, but there was the danger that such an arrangement, no matter how carefully it sought to protect the public interest, would be criticized as the use of the prestige, resources, and "perhaps even the armed forces" of the United States to provide profit for private capital. It would be difficult to calculate the amount of oil in the ground and so prevent the government from being left with little oil despite large profits to the company. And without control of the production and distribution of the company's oil, the government would be in the position of a junior partner with large oil companies with international ties and that, in their operations outside the United States, participated in quota and cartel systems.[29]

President Roosevelt approved the report, and on June 30, 1943, the Petroleum Reserves Corporation (PRC) was established with authority "to buy or otherwise acquire reserves of proved petroleum from sources outside the United States, including the purchase or acquisition of stock in corporations owning such reserves or interests therein." The president selected the stock purchase plan as the means of obtaining an interest in the oil of Saudi Arabia, and named Ickes head of the new corporation after Hull declined to serve. Ickes and Feis were chosen to oversee the negotiations with the owners of CASOC. Ickes quickly moved to strengthen the government's position in the negotiations by hiring former Under Secretary of the Interior and Texas oil lawyer Alvin J. Wirtz to represent the PRC in the negotiations. He also took over the responsibility for refinery construction in Saudi Arabia which the military had recommended; he felt that the project was so large—plans called for a refinery with a throughput of 100,000 barrels per day costing approximately $100 million—that it was certain to influence the company's attitude in the stock purchase negotiations.[30]

Negotiations for the Purchase of CASOC

On August 2, Ickes informed Texas Company head Rodgers and SOCAL President Collier that, in the interest of national security, the government wanted to buy CASOC. The government would reimburse their investment to date, and would pay them an overriding royalty on future production. He also told the "surprised and shocked" oilmen that the outcome of the purchase negotiations would have a bearing on the refinery project planned for Saudi Arabia. The oilmen had not expected the government to ask for all of CASOC, and Ickes's bold proposal "literally took their breath away."[31]

Rodgers and Collier recovered quickly, and for the rest of the meeting and at subsequent meetings argued that, if the government was concerned about oil for national security, the best course would be to support the rapid development of Saudi Arabia's reserves. Saudi Arabia was distant from the United States and probably impossible to defend in time of war. Development of its oil, however, would lessen the drain on nearby and defensible reserves in the Western Hemisphere. If the government felt that it had to have some reserves in Saudi Arabia, they were willing to consider long-term supply agreements or other arrangements of more moderate character. Rodgers and Collier claimed that their concession was so valuable as to defy measurement, and was in a sense a "reward" for the time, money, and effort their companies had expended in their worldwide search for oil. They planned to use Saudi Arabian oil to supply a refining and distribution network they were setting up in the Eastern Hemisphere and to support their position worldwide. Remuneration, therefore, would have to be generous and would have to include a high overriding royalty. The oilmen also questioned whether Ibn Saud would regard U.S. government ownership as a threat to his independence, pointed out that the move might provoke similar actions by the British and the Soviets, and argued that it would make other governments hesitant about granting concessions to U.S. companies. Finally, they warned that government ownership would be politically unpopular in the United States, where it would be seen as government entry into the oil business.[32]

The companies' adamant opposition to selling out led to a reassessment of the government's position. After the first round of meetings, Feis recommended that the PRC try for something less than 100 percent ownership, noting that he believed that the companies might be induced to sell a minority interest on reasonable terms. Davies, who exercised great influence on Ickes on oil matters, told his chief on August 16 that it had been a mistake to insist on 100 percent ownership, and that the government should try for no more than the simple majority interest that the British government held in AIOC. Davies also warned Ickes that he could find himself in "bad odor" with the oil industry since he had assured the industry that there would be no attempt at government control during the war. Even if this policy was maintained on the domestic front, the industry would regard the injection of the government into Arabian oil matters as an indication of a trend. After Bernard Baruch and James F. Byrnes gave him similar advice, Ickes met with President Roosevelt on August 30 and gained the president's approval to reduce the government's request to a minimum of 51 percent ownership. The PRC Board approved this change two days later.[33]

On September 8, Wirtz offered to lower the PRC's request from complete ownership to a controlling interest, to leave the operation of the properties in the hands of the present owners, and to assure SOCAL and the Texas Company of sufficient oil from the concession to carry out their expansion plans. Rodgers and Collier demurred, but indicated that they might consider selling the government a minority stock interest. Turning to the refinery project, which was already underway, the oilmen argued that, though the government would have to advance most of the money for the project, CASOC should retain ownership and be allowed to repay the government's advances over a long period at low rates. Although Feis supported the oilmen, Wirtz insisted that the government ought to have an interest in the refinery as well as the concession.[34]

Recognizing that the urgent military necessity of proceeding with the refinery project was putting the companies in a strong bargaining position, Ickes met with J. Frank Drake, president of the Gulf Oil Corporation, on September 10, and suggested that the government might be interested in acquiring stock in the concession Gulf shared with AIOC in Kuwait. Ordinarily a foe of government involvement in the oil business, Drake had already been in touch with the State Department about government assistance in freeing his company from the restrictions imposed on it by AIOC. Apparently concerned that his answer to Ickes's inquiry might influence the government's willingness to aid his company, he told Ickes that Gulf would consider the idea carefully. On September 24, Drake informed Ickes that Gulf was ready to "get down to cases" and work out a deal. In subsequent conversations, however, Drake asked for several hundred million dollars for sharing Gulf's half of the Kuwait concession with the government.[35]

Meanwhile, after gaining the support of the PRC Board to continue negotiations, Ickes attended the next meeting with SOCAL and Texas to demonstrate that the government "meant business." Ickes first emphasized the advantages of partnership with the government. All the capital needed to develop the property fully would be available, including funding for the construction of a pipeline from the oil fields to the Mediterranean. More important, the ability of the government to protect the concession remained in doubt as long as it remained a purely private interest. On the other hand, if "national safety" made it essential for the government to secure control of sufficient reserves to meet its future needs, he was "certain that the Government would seek sufficient power, legislative or otherwise, to safeguard itself." Fearing that if they did not agree to some sort of deal the government might try to take over CASOC as a war measure, Rodgers asked if Ickes would accept a one-third interest, provided an acceptable basis of valuation could be agreed upon. Ickes declined the offer, but agreed to consider a second offer by Rodgers whereby control over the oil produced would be in proportions different from stock ownership. Ickes assured the oilmen that regardless of how the stock was divided, the PRC was willing to let them operate the concession as a private commercial enterprise, reserving to the government only the right to insist on certain safeguards to protect the national interest.[36]

The PRC negotiators and the companies reached a tentative agreement on September 25 that would allow the PRC to acquire a one-third interest in CASOC for a price equal to the amount SOCAL and Texas had invested in the company. Before the rest of the details could be worked out, the Texas Company negotiators abruptly broke off negotiations charging that the proposed agreement was unfair since Texas had actually put twice as much cash into CASOC than SOCAL had. Infuriated, Ickes went directly to Rodgers and threatened to build the refinery elsewhere unless negotiations were resumed. This tactic worked, and on October 14, Ickes informed the PRC Board that an agreement was near. The terms called for the PRC to pay $40 million for a one-third interest in CASOC. Three of the company's nine directors would be from the government, and the government would have a preemptive right to all of CASOC's production in time of war or other national emergency and to 51 percent at all other times. The government could also block sales of oil or oil products to third parties if the national interest required. Otherwise the company would continue as a normal commercial operation. The PRC would advance CASOC the funds for the refinery and be repaid out of the company's future earnings minus "war costs" for the excess capacity beyond peacetime needs. Ickes emphasized that the deal gave the government what it was after—access to oil for national defense and control over the company's dealings with foreign governments and nationals.[37]

The "war costs" issue, however, led to a breakdown of negotiations. The Texas Company representatives insisted that at least 40 percent of the refinery's cost should be treated as "war costs" and not charged to the company. This demand added at least another $40 million to the government's costs and angered Ickes, who ordered negotiations terminated on October 15. Historian Irvine Anderson has argued that SOCAL and Texas were ready to make a deal, and that the falling out over war costs was due to Wirtz's failure to include these war costs in the government's offer of $40 million. It is not clear from the record, however, whether Wirtz was guilty of a negotiating error, or whether the Texas Company negotiators, in Ickes's words, "pulled another rabbit out of their hats," thereby delaying a decision while work on the refinery continued. In any event, the PRC Board approved Ickes's action on November 3, and resolved to cancel the refinery project in Saudi Arabia as well. Commodore Andrew F. Carter and General Walter B. Pyron insisted that government participation in foreign oil reserves was necessary to ensure U.S. access to the oil the nation needed, and convinced the board that the PRC should continue to consider an arrangement with CASOC at a reasonable price, as well as continue negotiations with Gulf for an interest in Gulf's Kuwait concession.[38]

Ickes later claimed that he broke off negotiations because SOCAL and Texas "were just pollyfoxing us and stringing us along," and had begun to lobby in Congress against "any deal of any sort." "I thought that the time had come when I wanted to call Mr. Rodgers's bluff, and I did call it." Moreover, word of the PRC's plans had leaked out, and the oil industry had begun mobilizing in opposition to government involvement in the oil business. The same day that he broke off talks with SOCAL and the Texas Company, Ickes had a long conversation with John A. Brown, president of the Socony-Vacuum Oil Company. Brown had learned of the negotiations and told Ickes that government ownership would stir up widespread opposition in the United States as well as encourage other oil-producing countries to enter the oil business. Brown maintained that from a national security standpoint government ownership was unnecessary, since U.S. oil companies would always cooperate with the U.S. government, especially in times of national emergency. The prospect of opposition from the oil industry, on whose support he depended for the functioning of the Petroleum Administration for War, coupled with the delaying tactics and high demands of SOCAL and the Texas Company, no doubt convinced the pragmatic Ickes that little was to be gained, and much to be lost, by continuing to press for government participation in CASOC.[39]

Brown's warning about industry opposition to government participation in the oil business was borne out on November 5, when the Foreign Operations Committee of the Petroleum Industry War Council (PIWC) released a report entitled "A Foreign Oil Policy for the United States." The basic thrust of this "program of principles and policies" was that the oil business should be left to private enterprise and that the government's role should be limited to providing an environment in which private enterprise could expand its operations. Concrete suggestions to achieve these ends included a relaxation of U.S. antitrust laws for foreign ventures; strong diplomatic support for U.S. companies abroad and encouragement of the Open Door; formalized industry-government cooperation; and coordination of the oil policies of the United States and other nations through the negotiation of an "International Oil Compact," modeled on the Interstate Oil Compact. As for national security, the PIWC argued that oil in the hands of U.S. nationals was "equally available for national security" as oil owned by the U.S. government.[40]

Ickes tried to convince the PIWC that a measure of government participation was necessary to ensure public support for whatever measures might be needed to protect U.S. access to foreign reserves, but his arguments fell on deaf ears. On December 9, the PIWC passed a resolution stating that the U.S. government "should under no circumstances acquire title or ownership or directly or indirectly engage in foreign oil exploration, development or operation." Although SOCAL, Texas, and Gulf dissociated themselves from these actions of the PIWC, negotiations with the first two on government participation were never resumed, and Drake continued to insist on such a high price that nothing could be worked out with Gulf.[41]

State Department Counterattack

While Harold Ickes and the Navy were busy transforming the Petroleum Reserves Corporation into something quite different from what Max Thornburg had envisaged, the State Department began examining what might be accomplished through an agreement with the British on Middle East oil. Foremost among U.S. objectives was the preservation of the U.S. concession position, closely followed by the removal of the various restrictions that, in the eyes of the United States, hampered development of the Middle East's vast reserves. In a May 1943 paper, Thornburg argued that the Red Line Agreement, which prohibited the companies making up the Iraq Petroleum Company (IPC) from engaging in oil development within the area of the old Ottoman Empire except as part of IPC, should be abolished as an "obsolete relic of a Secret Society which never had any other purpose than to tie the hands of its minority members." Thornburg also recommended that the 1928 As-Is (or Achnacarry) Agreement, which, in effect, divided world markets among the major companies, be terminated and that whatever agreements were necessary to stabilize prices and markets be openly arrived at and have the endorsement of responsible government agencies.[42]

The British were also interested in reaching an agreement on the development of the Middle East's vast resources, though for somewhat different reasons. In August 1943, Basil Jackson, AIOC's representative in the United States, told James Terry Duce, head of the Petroleum Administration for War's Foreign Division and a former official of CASOC, that AIOC was concerned about "such enormous quantities of oil overhanging the markets of the world." AIOC was determined to expand its markets after the war and was ready to cut prices to do so. For expansion to take place in an orderly manner, the U.S. and British governments would have to make arrangements to regulate production in the Middle East. According to Jackson, government action was needed because the companies on their own were not able to reach an agreement. There were too many conflicting personalities, and the U.S. companies were restricted by the antitrust laws.[43]

Jackson reported to London after the meeting that the time was "most opportune" for talks with the United States. U.S. oilmen were well aware of the "economic and political evils" of uncontrolled competition, but they were reluctant to act until the U.S. government had clarified its position on the type of cooperative agreement that would be necessary. Neither the U.S. government nor U.S. companies had formulated clear policies on foreign oil, but there were signs that the United States might be planning to take unilateral action to look out for its interests in the Middle East. Thus, it was imperative that talks on cooperation begin soon.[44]

U.S. oil companies were concerned about the situation in the Middle East for similar reasons. At the beginning of September, the president of Socony-Vacuum wrote State Department Economic Adviser Herbert Feis that intergovernmental agreements on Middle East oil were needed to assure "orderly development" of the region's resources. Middle East reserves were at least 18 to 30 billion barrels, far more than present markets east of Suez could absorb. Middle East oil would soon form part of Europe's oil supply, and the extent to which it did so would reduce the drain on Western Hemisphere reserves. To manage this transition in an orderly fashion would require agreements among governments, private producers, and private distributors on the rate and source of production from existing and new fields. Without such agreements the tremendous excess of supply over any existing or probable demand would result in price wars that could reduce the income of the producing countries "almost to the vanishing point," and result in the wasting of irreplaceable natural resources and needless duplication of capital investments.[45]

Drawing on these thoughts, the department's Committee on International Petroleum Policy, on September 16, proposed the establishment of a joint petroleum board to coordinate U.S. and British policies on Middle East oil matters. The board would make recommendations to the two governments on oil policy and help ensure that Middle East oil would be developed in accord with "sound conservation practices." The committee also recommended that the two governments agree to include other nations and that they work toward an international agreement to deal with oil on a worldwide basis. The State Department believed that cooperation with the British would provide stability for U.S. oil interests in the Middle East since it would remove British incentives for undermining the U.S. position. Without an agreement, there could be sharp conflict over marketing of Middle East oil. Markets, not concessions, were the basic problem since all concession holders in the Middle East had more oil than could be produced in the life of their concessions. An agreement with Britain would not only eliminate the restrictions hemming in the development of U.S.-owned oil, but could also be used to ensure adequate reserves for a postwar security system.[46]

A tour of the Allied war zones by five U.S. senators in the summer of 1943 underlined the potential for oil matters to lead to disharmony in Anglo-American relations. On their return to the United States, the senators charged that the British were husbanding their Middle East oil reserves to "win the peace" while the United States was exhausting its reserves "oiling" the Allied war effort. PAW and the State Department quickly produced reports countering the charges, and President Roosevelt held a press conference to assure the nation that the situation was "well in hand." Allied war decisions were made on the basis of how quickly and efficiently oil could be supplied to where it was needed. With nearly three-quarters of world productive capacity and 70 percent of available refinery capacity, the United States naturally supplied most of the oil used by Allied forces. Despite its great potential, the Middle East, where most of Britain's reserves were located, lacked facilities to produce more than around 5 percent of total world output. Moreover, Middle East reserves had been largely cut off from the major theaters in the early years of the war because of Axis control of the Mediterranean and a shortage of tankers. Despite the administration's quick handling of the matter, the incident showed how easily oil matters could feed traditional Anglophobia and strain relations with Britain.[47]

The formation and activity of the PRC increased British concern over U.S. intentions regarding Middle East oil. Harold Wilkinson, the British government's petroleum representative in the United States, summarized these concerns when he noted his fear that the PRC would be a "Washington Santa Claus bidding for . . . concessions with the lush money bags of the Treasury and Fort Knox." In late October, Geoffrey Lloyd, British secretary for petroleum, recommended to Prime Minister Churchill that the British government invite the United States to discuss postwar oil policy. There was more than enough oil in the Middle East for everyone and British policy should "aim at an orderly development with considerable intertwining of British and American interests." Such a strategy would help assure both nations of access to the oil they needed, minimize the potential for disagreements, and "might also assist us in diverting American minds from the unduly simplistic plan of helping themselves by taking from us." The Foreign Office also favored greater cooperation between the United States and Britain in the Middle East. Expansion of U.S. oil interests would give the United States a "highly strategic entanglement" in the Middle East, and thus provide a counterweight to Soviet influence. In addition, the Foreign Office felt that cooperation on oil matters was essential to overall collaboration with the United States on economic matters. Churchill, however, was reluctant to take the initiative, and decided that Britain should await a formal approach by the United States.[48]

The State Department was eager to begin talks with Britain on oil. Outmaneuvered by Ickes on the PRC, the department began to see an agreement with the British as a means of reasserting its leadership in foreign oil policy. The State Department argued that government partic-ipation in U.S. companies holding foreign reserves was a "highly questionable, not to say dangerous step." The best way to increase U.S. security was through the "orderly development" of Middle East oil. This would in turn reduce the drain on Western Hemisphere, including U.S., reserves. From the U.S. viewpoint, any accord would also have to deal with the various restrictive agreements and governmental practices that hampered the operations of U.S. companies in the Middle East. State Department officials felt that the resources of the area were "too important for their development to be permitted to remain subject to existing burdens," but they recognized that a solution would have to be found to the "fundamental problem of the disposition of larger quantities of Middle Eastern oil without disorganizing the markets for that oil." The British had been controlling distribution by favoring Iran while holding Iraq and Kuwait back. Any agreement with the British to end restrictions on U.S. companies would therefore have to provide for a new method of control.[49]

These goals could be achieved only through cooperation with the British, who were firmly entrenched in the Middle East, politically as well as economically. In the State Department's view, it was beyond the power of U.S. companies to remove the restrictions placed on them by the British, so it was necessary for the U.S. government to intercede with the British on behalf of the companies. Without an agreement there could be a commercial struggle over oil that would endanger U.S. interests in the Middle East. In contrast, an agreement would allow the United States to consolidate its position in an area hitherto dominated by the British and to end British commercial and political restrictions on U.S. companies. The British could be persuaded to go along with the United States because of U.S. financial power, control of oil tankers, and generally strengthened world position. Discussions should begin soon because of indications that the British were "utilizing present developments to achieve long-range ends."[50]

Ickes's problems with the PRC gave the State Department an opportunity to regain the initiative on foreign oil policy. On December 2, Secretary of State Hull informed the Foreign Office that the United States would welcome informal and preliminary talks with Britain on Middle East oil, and recommended that because of the importance of the subject the talks be undertaken promptly. Hull informed President Roosevelt of his action on December 8, and emphasized that the State Department should be in charge of the talks. On December 14, Wallace Murray, the department's senior Middle East specialist, recommended that the department advise the PRC that no further negotiations regarding U.S. government participation in U.S. companies be held until conversations with the British on Middle East oil could be held. An agreement with Britain could "accomplish on an international basis ... the purposes the Petroleum Reserves Corporation's negotiations with the oil companies were designed to further by unilateral action." Cooperation with Great Britain on Middle East oil and a postwar security system based on Atlantic Charter principles would protect U.S. interests, so government ownership was not necessary. Accordingly, Hull recommended to the president that negotiations by the PRC looking to government participation in companies holding foreign oil reserves be held in abeyance, while negotiations on oil were conducted with the British.[51]

Ickes strongly disagreed with Hull's recommendation and argued that government participation would strengthen the U.S. position in any negotiations with the British. Moreover, Ickes thought he should be in charge of any negotiations. The State Department could not handle the job: "They are a bunch of cookie pushers. And now they are so befuddled they don't even know how to push cookies." While conceding that he was glad the State Department had "started the ball rolling" with the British, Ickes wrote the president that he should be included in any negotiations. He was head of PAW and the PRC, and had the best oil staff in town. Moreover, "in a sense, this is my baby."[52]

Tactics as well as jurisdictional issues were involved. Although Hull and Ickes agreed that talks should be held in Washington and should lead to an agreement along the lines of the State Department's September draft, Hull wanted exploratory talks among "technicians" to precede a cabinet level conference. Ickes, in contrast, stressed the need for fast action, and proposed a cabinet-level conference, with himself leading the U.S. negotiating team. Fearing that foreign oil could become a partisan issue in an election year, Roosevelt decided in favor of Ickes and quick action. He named his petroleum administrator chairman of the U.S. delegation, which was also to include the under secretaries of state, war, and navy, and the deputy director of the War Production Board. Hull, however, dug in his heels and forced the president to change his decision. Hull, not Ickes, would lead the cabinet-level delegation, though Ickes would be included as co-chairman. Roosevelt sided with Ickes on the matter of the PRC. On January 10, the President asked Hull and Ickes to "get together and straighten out this problem." Though negotiations with the British were to be under State Department control, Ickes was to continue his negotiations with the companies to see where the United States stood before taking the matter up with the British. "Time is important, because after the war the American position will be greatly weaker than it is today. Can't we agree on a policy and a method of putting it into effect?"[53]

Divisions within the government complicated efforts to develop and implement policy on Middle East oil. Underlying the conflicts over public policy, however, were divisions within the oil industry. The impact of the structure and power of the U.S. oil industry on foreign oil policy became clearer as oil issues emerged from the confines of intragovernment debate into public forums in 1944.

Chapter 3

Obstacles to a Liberal Petroleum Order

Ickes and the military advocated direct government ownership of foreign oil reserves as a means of assuring the United States of the oil it would need in the future. After the demise of the purchase plan they proposed a government-owned pipeline from the Persian Gulf oil fields to the Mediterranean to protect U.S. interests and provide the tangible stake in Middle East oil necessary for public support. This plan also ran aground because of stiff opposition from the oil industry. As an alternative to government ownership, the State Department proposed an oil agreement with Great Britain to regulate the development of Middle East oil and lessen the drain on U.S. and Western Hemisphere reserves. Although the British were willing to cooperate with the United States, the domestic U.S. oil industry proved less amenable to designs for a liberal world petroleum order.

The PRC Pipeline Project

In November 1943, the Petroleum Reserves Corporation had sent a special Technical Mission to the Middle East to survey the region's oil resources. The mission's report in early 1944 painted a glowing picture of the region's oil future: "The center of gravity of world oil production is shifting from the Gulf-Caribbean region to the Middle East—to the Persian Gulf area—and is likely to continue to shift until it is firmly established in that area." The report set the indicated and proved reserves of the region at 26 billion barrels, with 9 billion barrels in Kuwait and at least 5 billion barrels in Saudi Arabia. U.S. proved reserves at this time were just over 20 billion barrels, and world reserves approximately 50 billion barrels.[1]

Commodore Andrew F. Carter, executive officer of the Army-Navy Petroleum Board and a member of the PRC's board, had accompanied the Technical Mission to the Middle East and consulted with oil company officials in the Middle East and in the United States. In an effort to regain the initiative for the PRC in oil matters, Carter recommended on January 17, 1944, that the PRC construct a pipeline from the Persian Gulf oil fields to the eastern Mediterranean. In return, the government would receive rights to a specified amount of oil at a fixed and favorable price, and an option on a billion barrels of oil for the life of CASOC's concession. Carter argued that this proposal would ensure that Saudi Arabia's oil would not "by one technicality or another, slip through the fingers" of the U.S. interests. Carter viewed his plan as a compromise between those advocating government ownership and those who wanted government involvement limited to backing the companies because it would preserve the principle of a government "equity" in the Persian Gulf area, while avoiding a direct clash with opponents of government participation in the oil business.[2]

Ickes liked Carter's idea. He felt that a government-owned pipeline was "practically as good as an interest in the oil concession" as a means of assuring a measure of U.S. government control over the region's oil resources. "If we build the pipeline . . . and run it, it will mean that the government will have a larger measure of control than it has ever had of any private industry, because without transportation, the oil is worthless." Moreover, the pipeline would give the United States a concrete interest in Middle East oil, thus making it politically possible for the United States to fight for its national interests as the British government fought for British interests. In addition, the project would help block further British encroachment on U.S. interests in the Middle East and strengthen the U.S. hand in negotiations with the British by giving the United States a "weapon with which to beat the British over the head." Ickes was also not unaware that the project would strengthen the PRC's hand in its fight with the State Department.[3]

Ickes's deputy, Ralph K. Davies, however, opposed the idea and warned Ickes that it was "certain to raise hell in the industry and on the Hill . . . and quite likely with the State Department." Davies felt that the plan was "half-baked," and predicted that the oil industry would view it as the entering wedge of government involvement in the oil industry. The whole industry, not just the big companies, would join the fight. To make matters worse, the central purpose of the plan—"American protection against British connivance"—could not be used to defend the plan without stirring up prejudices against an important ally in the midst of a war. The result would be defeat for Ickes, and the State Department would take control of foreign oil policy. Rather than risk defeat, Davies recommended that the problem of British competition be dealt with through "forceful" conversations with the British leading to an Anglo-American understanding on oil.[4]

Ickes ignored Davies's warnings. To give SOCAL and Texas extra incentive to cooperate, he dropped his objections to the refinery the Arabian partners were planning. The companies were eager to cooperate. Government construction of the pipeline would save them the expense of building one themselves, provide low-cost access to European markets, and enhance the security of their investments by giving the United States a concrete interest in the stability of the Middle East. U.S. government involvement was necessary, ARAMCO Vice-president James Terry Duce later explained, because "the partners of everybody else who operates in the Near East are the government of Great Britain and the government of France." After a week of hectic negotiations an agreement was reached on January 24. According to the terms, the PRC would build, own, and maintain an oil pipeline from the oil fields in Saudi Arabia and Kuwait to an eastern Mediterranean terminus. Charges for the use of the pipeline were to be sufficient to cover maintenance and operating costs and to amortize the entire investment over twenty-five years. In return, the companies were to set aside a crude oil reserve of 1 billion barrels for use by U.S. military forces, with the government having the option to buy this oil at 25 percent below market prices. To insure that the oil produced by the companies would be sold in conformity with U.S. foreign policy, the agreement stipulated that the companies had to give the State Department prior notice of sales to foreign governments and that no sales of petroleum or petroleum products would be made by the companies to any government or the nationals of any government when, in the opinion of the State Department, such sales "would militate against the interests of the United States."[5]

The plan won wide support within the government. The Joint Chiefs of Staff unanimously endorsed the project on January 25, noting that in view of the inadequacy of domestic oil reserves and the potential of the Middle East "the assurance that a portion of the oil in place will be maintained as a military reserve is of utmost importance to the U.S. Army and Navy." James F. Byrnes, director of the Office of War Mobilization and Roosevelt's "assistant president," wrote the president that the project would protect U.S. interests in Saudi Arabian oil without involving the government directly in the oil business, and should be supported. The State Department, however, resisted the project and insisted that the pipeline and other facilities be turned over to private industry after the war. A compromise was reached whereby the PRC Board agreed to review the question of the pipeline's ownership after the war. This satisfied Secretary of State Hull, and on January 31, he gave his approval to the project. President Roosevelt gave his "OK" on February 3, and three days later details of the agreement were released to the press.[6]

A storm of opposition greeted the announcement of the pipeline project. Far from bringing "a favorable response from 10 Downing Street," as Ickes later claimed, the project almost scuttled the proposed oil talks. British reaction, both in the press and in Parliament, was bitter, and the government was embarrassed by questioning in the House of Commons that revealed that it had not been consulted about U.S. plans. Particularly galling to the British was the fact that the United States had continually refused to release materials to them to increase the capacity of the pipelines from the Iraq Petroleum Company's fields to the Mediterranean. It was clear that the United States was planning to increase the production of U.S. oil in the Middle East at the expense of Britain, and it was suspected in some quarters that the pipeline plan was only part of a broader plan of U.S. penetration into the Middle East. The leading critic of collaboration with the United States, Lord Privy Seal Beaverbrook, a self-made millionaire newspaper magnate, argued on February 11 that "oil is the greatest single post-war asset remaining" to Britain, and it "should refuse to divide [its] last resource with the Americans."[7]

On the other hand, some officials in the Foreign Office and in the British Admiralty and Chiefs of Staff saw advantages for Britain if the United States became more involved in the Middle East. Foreign Secretary Anthony Eden reminded his cabinet colleagues that Britain depended on the United States for wartime oil supplies and for equipment necessary for its oil industry, and that many other economic questions were pending between Britain and the United States. "In these circumstances to deny the United States Government an opportunity even to discuss oil problems would almost certainly prejudice the development of our oil industry after the war . . . and would also play into the hands of all enemies of Anglo-American cooperation."[8]

Eden's views won out, but the British insisted that the talks be worldwide in scope and that they be held in London. They would agree to talks on the Middle East alone only if the discussions were of an exploratory nature and on the technical, as opposed to the ministerial, level. The British also insisted that existing concession rights be excluded from any discussions. On February 16, the influential Truman Committee had issued a report that complained that the United States had borne the burden of "oiling the war," and suggested that the British might transfer some of their concession rights to the United States as payment for lend-lease assistance. Oil was a leading source of foreign exchange and one of Britain's most important overseas assets, and the British had no intention of giving up any concession rights.[9]

These opposing points of view clashed in a series of exchanges between President Roosevelt and Prime Minister Churchill which laid the basis for an eventual agreement on Middle East oil. On February 20, Churchill expressed his desire to avoid a dispute over oil but noted apprehension in Britain that the United States was seeking to deprive Britain of its oil assets in the Middle East. Great Britain, he added, did not like to be pressured into talks. Roosevelt replied two days later that because of reports from the Middle East the United States was concerned that the British were trying to "horn in" on U.S. holdings in Saudi Arabia. The conflict demonstrated the need for high-level discussions. Two days later,

Churchill replied that he did not like the way the president's telegram preempted discussion, and that the cabinet was "very much concerned at the apparent possibility of a wide difference opening between the British and United States governments." Roosevelt smoothed over the differences in a private telegram on March 3 in which he assured the British leader that "we are not making sheep's eyes at your oil fields in Iraq or Iran." Churchill, on March 4, thanked Roosevelt for his assurances and disclaimed any intention of seeking to "horn in" on the U.S. position in Saudi Arabia. Churchill pointed out, however, that though Britain was seeking no advantage from the war "she will not be deprived of anything which rightfully belongs to her after having given her best services to the good cause—at least not so long as your humble servant is entrusted with the conduct of foreign affairs." Accordingly, he insisted that the talks start on the expert level, and, as he informed the president on March 6, that there be "no question . . . of any transfer of existing rights or properties in oil." Roosevelt agreed, and preliminary talks at the expert level were scheduled for mid-April in Washington, with higher-level talks to follow.[10]

The U.S. oil industry proved harder to assuage. Already up in arms about government involvement in the oil business, the PIWC had, on February 2, passed a resolution calling for the abolition of the PRC. Joseph Pogue, oil economist of the Chase National Bank and a prominent industry spokesman, charged that the pipeline plan put the government into the oil business, and complained that the companies involved had been forced to pay price rebates and to surrender freedom of managerial control for protection to which they were entitled as U.S. citizens. In addition, the project would involve the United States in "international complications and possible future armed conflict." The PIWC condemned the project on March 1, and distributed over 90,000 copies of a lengthy critique of the project prepared for the Independent Petroleum Association of America (IPAA) by George A. Hill, Jr., president of the Houston Oil Company. Hill denounced the proposed plan as "fascist" and charged that it represented the "surrender of the freedom of the American system of free enterprise that we are now fighting to preserve."[11]

These sentiments were genuine, but behind them lay the issue of favoritism. Many feared that the pipeline would give the companies involved an unfair advantage over their competitors. Jersey and Socony, which shared in the existing pipeline from the Persian Gulf to the Mediterranean, were concerned that the pipeline would put Arabian oil into the European market in competition with their Caribbean and Iraqi oil. (At this time, oil from the Caribbean and the Gulf of Mexico accounted for approximately 73 percent of the European market whereas oil from Iran and Iraq made up around 23 percent.) Domestic producers feared that the U.S. market could be flooded with foreign oil as Arabian oil took away European markets for Caribbean oil. In general, industry analysts feared that the pipeline would lead to disequilibrium in world oil markets as it made too much oil available too soon for the European market to absorb. Thus, interest as well as ideology motivated industry criticism of the pipeline plan.[12]

Determined opposition to the project also arose in Congress, where the anti–New Deal conservative coalition was at the peak of its effectiveness. On January 21, Senators E. H. Moore of Oklahoma and Owen Brewster of Maine had introduced a resolution calling for the abolition of the PRC. Brewster, a long-time critic of the New Deal, immediately attacked the plan as a "radical reorientation of our foreign policy." Moore, an independent oilman and ex-Democrat, accused "White House planners" of "visualizing the socialization of the oil business and of other basic industries." Later, in a debate with Ickes on the radio program "Town Meeting of the Air," Moore voiced the fears of the oil industry and business in general when he stated that "every phase of American private enterprise is fearful that this bold move is the beginning of an offensive on all business."[13]

Press comment was also largely negative. *Newsweek* ran a story that repeated almost verbatim criticisms of the project circulated by the PIWC. Other papers seized on the theme of government entry into business and the implications of growing involvement in the Middle East. The liberal daily *PM*, in contrast, complained that the deal promised hundreds of millions of dollars in profit to the companies while risking the use of U.S. military power to protect private investment. According to *PM*, the project was a "vicious and unjustifiable sellout to the oil companies that will endanger American peace and the economic future of the world." On the other hand, Roscoe Drummond of the *Christian Science Monitor* compared the pipeline to the Panama Canal, and argued that rather than putting the government into the oil business it was a method of keeping the government out of private business by enabling the companies to transport the oil directly to where it was most needed.[14]

More measured criticism came from Herbert Feis, who had left the State Department at the end of October 1943. In March, Feis published a study entitled *Petroleum and American Foreign Policy* in which he warned against the United States becoming too deeply involved in the "maze of interests, economic and political, that existed in, or centered on, the Middle East." Speaking before the Foreign Policy Association on April 15, Feis returned to his plan of an international agreement on oil and declared that the pipeline plan was fraught with dangers to world peace unless other countries were brought in.[15]

Carter and Ickes defended their brainchild vigorously. Carter felt that the oil industry was unduly concerned about the "ghost of governmental entry into the oil business." He explained to the head of the PIWC that the PRC was only concerned with oil matters outside the United States.

It was not intended to compete with private business, but rather to promote private control of the oil industry. Ickes also denied that the plan would constitute an "entering wedge" of government control. He told the head of the Independent Petroleum Association that it was "designed and intended" to assist the U.S. companies operating in the Middle East because the "future security and welfare" of the nation were dependent on the continued availability of Middle East oil. Ickes pledged to oppose any efforts to impede or prevent completion of "this vital undertaking." Secretary of the Navy Frank Knox also strongly defended the project as a "military necessity" and criticized the "selfish oil companies" that opposed the plan.[16]

The government position, however, was far from unified. The State Department was well aware that the project strengthened Ickes's hand in foreign oil matters and there was a strong measure of opposition to the project throughout the department. In addition, the controversy aroused by the project brought to the surface latent tensions between Ickes's top advisers Deputy Petroleum Administrator Ralph K. Davies and Under Secretary of the Interior Abe Fortas, Ickes's thirty-four-year-old "administrative alter ego." On April 8, Fortas withdrew an earlier recommendation that all petroleum activities be consolidated under Davies. Referring to Davies's "intellectual and moral commitments to the oil industry," Fortas charged that Davies would not support any program which was likely to meet with the opposition of the industry. Such a person, Fortas concluded, should not be Ickes's principal adviser on oil policy. Ickes, however, backed Davies. To compound matters, Carter, flush with success at coming up with the pipeline scheme, mounted a drive to displace Ickes as head of the PRC. Though Ickes was able to fight off this challenge, it distracted him from more serious business and split the ranks of the project's defenders.[17]

As a result of the intense opposition the project never came to fruition. "By seizing on the issue of government versus private ownership," an astute analysis in *Fortune* magazine pointed out, the pipeline's critics "caused the project to be thrown into the furious arena where the Administration's whole approach to domestic economic affairs is being debated." A Special Senate Committee Investigating Petroleum Resources, set up on March 13 in response to a motion by Senators Moore and Brewster, held eleven executive session hearings on the pipeline plan between March 31 and June 13. The committee was packed with critics of the plan and pressed Ickes to promise that the PRC would not enter into any binding agreement on the project without giving the committee thirty days' notice. Ickes refused, pointing out that the PRC was operating legally and it would set a bad precedent to ask for congressional approval before taking action it was entitled to perform. President Roosevelt, however, acquiesced after the committee's chairman threatened to hold open hearings on the pipeline and on negotiations with the British. On June 12, the president wrote the committee that although he could legally put through a contract without notice, he was asking Ickes to give the committee thirty days' notice before entering into any final contract. After Ickes failed to get the Joint Chiefs of Staff to declare the project a military necessity, the plan was postponed until after the completion of talks with the British. It was never revived, and in 1945 SOCAL and Texas decided to go ahead with the pipeline on their own.[18]

The Anglo-American Oil Agreement

Meanwhile, preparation for oil talks with the British had continued. In early April, an Inter-Divisional Petroleum Committee set out a "Foreign Petroleum Policy of the United States" which called for a "broad policy of conservation of Western Hemisphere petroleum reserves . . . in the interests of hemispheric security," to be made possible by the expansion of Middle Eastern production. U.S. policy should seek to assure the United States of "a substantial and geographically diversified holding of foreign petroleum resources in the hands of United States nationals." According to the paper, this policy "would involve the preservation of the absolute position presently obtaining, and therefore vigilant protection of existing concessions in United States hands coupled with insistence upon the Open Door principle of equal opportunity for United States companies in new areas." To effectuate these policies, the committee endorsed the negotiation of an agreement with the British on the broad principles governing petroleum development and distribution. The agreement should provide for the establishment of a joint Anglo-American petroleum commission to regulate the production and distribution of Middle East oil. This commission could schedule exports, protect the interests of the producing countries, abolish the existing political and commercial restrictions on the operation of U.S. companies, and look out for the security needs of the two countries.[19]

The idea of a petroleum commission to oversee the development of Middle East oil was at the heart of the draft "Memorandum of Understanding" presented by the U.S. negotiating team to the British delegation when oil talks began on April 18, 1944. The State Department had originally envisaged the commission as a means of facilitating the prompt and orderly expansion of Middle East production in order to conserve Western Hemisphere and especially U.S. reserves. Under the influence of Ralph K. Davies, who was appointed vice-chairman of the U.S. group, the focus of the commission was shifted to restraining and allocating Middle East production during the period following the war, when worldwide productive capacity was expected to exceed demand by a large margin. Davies explained to the British on April 20 that production in 1946 would exceed 1938 production by 37 percent whereas demand would probably be the same because of the effects of the war. To bring supply into line with demand, production in selected areas would have to be cut back to "tolerable levels." Production could be cut back in Iran, Iraq, and Venezuela, held steady in the Netherlands East Indies, but allowed some "necessary expansion" in Saudi Arabia and modest increases in Kuwait and Qatar. By 1950 world demand would have recovered sufficiently to assimilate expanded production in the Middle East: Western Hemisphere production and consumption would be in balance, and markets outside the Western Hemisphere would be supplied from the Middle East.[20]

Although the British could not have been pleased with the way Davies proposed slicing the pie, they were well aware that cooperation in some form or another was needed to prevent the negative economic and political impacts of unbridled competition. Therefore, the only change they proposed was to limit somewhat the power of the petroleum commission, which in the U.S. draft was given limited executive authority to "implement and execute" its recommendations once they had been approved by the two governments. The British group felt strongly that it would be inadvisable to give the commission more than advisory and consultative functions, arguing that the commission's authority would, in any event, be based on the confidence the oil industry had in its decisions. The U.S. delegation agreed to this change and to deleting specific mention of the Middle East from the description of the commission's functions.[21]

The Memorandum of Understanding agreed to on May 3 provided for the establishment of a Joint Petroleum Commission composed of representatives of the two nations. Its chief function would be to guide the operations of the international petroleum industry by preparing estimates of demand for petroleum, and by suggesting the manner in which the estimated demand could best be met by "production equitably distributed among the various producing countries" in accordance with such criteria as "available reserves, sound engineering practices, relevant economic factors, and the interests of producing and consuming countries." The commission would also deal with short-term problems of joint interest relating to the production, processing, transportation, and distribution of petroleum on a worldwide basis. The commission was to make reports and recommendations to the two governments on how to further the "efficient and orderly" development of the international petroleum industry, and the two governments agreed that "upon approval of the recommendations of the Commission they will endeavor, in accordance with their respective constitutional procedures, to give effect to such approved recommendations and, whenever necessary and advisable, to ensure that the activities of their nationals will conform thereto."[22]

Two areas of major disagreement had to be reconciled. The U.S. draft called for respect of existing valid concession contracts and acquired rights and observation of the principle of equal opportunity in the acquisition of new rights. The British, still upset over the way the United States had settled unilaterally with Mexico in 1941, wanted this clause strengthened to state that the two governments would not merely respect all valid concession contracts but would also support them and see that they were observed. Although this recommendation would apply only to such cases as Mexico, Venezuela, or Iraq where companies of both nations were involved, the U.S. delegation felt that it would involve commitments of such scope and duration that it raised questions of general foreign policy beyond the delegation's authority. Therefore the British group agreed to let the language of the U.S. draft stand until the question could be addressed at the cabinet-level talks.[23]

The British also felt that there should be appropriate recognition of their position as a nation entirely dependent on imports for their oil supplies, and proposed that the Atlantic Charter principles of the agreement be qualified not only by national security, as proposed by the United States, but also by "industrial and commercial well-being." The head of the British delegation pointed out that the "position of the United States as a very large producer with ample indigenous supplies was a sheltered position." Britain wanted to assure itself of an adequacy of supplies comparable to that enjoyed by the United States because of its natural endowments. Although the oil industry representatives advising the U.S. delegation were inclined to agree to the British request, the State Department vigorously opposed it, pointing out that the British proposal was contrary to the principle of equal access and could have adverse effects in other commodity areas. The U.S. delegation was able to convince the British to withdraw their reservation by assuring them that the United States understood that Britain might want to draw its supplies from its own companies. On the basis of this assurance, which was incorporated into the official minutes, the British group accepted, for the time being, the language of the U.S. draft.[24]

Although the British delegation agreed to the U.S. draft Memorandum of Understanding, opposition to the agreement within the British government delayed the selection of the British cabinet-level delegation. In addition, Prime Minister Churchill preferred to postpone the final talks until after the U.S. elections. President Roosevelt, however, was anxious about the political impact of delay, and wrote Churchill on June 7 that he hoped that the British would send their delegation soon, "as the situation is becoming embarrassing." On June 16, the day after the Special Senate Committee Investigating Petroleum Resources announced that the pipeline project had been deferred until after the completion of Anglo-American discussions, the War Cabinet authorized ministerial talks for the completion of the oil agreement. In a move designed both to quiet critics of an agreement and to ensure the best possible deal for Britain,

Churchill named Beaverbrook head of the British delegation. Three members of the ministerial oil committee were sent along to keep Beaverbrook in line. Although Beaverbrook had been a leading opponent of an agreement, his political fortunes had been declining, and, in the opinion of the U.S. Embassy in London, "the Beaver could not afford to return without an agreement."[25]

Opposition to the agreement was already forming within the United States. In order to ensure oil industry support for the agreement, the State Department had conferred with a cross section of industry leaders before the talks with the British, and had selected ten oilmen to advise the U.S. delegation. Three members of this group were allowed to attend meetings, and the larger group had access to all drafts of documents and participated in the formulation of the U.S. position. Their participation failed to gain the industry's support, however. Suspicious of the government's motives because of the activities of the Petroleum Reserves Corporation, many oilmen feared that adoption of the agreement presaged a body of special restrictive legislation curtailing the power of the oil industry. The industry advisers, especially the representatives of companies with little or no foreign production, wanted to insulate the United States completely from the coverage of the agreement. They were also concerned about the powers of the proposed petroleum commission, which they feared could exercise extensive control over the oil business. To protect itself, the industry wanted a "full vote and voice" on the proposed commission. George A. Hill, Jr., president of the Houston Oil Company, who had emerged as a prominent spokesman for the independent oil companies with his pamphlet against the pipeline, feared that the agreement would obligate U.S. nationals to conform to its provisions and that it would provide the U.S. government with authority to enforce those provisions. Although the State Department pointed out that the agreement gave the government no new powers, the oilmen were not satisfied.[26]

Industry's concerns were shared by Congress, which was also suspicious of the government's intentions because of Ickes's handling of the PRC. To calm these fears, the State Department made draft documents available to leading senators. Senator Tom Connally of Texas, head of the Foreign Relations Committee and ever sensitive to his state's leading industry, was concerned that the petroleum commission would have jurisdiction within the United States. "I want to cooperate in this oil program, but I don't want to see us turn control or dominance of our oil industries here at home or even throughout the world over to some international body that will gyp us." Senator Owen Brewster, when shown the agreement, remarked that it amounted to an "international governmental voluntary cartel controlling petroleum production." The senators' main concern, however, was that they have a voice in determining foreign oil policy. To this end they argued that any agreement with Britain on oil would require ratification by the Senate as a treaty. The State Department, on the other hand, felt that an agreement that merely enunciated broad principles and established an advisory petroleum commission hardly justified treaty procedure, and should be put into force as an executive agreement. Here the matter stood when the British came to Washington in late July to negotiate the final agreement.[27]

Talks began in Washington on July 25. Because of ill health, Secretary of State Hull participated only in the opening ceremonies, and consequently Ickes was left in charge of the U.S. group. To everyone's surprise the talks went well. The British readily agreed to some minor changes in phrasing proposed by the United States to meet the objections of the industry advisers. For their part, the British proposed three substantive changes to the draft Memorandum of Understanding. Two of these—that valid concession contracts not only be "respected" but "supported" and that invitations to other nations to participate in the negotiation of an international petroleum agreement be postponed until after the war— were withdrawn after the United States raised objections. The U.S. delegation argued that any commitment to support or even to collaborate on matters of concession rights entailed obligations toward third countries that could easily be misunderstood. The United States opposed postponing the multilateralization of the agreement because it had publicly stated its "willingness to discuss petroleum problems of mutual interest with any government at any time."[28]

The third British amendment, though ultimately withdrawn, caused serious difficulties and threatened to prevent an agreement. In order to provide itself with "safeguards" that might be required if Britain found itself in foreign exchange difficulties after the war, the British delegation proposed revising the agreement to recognize "the right of each country to draw its consumption requirements, to the extent that may be considered necessary, from the production in its territories or in which rights are held by its nationals." Britain was likely to find itself in an adverse foreign exchange position after the war, and wanted to have the right to deal with this situation as it saw fit, even to the extent of excluding dollar oil from the United Kingdom and procuring supplies exclusively from British-controlled sources. Without some such controls, Britain might have to expend some $100 million a year on oil which could have been obtained from sterling sources. Britain did not claim this right as a principle, but only as a safeguard, and would return to free trade in oil as soon as its foreign exchange position permitted.[29]

The U.S. delegation firmly rejected the British argument. Davies argued that the British proposal was inconsistent with the spirit of the agreement, which was intended to remove restrictions on world oil development. Exclusion of dollar oil from British markets would have a disruptive effect on world oil markets and lead to severe commercial rivalry, making the achievement of order and stability in the international oil trade impossible. Davies also pointed out that the agreement was dependent on the voluntary cooperation of the oil industry, and predicted that U.S. oil companies, which could be frozen out of sterling markets by the British proposal, would not cooperate if the British amendment were adopted. The petroleum commission would see to Britain's exchange position, since it would be a "relevant economic factor" to be considered in the allocation of output.[30]

With the outcome of the talks threatened by this impasse, Ickes put pressure on Beaverbrook by arguing that British insistence on their position could lead to "serious repercussions" in the United States which could determine the outcome of the upcoming election. Postponement of the matter would not help, since it would make the agreement a partisan political issue. Faced with these prospects, the British agreed to withdraw their amendment in exchange for a statement in the official minutes to the effect that Britain might be forced "to take into account the exchange it could lose or gain by the purchase or sale of petroleum in deciding sources from which the petroleum it required should be drawn." The United States "took note" of the British reservation, and cautioned that such action would be inconsistent with the purpose of the agreement. Both countries pledged to try to work out such problems through the proposed petroleum commission before taking unilateral actions. If no solution could be found, either country could terminate the agreement on three months' notice. With this hurdle cleared, the final agreement, little changed from the May 3 Memorandum of Understanding, was signed on August 8, 1944.[31]

Declining power led Britain to accept an agreement that reserved a privileged position for the U.S. domestic oil industry, while exposing all of Britain's oil production, which was in other countries, to the competition of the powerful U.S. international oil companies. Domestically, however, it was not possible to bridge the divisions in the U.S. oil industry. Although initial industry comment was favorable, widespread opposition to the agreement quickly developed, especially among independent producers. In an open letter to the Senate Foreign Relations Committee on August 21, J. Howard Pew, president of the Sun Oil Company and a vigorous opponent of government intervention in oil matters, denounced the accord as the first step toward a "super-state cartel." Pew did not like the vagueness of the agreement's wording, which he charged was "just as innocuous or as vicious as its administrators chose to make it." It could provide a "framework within which there could be developed unlimited government control of the petroleum industry in its domestic operations as well as in the foreign field." He felt forced to the conclusion that the agreement was "a deliberate attempt to place the American petroleum industry under the control of the Federal Government." Echoing Pew's charges, Ralph T. Zook, president of the Independent Petroleum Association, warned the Interstate Oil Compact Commission that the agreement represented yet "another path to Federal control over the oil industry."[32]

George A. Hill, Jr., predictably, was in the forefront of the agreement's critics. Hill's opposition centered on the argument that as a treaty the agreement would give the federal government new powers over the oil industry. Concerned about other important matters before the Senate, the State Department had bowed to Senate pressure and submitted the agreement in treaty form to the Foreign Relations Committee. The committee's assistant counsel, Henry Fraser, had convinced the senators that treaty status was required since the agreement, in his opinion, obligated the United States to take actions that would require congressional approval but that were beyond the powers delegated to the Congress by the Constitution. A treaty, however, could serve as the basis for legislation beyond the constitutional competence of Congress. Hill seized on this point to charge that the "subtle and ingenious device of an international treaty" was being used to give Congress control over the oil industry. The proposed treaty was a "design for the progressive germination of cartels under government blessing."[33]

Behind the rhetoric were fears that the agreement would allow cheaper oil from the Middle East and Latin America to flood the United States, thereby ruining domestic producers. This was not an unrealistic assessment of the operating implications of the agreement. Recommendations of the petroleum commission concerning the volume of oil in international trade would have an effect on U.S. domestic production. Any volume of oil in excess of that required by markets outside the United States would, because of lower costs, force its way into U.S. markets, thereby lowering prices and/or the aggregate level of marketable domestic production. In either case, companies without access to cheap foreign oil would be at a competitive disadvantage vis-à-vis companies with such access.[34]

The major oil companies, in contrast, were inclined to support the agreement. Oil analyst Joseph Pogue of the Chase National Bank pointed out that some sort of coordination was needed to facilitate the "orderly and efficient development" of the world's oil resources in order to avoid both shortages and surpluses "with their train of wasteful consumption and disinvestments." This need was especially acute in the Middle East where, according to the report of the PRC Technical Mission, there was "productive capacity of as much as four times its probable market outlet." As it became evident that they would not control the proposed petroleum commission and that the agreement would not change the antitrust laws, however, the major companies' enthusiasm for the agreement waned. The ending of British restrictions, a key goal of the majors, could be achieved in other ways, and the remaining advantages of the agreement were not great enough to justify open conflict with the independents. The majors did not come out against the agreement, but their increasingly lukewarm support was tantamount to mild opposition.[35]

Industry opposition presented the agreement's supporters with some hard choices. Ickes concentrated on preventing the Petroleum Industry War Council from condemning it outright, a position that would have been politically damaging to the president as well as to Ickes. With Davies's help, he convinced the PIWC to defeat a resolution by Ralph Zook denouncing the agreement, and instead to appoint a committee to suggest possible revisions. This tactic prevented further action until after the elections. In late November, Ickes recommended to the president that the agreement be withdrawn from the Senate and revised to meet oil industry criticism.[36]

The State Department came to a similar conclusion. Although oil analyst John Loftus, a Johns Hopkins–trained economist, argued that industry dissatisfaction was so extensive that any revision to accommodate it would result in an entirely different as well as a "meaningless and ineffectual" document, the department decided in early November that the agreement should be withdrawn from the Senate, revised to meet oil industry criticisms, renegotiated with the British, and then put into force as an executive agreement. Assistant Secretary of State Dean Acheson met with Senator Connally on December 1 and explained that the State Department felt that the agreement should be withdrawn to eliminate the misconception that it covered U.S. production, gave executive power to the proposed petroleum commission, and extended the powers of the federal government over the oil industry. Pressed by Connally, Acheson assured the suspicious senator that the department had no intention of withdrawing the treaty just to make some slight changes and then to bypass the Senate by effectuating it as an executive agreement.[37]

Acheson's assurances undercut the department's strategy. The following morning, Connally issued a statement to the press predicting that the agreement would not be ratified in its present form. The proposed treaty was "unfair to the American oil industry" and "not necessary to the general welfare." On December 6, the PIWC endorsed a revised version of the agreement drafted by its Committee on National Oil Policy. The PIWC redraft specified that the rulings of the proposed petroleum commission were advisory and not binding, that each nation retained the right to limit imports of petroleum products, and that the terms of the agreement applied only to the international petroleum trade and not to either nation's domestic petroleum industry. The oil industry version also provided antitrust immunity to the oil companies for actions taken pursuant to recommendations of the petroleum commission. Faced with these actions, Acheson met with Connally the day after Christmas and got his permission to withdraw the agreement. Accordingly, President Roosevelt formally requested the Senate to return the treaty in early January, so it could be revised "to remove grounds for misunderstanding."[38]

Revision, Renegotiation, and Rejection

After the withdrawal of the Anglo-American Oil Agreement from the Senate, Ickes moved immediately to take control of the revision process. His old rival, Secretary of State Hull, had been forced to retire at the end of November 1944 because of poor health. Hull's successor, former head of U.S. Steel Edward R. Stettinius, Jr., was preoccupied with the problems attendant to the ending of the war and, in addition, seemed less inclined than Hull to guard the department's prerogatives. When the President's Committee on Oil, which had been formed to oversee negotiation of the agreement, met on January 9, 1945, Ickes convinced the committee to authorize State Department oil adviser Charles Rayner and Ralph K. Davies to meet with oil industry representatives and work out a revised agreement. Two days later, Ickes met with Richard Law, minister of foreign affairs in the British cabinet, and informed Law of his plans to revise the agreement. Law told Ickes that, though the British government could not accept an oil industry proposal, it would accept amendments to the original agreement to meet industry criticisms. Ickes met with the National Oil Policy Committee of the PIWC on January 15, and warned the oilmen that unless they cooperated in revising the agreement, he would "take this case to the people by a series of speeches beginning in Texas." Working closely with Davies and Rayner, the PIWC committee put in "practically three days and three nights" from January 15 to 17 and produced a draft revision.[39]

Because of the Yalta Conference, little action was taken until February 22, when Ickes and Davies presented the revised agreement to a joint gathering of the President's Committee on Oil, the Senate Foreign Relations Committee, and the Foreign Oil Policy Committee of the PIWC. Everyone present, with the exception of the State Department representatives, seemed to favor the revision. The next day, Senator Connally praised Ickes's efforts to involve the Congress and the oil industry in the revision, noting that this approach would keep him from "getting kicked in the pants like before."[40]

The State Department did not like the revised agreement and tried to hold Ickes off until Secretary of State Stettinius returned to the country. The department feared that treaty status for the agreement would set a bad precedent for other international economic agreements. Treaty status was also linked to the industry's insistence on antitrust immunity, since only a treaty could override U.S. laws. While sympathetic to the oil industry's desire for protection on this score, the department felt that the provision contained in the revised agreement was "utterly inappropriate" in an international agreement and was "without precedent in antitrust policy." All the important antitrust exemptions had been part of legislation under which the government maintained some sort of continuing control over the industry concerned. In this instance, however, one provision of the treaty committed the government to a policy of antitrust immunity for the oil industry, while another disavowed any intention to impose government control. Such an exemption could have "far-reaching" effects on the Department of State's efforts to combat cartels.[41]

There were other problems as well. The clause guaranteeing the right of each nation to regulate its oil imports not only cast doubts on the U.S. commitment to reduce trade barriers, but was also totally at odds with the agreement's original purpose of reducing the drain on U.S. reserves. The emphasis of the agreement had been shifted from facilitating the development of Middle East oil to the "orderly development" of the international petroleum trade. The shift, arrived at in part to exclude U.S. domestic production from the agreement's coverage, could be seen as promoting "orderly marketing, avoidance of a surplus of petroleum on world markets with a depressing effect upon prices, and the other objectives of an international petroleum cartel arrangement." In light of these objections, and because some sort of Anglo-American understanding on oil was "essential" to protect the U.S. position in the Middle East, the State Department wanted to substitute a joint Anglo-American declaration of principles for the agreement. A joint declaration would contain the important principles, be easy for other nations to accede to, and would avoid the treaty issue.[42]

Ickes attempted to bypass the State Department by going straight to President Roosevelt. Complaining of State Department obstruction, he argued that the antitrust immunity provision was necessary to gain industry support, and was also not unreasonable because of the oil industry's "painful recollection" of recent antitrust suits. According to Ickes, the clause was not self-executing and would require legislation to designate the agency that would grant the immunity as well as establish the standards for such grants. Ickes wanted the president to bring the State Department into line and to persuade the Justice Department to accept the antitrust immunity clause contained in the revision.[43]

President Roosevelt was not persuaded, forcing Ickes to meet with Attorney General Francis Biddle in early April to discuss the Justice Department's objections to the antitrust provision. The Justice Department proposed to substitute an elaborate clause that specified that legislation was necessary to grant antitrust immunity, provided for continuous government supervision of any action taken under a grant of immunity, and stated that the immunity would not involve the power to fix prices or allocate markets. Ickes did not like the Justice Department proposal, but Stettinius, who had finally returned to Washington, agreed to the Justice substitute. The Foreign Oil Policy Committee of the PIWC, however, after struggling with the new language for several days, decided to drop its insistence on antitrust protection altogether rather than tie itself to the specifics contained in the Justice Department proposal. With this matter out of the way, agreement was quickly reached on a final draft of the revision. To meet State Department objections, the import restriction clause was weakened, and the agreement was made more open to accession by other countries. The new version was cleared with the appropriate Senate committees in May.[44]

President Roosevelt died on April 12, and it was not until early June that the revised agreement was presented to President Harry S. Truman. Then it was decided to wait until after the British elections in July before embarking on renegotiation. Further delay resulted from Ickes's refusal to lead the U.S. delegation until he was assured that he would not be replaced at Interior. Thus it was not until mid-September that a U.S. delegation, with Ickes at its head, arrived in London to renegotiate the Anglo-American Oil Agreement. Ickes brought along six oil industry advisers, who acted as "wet nurses . . . self-charged with the responsibility of seeing to it that I neither nationalized the oil industry nor internationalized it." Significantly, almost all of the advisers represented companies whose operations were mainly within the United States, in other words, the industry segment that had opposed the original agreement. Although the advisers were excluded from the actual negotiations at the insistence of the British, Davies assured them that "we won't change the agreement in any way, shape or form without sitting down with you and telling you what it is all about." The advisers were treated with the utmost deference, and even accompanied the delegation on a "junket" to Europe after the completion of the negotiations.[45]

Negotiations with the British went smoothly. The British elections had seen Churchill's wartime coalition replaced by a Labour government, but oil policy, like most aspects of British foreign policy, changed little under the new government. The British were chiefly concerned about being able to utilize oil policy as a means of meeting prospective foreign exchange difficulties. Their desire in this regard caused much less difficulty than before because the import limitation clause, inserted at the insistence of U.S. independent oil companies, gave Britain the right to restrict imports of dollar oil into its home market. The U.S. delegation, while admitting the right of the British under the terms of the agreement to take such action, argued that exclusion of dollar oil from sterling markets would destroy the objective of orderliness in world markets and thus undermine the basis of the whole agreement. As before, both sides noted their positions on the issue in the minutes of the final session.[46]

The British were not so comfortable with some of the other changes proposed by the United States. They had no objection to limiting further the powers of the proposed petroleum commission, and were in fact quite in accord with this change. They were less pleased with the total exclusion of domestic oil from the agreement's coverage. U.S. domestic production was huge, whereas domestic oil production in Britain was minuscule. Thus to ensure equal treatment either the domestic oil industry of both countries would have to be brought within the scope of the agreement or oil required by either country from production controlled by its nationals, wherever its source, should be excluded. Otherwise, the U.S. proposal would mean that British oil from Iran would be accessible in international trade to all countries on a competitive and nondiscriminatory basis, whereas oil produced within the United States would not. The British also did not like making the agreement open to any nation to accede to by declaration. The international oil industry was not a general concern like matters of international finance, and the immediate accession of other countries would be "impracticable."[47]

These differences were settled without great difficulty. After conferring with its oil industry advisers, the U.S. delegation rejected the British arguments concerning the exclusion of domestic oil from the scope of the agreement. Neither the U.S. oil industry nor the U.S. Senate would agree to any document that entailed any degree of control over the domestic oil industry. Excluding British-controlled foreign oil from the agreement's coverage would seriously undermine the objective of free trade in the international oil business. The United States wanted to make the transition to a multilateral agreement easy to avoid the appearance of an Anglo-American "freeze-out," but the U.S. delegation agreed to revise the wording of the agreement to put the extension of the agreement to other countries on essentially the same basis as in the earlier agreement. With these and a few minor matters settled, the final agreement was signed on September 24, 1945.[48]

Having kept the U.S. oil industry and the Senate fully informed at all stages of the revision process, Ickes felt confident that the revised agreement would have little difficulty in securing Senate approval. The Independent Petroleum Association of America endorsed the agreement on October 17. Ralph T. Zook, the association's president and one of the first agreement's fiercest critics, was now convinced that the agreement was needed to prevent the importation of large amounts of cheap foreign oil which would eliminate all but the "flush" oil producers in the United States. In a conciliatory gesture a week later, Ickes promised the PIWC that when the agreement was ratified he would "liquidate" the Petroleum Reserves Corporation. The PRC's purpose, Ickes claimed a great deal less than truthfully, had been to get the British to talk about the oil situation. Congress also seemed to be pleased with his handiwork, and Ickes expected quick approval after President Truman submitted the revised agreement to the Senate on November 1.[49]

The State Department had never been happy with the revised agreement, and the prospect of its passage raised concerns that control over key aspects of foreign oil policy could be vested with Ickes. State Department economist Claire Wilcox warned Assistant Secretary of State for Economic Affairs William L. Clayton in late January 1946 that it was imperative for the department to assert at a high level its interest in the agreement. Wilcox enclosed draft testimony, prepared by Petroleum Division head John Loftus, which explained that an international agreement on petroleum was needed because of the strategic character of oil, the resulting tendency for governments to intervene in oil matters, and the "tendency latent in the organization of the industry towards private restraints and agreements." These characteristics of the oil industry hindered implementation of a "liberal expansionary economic foreign policy" as desired by the Department of State. Consequently, if such objectives of U.S. public policy as equality of access, freedom of competition, and nondiscrimination were to be realized, a public international agency was needed to oversee the industry's operations and to make recommendations for public policies to free the international oil trade from all barriers and restrictions, whether public or private. Wilcox recommended that Clayton make a strong public statement on these public policy aspects of the agreement "for the sake of the record and in view of the misuse to which the petroleum agreement and the consultative machinery to be established thereunder may be put."[50]

Winter recess, and more pressing business, prevented the Senate from taking up the agreement immediately. In the meantime, opposition to the agreement had developed. Sinclair Oil Corporation, which had little foreign production, led the opposition, claiming that the agreement would give the federal government control of the U.S. oil industry. Unconvinced by the arguments of Zook and other "insiders," independent Texas producers again charged that the accord would allow the United States to be flooded with cheap oil from foreign countries. The University of Texas and other eleemosynary institutions in the state dependent on oil royalties joined in the chorus of opposition. The East Texas Oil Association distributed a pamphlet *Watch Your Purse Uncle Sam*, which depicted Britain as a "greedy, selfish nation," and asked "would free American enterprise survive if the British lion once set his cartelized fangs deep into the throat of the oil industry of the United States?" The agreement was a "cunningly worded instrument" designed to give the British and a few integrated oil companies control of the world's oil supply.[51]

Just as the opposition was getting underway, the agreement lost its major supporter; Ickes resigned as secretary of the interior on February 12, 1946. Ickes and Truman had never been close. When the independent

Ickes publicly opposed the president's nomination of oilman and Democratic party treasurer Edwin W. Pauley as under secretary of the navy,
the president reacted angrily and left Ickes with little choice but to resign.
Davies, too, was on his way out of the government as the Petroleum
Administration for War was winding down its activities. Ickes's successors
at the Department of the Interior had not been involved in oil matters
and were not personally committed to the agreement. They were reluctant
to support it strongly, especially after the department's solicitor warned
that it had the potential to foster cartels and could be used as a "club"
against small countries trying to revise unfavorable and morally dubious
concession contracts.[52]

Ickes's departure relieved State Department fears that he would control
the proposed petroleum commission. Nevertheless, the department still
had major misgivings about the agreement. Successive drafts of the
agreement had progressively departed from the department's original
objective of stimulating Middle East production in order to reduce the
drain on Western Hemisphere reserves. Wilcox complained to Clayton on
February 19 that the revised agreement was the "agreement of a faction
of the oil industry out of Ickes by Davies." It was "either dangerous or
useless. . . . If employed as a cover to set up a cartel to allot quotas and
fix minimum prices, it is dangerous. If not so employed, it is useless."
Moreover, the situation had greatly changed since negotiations had begun
in 1944. The altered balance of power in the Middle East made it unlikely
that the British would oppose U.S. commercial expansion. U.S. companies
were expanding their production in the Middle East, and whatever needed
to be done to help them could be done just as well without an agreement.
The Soviets and small oil-producing countries would view a bilateral
understanding between the United States and Britain with suspicion, while
U.S. independent oil companies would oppose it as a cartel. The Petroleum
Division recommended that the department either "kill" the agreement
or assert its own interest in it. The easiest way to kill it would be to have
hearings on it postponed indefinitely since time would work against
ratification. The department could also "damn it with faint praise" or
support it in general while pointing out its negative aspects. If the
department decided to support the agreement, it should assert its interest
in it and "seize the reins" on the U.S. side of the proposed petroleum
commission.[53]

In early March, the State Department requested that hearings on the
oil agreement be postponed, ostensibly so as not to jeopardize the prospects
of the vital loan to Great Britain which Congress was considering. The
British were reluctantly brought to agree to the postponement. Senator
Connally, faced with a primary election in July, was more than happy to
delay consideration of the agreement.[54]

Epilogue

Although the British loan was passed in July, no action was taken on the oil agreement for almost a year. Opposition to the agreement increased over time. Petroleum Division head John A. Loftus inadvertently contributed to this result in August 1946, when he implied on a radio program that the proposed petroleum commission might have more than advisory powers and seemed to support international control of the oil industry. In October, the IPAA reversed its stand and passed a resolution condemning the agreement. The American Petroleum Institute agreed to continue its support of the agreement only after the State Department accepted a reservation that provided that the proposed petroleum commission had "no right or authority to regulate or control the foreign operations of American nationals, nor in any way to regulate or control the domestic industry of the United States." Meanwhile, the Texas independents kept up a barrage of criticism of the agreement.[55]

In June 1947, Senator Arthur Vandenberg of Michigan, who had replaced Connally as chairman of the Foreign Relations Committee, set hearings on the Anglo-American Oil Agreement. Vandenberg felt confident that the agreement would be approved and expected the hearings to last only a few days. Connally, however, kept the committee tied up for almost a month with a parade of opposition witnesses who repeated the by now familiar arguments against the agreement. In contrast to the vociferousness of the treaty's opponents, the witnesses in favor of the treaty made a "perfectly wretched showing . . . weak, inconclusive and lacking any aggressiveness." Ickes watched in dismay from the sidelines as Vandenberg failed to call him to testify for fear of harming the agreement's chances. Nevertheless, the committee reported favorably on the treaty on July 7, with only Connally registering dissent. Despite the committee's favorable report, the agreement never came to a vote. No one, it seemed, cared enough about it to work for its passage. President Truman had never felt strongly about the agreement, and was, in any event, tied up with the fight over tidelands oil. Ickes and Davies were no longer with the government, and the State Department was tepid in its support, if not secretly hostile. Most importantly, the major oil companies had worked out their own salvation and no longer needed the agreement. The treaty was re-referred to the Foreign Relations Committee at the beginning of the Eighty-first and Eighty-second Congresses before being returned to the president on July 5, 1952.[56]

The attempt at a liberal solution for the problems facing U.S. foreign oil policy failed because of the structure of the U.S. oil industry. The State Department's strategic goal of promoting increased Middle East production in order to conserve Western Hemisphere oil reserves was compatible with the interests of the international oil companies, all of which held reserves in the Middle East. Domestic producers, however, feared that the proposed petroleum commission would allow the U.S. market to be flooded with cheap foreign oil. Although it was paranoid to believe that the federal government was plotting the destruction of the domestic oil industry, the State Department's strategic goal conflicted with the short-term financial interests of domestic producers. With more than twenty states having some sort of oil industry, the concerns of domestic producers were heard in Congress. This opposition forced the State Department to look for other means to secure the nation's stake in foreign oil.

Chapter 4

Return to Open Door Diplomacy

While it was negotiating an international accord on oil with the British, the State Department had also been working to create a favorable environment for U.S. oil operations in Iran, Mexico, and Saudi Arabia. Increasing concern over securing access to Middle East oil transformed U.S. policy toward Iran from relative indifference to deep concern for Iranian independence and territorial integrity. State Department policy toward Mexico focused on reversing nationalization of the Mexican oil industry and reintegrating Mexican oil into the privately controlled world oil economy. In Saudi Arabia, U.S. policy concentrated on enhancing U.S. influence, preventing the Saudis from becoming dependent on the British, and stabilizing the desert kingdom's finances.

Oil and U.S. Policy in Iran, 1941–1945

At the same time that they were negotiating an oil agreement, the United States and Britain were engaged in a scramble for concession rights in Iran. Complicating matters was the presence of their wartime ally, the Soviet Union, which had not been invited to the oil talks.

In August 1941, forces of Great Britain and the Soviet Union had invaded Iran. Strategically located and rich in oil, the ancient Middle Eastern nation had become vital to the anti-Hitler coalition when the Germans attacked the Soviet Union in June 1941. After signing a mutual assistance agreement on July 12, the British and the Soviets gave Iran an ultimatum that they knew Iran could not meet: that the nearly 2,000 Germans then resident in Iran be expelled. Iranian leader Reza Shah had increasingly turned toward Germany in the 1930s to counterbalance British and Soviet influence. With German armies driving toward the Caucasus, the British and Soviets invaded Iran on August 25. Iranian resistance quickly collapsed, and on September 16, Reza Shah was compelled to abdicate in favor of his eighteen-year old son, Mohammad Reza Pahlavi. Iran was divided into a northern zone of Soviet occupation and a southern zone under British control. In a tripartite treaty drafted by the occupying powers, Britain and the Soviet Union pledged to respect the territorial integrity, sovereignty, and the political independence of Iran, and promised to remove their forces from Iranian territory no later than six months after the end of hostilities with Germany and its associates.[1]

Although the United States declined Iranian requests to dissuade the British and Soviets from attacking, and to cosign the treaty governing the occupation, the United States kept a close eye on the activities of its Allies in Iran. In this it was aided by the Iranians themselves, who zealously reported any transgressions by the occupying powers. Soviet violations of Iranian sovereignty, in particular, were highlighted, and many in the State Department soon became convinced that the Russians were planning to "Sovietize" their zone, if not all of Iran. Whatever Soviet intentions, their policy of treating their zone as an autonomous area, distinct from the rest of the country, and of restricting access to it elicited suspicions. In contrast, U.S. relations with the British in Iran were more cordial, though there were problems concerning a British request for lend-lease material to build an oil pipeline and over British handling of food distribution and transportation. The pipeline issue raised fears that the British were planning to exclude U.S. interests from participating in Iranian economic affairs. In November 1942, the U.S. minister in Iran voiced a suspicion that the British wished "to entrench themselves solidly in Iran for fear [that] ever increasing American influence and activity here may have adverse effects on their long-range interests."[2]

Iran was declared eligible for direct lend-lease aid in March 1942, and in September the Combined Chiefs of Staff assigned the United States direct responsibility for the Persian Corridor supply route through Iran to the Soviet Union. During the next year, the U.S. Army's strength in Iran increased from fewer than 500 to nearly 30,000 troops. Moreover, in response to Iranian requests, the United States sent a number of advisory missions to Iran. The United States was concerned that instability in Iran could interfere with the supply effort, and the State Department hoped that the adviser program would help the Iranians "see that their real interests lie on the side of the United States." The British, strapped for resources and concerned over growing Soviet influence and possible instability, supported the U.S. adviser program. As a high Foreign Office official noted in June 1943, "the success of the present American advisers in Persia is probably the best hope for establishing and maintaining the kind of Persia that suits our interests."[3]

Iran was economically and strategically important to Britain as its main source of oil, and because of its geographic position directly athwart the lines of communication of the British Empire. Iranian oil production in 1939 was over 78 million barrels, and the huge refinery complex at Abadan was the chief source of refined products east of Suez. The Soviet Union's interest in Iran was scarcely less vital, for Iran not only provided a "back door" to the Soviet Union, but did so near the Soviets' vital oil center in the Caucasus. The interests of the British and the Russians had traditionally led to conflict, but there was always the possibility of a compromise along the lines of the 1907 Anglo-Russian Agreement, which divided Iran into a northern and a southern sphere of influence.[4]

The United States initially became involved in Iran because of its importance as a supply route to the Soviet Union. As the United States became more interested in the oil of the Middle East, however, Iran came to be seen as a strategic buffer between the Soviet Union and U.S. oil interests in the Persian Gulf. In mid-August 1943, Secretary of State Cordell Hull pointed out to President Roosevelt that "it is to our interest that no great power be established on the Persian Gulf opposite the important American petroleum development in Saudi Arabia." Hull recommended that the United States adopt a "policy of positive action in Iran," mainly through the use of U.S. advisers, "so as to facilitate not only the United Nations war operations in that country but also a sound postwar development." Hull believed that the British would support or at least acquiesce in U.S. policy. The attitude of the Soviet government on the other hand was "doubtful," but the U.S. government "should be in a position to exert considerable influence if occasion should arise."[5]

At the Foreign Ministers Conference in Moscow in October, Hull managed to get British Foreign Secretary Anthony Eden to support a proposed declaration on Iran that included a promise of support for the "foreign advisers . . . working to improve conditions in Iran." The Soviets, however, refused to go beyond their commitments under the Tripartite Treaty. At the Tehran Conference, November 27 to December 2, 1943, Stalin agreed to a declaration that promised Iran some postwar assistance for its contribution to the war effort and pledged that the "Governments of the United States, the Union of Soviet Socialist Republics, and the United Kingdom are at one in their desire for the maintenance of the independence, sovereignty and territorial integrity of Iran."[6]

These plans were upset by what Herbert Feis later termed the "apparently unlimited quest of American and British interests for oil rights everywhere." In February 1943, the Iranian government, as part of its efforts to attract U.S. support, approached the Standard Vacuum Oil Company (Stanvac) about a possible concession in Iran. Stanvac, a joint subsidiary of Standard Oil of New Jersey and the Socony-Vacuum Oil Company, needed oil and decided to pursue the Iranian offer. Although the U.S. minister in Iran pointed out that efforts by U.S. oil companies to obtain concessions would lead the British and Soviets to suspect that U.S. policy toward Iran was "not entirely disinterested," this warning was not heeded. The department informed the U.S. legation in Iran in mid-November that "because of the importance of petroleum, both from the long-range standpoint and for war purposes, the Department of State looks with favor upon the development of all possible sources of petroleum."[7]

Meanwhile, the British government was encouraging Shell to seek a concession in Iran. Although British interests held a monopoly over Iran's known reserves through the Anglo-Iranian Oil Company (AIOC), the British government felt that unless Britain "kept its end up," U.S. interests would get all the oil that was left in Iran. Shell's representatives arrived in Iran in the fall of 1943, closely followed by Stanvac officials. In March 1944, the Sinclair Oil Company informed the Iranian government that it too was interested in a concession. To introduce some order to the process, the Iranians hired a U.S. petroleum consulting firm in early April to advise them on oil matters. The firm's principals, Herbert Hoover, Jr., and A. A. Curtice, fresh from a similar assignment in Venezuela, arrived in Iran in July in the midst of what was becoming a classic scramble for concession rights.[8]

Sinclair's entry into the race complicated State Department efforts to secure the concession for U.S. interests. With two U.S. companies competing, efforts on behalf of either would be to the detriment of the other, and could expose the department to charges of favoritism. To avoid this problem, Wallace Murray and the officials handling the issue in Iran, Chargé Richard Ford and newly arrived petroleum attaché John H. Leavell, recommended that the department urge the companies to consolidate their efforts. Sinclair agreed to cooperate with Stanvac, but only on the condition that it receive managerial and operating control as well as an equal interest. Otherwise, Stanvac with its interests in Iraq and elsewhere might at some point attempt to hold back Iranian development. Stanvac's owners, though in need of additional sources of petroleum, would not agree to control by Sinclair. The State Department, mindful of the furor Ickes had created with his pipeline scheme, hesitated to pressure the companies. Therefore, the department informed Ford that the principle of equal access and opportunity prevented the U.S. government from pressuring U.S. companies to join forces.[9]

While State Department attention was focused on competition with the British and between U.S. companies, another factor was being ignored. The Russians had traditionally objected to any other foreign power establishing a presence in northern Iran. A Soviet-Iranian treaty in 1921 had given up oil concession rights held by a Russian on the condition that the concession would never be ceded to a third power. The Soviets had objected to a U.S. oil concession anywhere in Iran in 1940, and in 1941, they had expressed a desire for their own concession in northern Iran. In February 1944, a Soviet Embassy spokesman informed a U.S. correspondent that the Soviet Union possessed prior rights to oil exploitation in northern Iran. In addition, two reports by the Office of Strategic Services (OSS) in the spring of 1944 warned that the Soviets would oppose any U.S. concession in northern Iran and were not likely to favor U.S. concessions anywhere in the country. The OSS believed that the only way the United States could proceed without arousing Soviet suspicions was to reach an agreement with the Soviet Union before making any moves. The State Department, nevertheless, neglected to consult with the Soviets, an omission highlighted by Soviet exclusion from the Anglo-American Oil Agreement negotiations.[10]

The scramble for oil concessions grew hotter as the summer wore on. On August 4, Ford telegraphed Washington calling for quick action "to gain this rich prize for American interests." He recommended that the companies send high-level executives with power to conclude deals on the spot and that the department grant them number-one air priorities. The situation changed dramatically on September 6 when the Soviet Union informed Iran that it was sending a mission to discuss oil concessions in the northern part of the country. A week later, a Soviet mission, headed by Vice-commissar for Foreign Affairs Sergei Kavtaradze, arrived in Tehran to begin talks. On October 2, Kavtaradze proposed that Iran grant the Soviet Union exploration rights in a 175,000-square-kilometer area in northern Iran for a period of five years, after which time the area would be reduced to 115,000 square kilometers. Kavtaradze also asked for mineral rights and the exclusion of all other foreigners from the concession area.[11]

The Iranian government, which had initiated the whole process as a means of countering Soviet influence, found itself faced with the choice of granting the Soviet request, thus increasing Soviet influence in northern Iran, or of incurring Soviet displeasure by refusing. The British also faced a quandary. Shell needed additional sources of oil, and some British officials feared that denying the Soviets a concession could be used by Iranian nationalists to undermine AIOC's existing rights. On the other hand, the British ambassador to Iran, Sir Reader Bullard, an "old-school British Middle Eastern specialist," believed that the Soviets would use any concession rights they obtained to establish hegemony over all Iran. Bullard and others in the Foreign Office argued that maintenance of Iran as a buffer state between the Soviet Union and the Persian Gulf was more important than a new concession for British oil interests. The best course was to urge the Iranian government not to grant any concessions. Newly arrived U.S. Ambassador Leland B. Morris, a veteran foreign service officer, recommended that the United States "either singly or cojointly with the British" oppose the Soviet request. If higher policy precluded action of this nature, the United States should encourage the Iranian government to postpone for six months or a year the granting of concessions.[12]

In consultation with representatives from the Majlis (Parliament), the Iranian government decided to postpone all oil concession negotiations until the end of the war. This decision was announced October 8. The announcement claimed that the decision had been made on September 2, before the Soviets had indicated their interest in a concession, but this was, as Ambassador Morris observed, "too apparently a desperate diplomatic lie." Although there was domestic opposition to the granting of concessions to any foreigners, the decision was clearly aimed at denying a concession to the Soviet Union. The motive was clear to the Soviets, and they reacted with tactics that seemed to Morris to "smack of Hitlerian methods:" The pro-Soviet Tudeh party was openly employed for the first time in the war to organize demonstrations against the Iranian government; traffic heading north into the Soviet-occupied zone was stopped; food shipments south from the Soviet zone were temporarily halted; and Kavtaradze and Soviet Ambassador Mikhail A. Maximov threatened Iranian officials with dire consequences for the affront.[13]

The United States and Britain, for obvious reasons, took the Iranian decision in stride. The Soviet reaction, however, caused alarm in Washington and London, and both governments sent protest notes to Moscow. The U.S. note, delivered on November 1, stressed that in light of the Tehran Declaration, the United States "would not be able . . . to concur in any action which would constitute undue interference in the internal affairs of Iran." Nevertheless, the United States, despite Iranian and British urging, went no further than to support the Tehran Declaration verbally. As Wallace Murray pointed out in early November, "we are in the midst of a war and . . . we cannot take any action which would interfere with the conduct of the war and with our vitally important relations with Soviet Russia."[14]

The Soviets were operating under similar constraints, and the situation began to improve in November, after the Iranian prime minister responsible for the postponement decision resigned. On December 2, the Majlis passed a law that prohibited any discussion of oil concessions with foreigners. Under the law, which had been sponsored by nationalist deputy Mohammed Mossadegh, all future decisions regarding oil concessions would be under the control of the Majlis. The law surprised everyone, and angered the Soviets who protested that it was obstructive. The Soviets, however, took no action, and a week later the Soviet oil mission, which had remained in Iran since September, returned home.[15]

In an analysis of the crisis in mid-December, Ambassador Morris pointed out that U.S. policy was partly responsible for what had occurred. Morris noted that if one considered the "extensive operations" of the various U.S. advisory missions, the activity of the U.S. oil companies, and the active interest in Iran manifested by the U.S. government itself, it would not have been hard for the Soviet government to conclude that the United States aimed at securing a permanent position in Iran which, in the Soviet view, could some day prove disadvantageous or even dangerous to them. Morris's view was shared by the U.S. chargé in the Soviet Union, George F. Kennan. Kennan believed that the Soviets had reacted to the scramble by British and U.S. oil companies for concessions in Iran not because they needed the oil but because they considered northern Iran, which bordered on their vital Caucasian oil center, so "essential to Russian security that no other great power should have even the chance of gaining a footing there." Although the crisis ended with the departure of the Soviet oil mission, it was clear that Soviet efforts to assure themselves predominant influence in this vital border area would continue.[16]

Stalemate in Mexico

After the high octane refinery fiasco, the State Department continued its efforts to reverse the nationalization of the Mexican oil industry. Mexico's potential as an oil producer was great, and as wartime demand continued to strain U.S. resources, attention once again focused on the nation's southern neighbor. In November 1943, the department reminded Ambassador George S. Messersmith of the "vast importance" of his negotiations with Mexico in view of the "rapid depletion of United States oil reserves." Reopening Mexico's reserves to the United States was "necessary to the future welfare of this country."[17]

Harold Ickes had also not lost interest in Mexico. Stymied in Saudi Arabia, Ickes turned to Mexico as a possible field of operations for his Petroleum Reserves Corporation (PRC). In late 1943, he suggested to sympathetic Mexican officials that the PRC could do exploratory work, furnish technical assistance, and provide financing for the Mexican oil industry in return for first call on the oil produced. The Mexicans were enthusiastic about the idea, as was President Roosevelt, who liked the concept of a "cooperative arrangement" with Mexico on oil development. In early January 1944, the president summoned Ambassador Messersmith to Washington to talk about the idea. What Ickes had in mind was U.S. assistance for the general development of the Mexican oil industry. President Roosevelt, according to Messersmith's record of their conversation, was thinking primarily in terms of limited assistance to Mexico for the development of a joint military petroleum reserve. Such assistance would be separate from the general question of Mexican oil policy, and Messersmith apparently received the president's approval to continue his efforts at convincing Mexico to reopen its oil industry to foreign participation.[18]

Since Mexican law strictly limited foreign participation in the nation's oil industry, U.S. efforts focused on finding a "formula" that would allow foreign companies to operate in Mexico without violating the law. Messersmith and Mexican Foreign Minister Ezequiel Padilla were able to agree that all subsoil rights were the property of the Mexican government and that all questions regarding labor should be under the complete authority of the Mexican government. To meet Mexican objections to foreign concessions, Messersmith repeated the proposal, first advanced in the spring of 1943, that Petróleos Mexicanos (PEMEX) contract with U.S. companies which would undertake development and production activities. The companies would recover their outlays out of oil produced, and thereafter the companies and PEMEX would share the oil produced on some agreed-on basis. Under pressure from ex-President Lázaro Cárdenas, the author of nationalization, the Mexican government decided that the U.S. proposal turned over too much responsibility to foreign interests. When Messersmith returned to Mexico in February 1944, the Mexican government presented him with a counterproposal which provided for closer PEMEX control over oil operations.[19]

Messersmith sent copies of the new proposal to the department's Petroleum Division as well as to Jersey and the Cities Service Oil Company. After the oil companies rejected the Mexican plan as not providing enough company control or enough reward for their efforts, the Petroleum Division worked out a substitute plan which called for U.S. companies to contract directly with the Mexican government on the basis of a 51-49 percent split of profits and operational control between the Mexican government and the companies respectively. Cities Service was willing to negotiate on the basis of this plan, but Jersey felt that it did not offer enough managerial control or opportunity for profit. In addition, Jersey felt that both plans involved too much "subterfuge." If Mexico wanted to attract foreign capital, it should write a petroleum law "the fundamental nature of which in itself constituted a clear invitation to foreign capital to participate in Mexico's petroleum industry."[20]

Taking his cue from Jersey, Messersmith told the Mexican government in mid-April that the United States did not want oil companies coming back into Mexico under conditions that would again lead to trouble. "Small and irresponsible companies" would not be able to carry through any responsible development and would only become a source of trouble. The major companies, in contrast, had a "deep sense of responsibility" and any operations they undertook would be financed from their reserves. The major companies, however, would invest their shareholders' money only if they were allowed adequate control over their operations and the opportunity to make a "reasonable profit." In Messersmith's opinion, Mexico needed the money it could get from oil exports, and PEMEX was not big enough to do the job that needed to be done. The world oil industry was going to remain in private hands, so Mexico would have to come to terms with the major oil companies. Instructions sent to Messersmith in early June echoed these sentiments: If Mexico wanted to improve its economy and develop its oil resources, it should "consider the wisdom of adopting a more enlightened attitude toward inviting foreign capital back to Mexico to assist in developing its petroleum industry."[21]

Meanwhile, Mexican officials had approached Ickes about the possibility of U.S. government loans to PEMEX for oil exploration and development. They proposed that the Mexican government would obligate itself for the money and the loan could be repaid in oil. Ickes presented this idea to President Roosevelt, who, according to Ickes, said that it had been his idea all along that the United States would help Mexico develop its oil and take payment in kind. Ickes informed the Mexicans of the president's response and suggested that they pursue the matter with the State Department. This turn of events brought Messersmith's negotiations with the Mexican government to a halt as Cárdenas put pressure on the Mexican government to pursue an oil loan. Although Messersmith assured Padilla that the U.S. government would never lend any foreign government money for oil exploration and development, Padilla insisted that the conversations regarding the reentry of U.S. companies could not continue until Mexico received a definite answer on the possibility of a loan. Messersmith urged the State Department to take the matter up with the president so that he could have an authoritative statement ruling out any U.S. government assistance to PEMEX to present to the Mexican government. If it did so, he was "not at all pessimistic" that the Mexicans would find a way to let U.S. interests back into Mexico.[22]

In response to Messersmith's pleas the Petroleum Division prepared a policy statement on the loan issue. The Petroleum Division doubted that a loan would help PEMEX, which it felt to be incompetent, but the most important consideration was the fear that such a loan would help PEMEX and thus contribute to the exclusion of private U.S. interests from any future participation in Mexican oil development. Moreover, the Petroleum Division argued that if, in the light of the Mexican expropriation of U.S.-held properties, the U.S. government were to make such a loan to the Mexican government, this action would greatly encourage other countries, particularly in Latin America, to nationalize their petroleum and other resources. The Executive Committee on Economic Foreign Policy agreed with this analysis and unanimously recommended that the Mexican government should be informed that the United States was not considering any such loan to Mexico and did not contemplate considering such a project in the future.[23]

These statements were apparently not sufficient, and Foreign Minister Padilla was sent to Washington to confer directly with President Roosevelt. Concerned that all go well, Messersmith urged the State Department to emphasize strongly that the United States should under no circumstances loan Mexico money for oil exploration and development. Although Padilla and Ávila Camacho understood this, the case had to be made as clear as possible so that they could use it against Cárdenas, who was blocking the development of a "reasonable and constructive" oil policy. The State Department sent the White House a memorandum on July 4 warning that a loan to PEMEX would exclude private U.S. companies from Mexican oil development, but Ickes managed to get in the last word. In a conversation with the president on July 7, Ickes discounted the State Department's objection that a loan would establish a bad precedent, and told Roosevelt that the State Department was trying to get the big companies back into Mexico.[24]

Padilla met with President Roosevelt on July 11. The president later claimed that he clearly distinguished between oil for commercial purposes and oil for government use. Padilla, however, got the impression that Roosevelt favored development of Mexico's oil resources through government to government cooperation. According to Padilla, Roosevelt told him that he did not have confidence in the oil companies and felt that the U.S. government should loan Mexico money so that Mexico could develop its oil on its own. The president also told Padilla that the State Department was "old fashioned" on the subject of oil and advised him to talk with Ickes. The amazed Padilla then went to see Ickes who also warned him about dealing with the oil companies, and told him that the U.S. government was willing to finance Mexican oil development and take payment in kind. Padilla pointed out that any loan would have to be long-term and at a very low interest rate, and that the Mexican government would have to retain control of all operations inside Mexico. Ickes replied that this was no problem and advised Padilla to prepare a proposal in writing. Ickes told Padilla that the State Department would handle the matter, but when Padilla spoke with Secretary of State Hull the following day, he was informed that no money was available for loans.[25]

President Roosevelt assured the State Department on July 19 that what he had proposed to Padilla was the development of joint military oil reserves in Mexico, not oil development in general, but the damage had already been done. Padilla explained to Messersmith on July 21 that since Cárdenas had no doubt learned about the president's promise of aid, the Mexican government had no choice but to pursue the matter of a loan. To make matters worse, Messersmith suspected that Phillips Petroleum Company, which had a contract with PEMEX to operate the 100 octane refinery being built with U.S. aid, was trying to use the contract as a means of getting into oil exploration and development in Mexico "through the back door." If the Mexicans were able to get aid from Phillips, it would lessen the pressure on them to change their general oil policy. In early September, Mexico requested a long-term, low-interest loan for oil development.[26]

In extensive talks with State Department officials in early October, Messersmith argued that it was "impracticable and undesirable" for the United States to loan Mexico funds for commercial oil development. Such a loan would set a bad precedent internationally, and would not help Mexico since PEMEX would waste any money given it. On the other hand, the State Department had a "primary responsibility for and an interest in" helping U.S. oil companies get into Mexico. Therefore, his conversations with the Mexican government should continue. Hull approved these recommendations on October 2, but it was not until mid-December that Messersmith was able to discuss the issue with President Roosevelt. According to Messersmith's record of the conversation, the president stated that Padilla had misunderstood him in July. Although the United States was ready to loan Mexico a small amount of money for development of a joint military oil reserve, loans for general oil exploration and development were "out of the question and not to be considered."[27]

On the basis of this conversation, Messersmith informed Padilla that a loan was "out of the question." He also advised Padilla that there was no point in Mexico talking with small oil companies since they did not have the necessary financial resources and would not be able to get money from banks or the U.S. government. The large companies, on the other hand, were ready to do business on an equitable basis. The official U.S. reply to the Mexican loan request, delivered in late January 1945, carried much the same message. The United States would not lend Mexico money for oil exploration and development. If Mexico wanted outside help it would have to make arrangements with private interests.[28]

Because of the onset of the year-long Mexican presidential election campaign, Messersmith decided not to press ahead with the conversations on oil policy until after the elections. In particular, he did not wish to harm the chances of Padilla, who was the conservative candidate, by forcing the foreign minister to take an unpopular stand on the emotional issue of oil policy. The State Department approved his decision to postpone conversations, and in October 1945, obtained President Truman's approval to continue its general oil policy toward Mexico. President Roosevelt's idea of a joint military reserve, which Messersmith had previously been instructed to pursue, was quietly moved to the back burner, and Messersmith was to discuss the subject only if the Mexicans brought it up.[29]

State Department policy toward Mexican oil failed in its major objective of reversing nationalization. On the other hand, the department was able to maintain pressure on Mexico to change its oil policy by preventing U.S. government aid to the Mexican oil industry and discouraging arrangements between PEMEX and independent U.S. oil companies. These policies favored the major oil companies, and in particular Jersey, for while Mexico remained off limits to them, the threat of oil development outside their control was minimized.

Search for Security

The failure of the plan to purchase Saudi Arabia's oil and the overwhelming opposition to the pipeline plan did not signal a diminution in U.S. interest in Saudi Arabia. On the contrary, the State Department and the military continued to look for ways to secure U.S. control of "one of the world's greatest prizes"—the oil of Saudi Arabia.[30]

U.S. lend-lease assistance, coupled with continued British subsidies, kept the Saudi Arabian government afloat during 1943. It was clear by early 1944, however, that more aid would be needed to meet Saudi Arabia's continuing financial problems. Oil revenues were not yet sufficient to balance the king's budget, as oil production in 1944 averaged only a little more than 20,000 barrels per day. In addition, the British were planning to reduce their subsidies. With stability returning to the Middle East, the British saw little reason to continue to bankroll what they viewed as an inefficient and wasteful regime. The United States, on the other hand, was reluctant to risk the possible consequences of not meeting the king's needs. On April 13, Secretary of State Hull recommended to the president that the United States consider assuming 50 percent of the subsidy to Ibn Saud. If Saudi Arabia was permitted to lean too heavily on the British for support, there was the danger that the British would demand a "*quid pro quo* in oil." President Roosevelt immediately approved the recommendation.[31]

The British were not ready to concede their predominant influence in Saudi Arabia, but they wished to avoid conflict with the United States. At talks held in London in mid-April, the Foreign Office was careful to assure the U.S. representative, Wallace Murray, that Britain had no intention "to undermine or prejudice American oil rights in Saudi Arabia." The British also agreed to begin discussions on a joint assistance program and a joint military mission. These instances of British cooperation failed to satisfy the State Department's Middle East hands, who viewed British efforts to impose some order on Saudi finances as part of a larger plan to control Saudi affairs. The U.S. representative in Saudi Arabia, James S. Moose, was convinced that his British counterpart was trying to undermine U.S. influence with King Ibn Saud. Moose warned that merely matching British aid was not enough. The United States should take over the entire problem of Saudi supply and finances until oil revenues substantially increased.[32]

Although a joint aid program was worked out in July, the Near East Division continued to press for a larger U.S. role in Saudi Arabia. In late September, William A. Eddy, who had replaced Moose earlier in the month, immediately proposed a ten-year, $55-million loan program to stabilize the economy and to finance development projects until oil revenues increased. Eddy, who has been called "the nearest thing the United States had to a Lawrence of Arabia," also felt that the United States should construct airfields and roads and provide military aid. These suggestions struck a responsive chord in Washington, and the Near East Division quickly developed plans for long-range assistance to Saudi Arabia. The

Saudis would need $57 million in assistance over the next five years. The War Department could provide some $20 million in return for airfield and overflight rights, while another $37 million could be provided to secure oil for Navy use. Saudi Arabia's oil reserves were "among the greatest in the world," and $57 million was a "relatively small investment on the part of this government to safeguard our national interest in the oil reserves of Saudi Arabia."[33]

The Army was interested in building an airfield in Saudi Arabia, and conversations with Navy officials revealed a great interest in Saudi Arabian oil. James V. Forrestal, who had succeeded Frank Knox as secretary of the navy after Knox's death in late April 1944, felt that it was "distinctly in the strategic interest of the United States" to encourage U.S. oil companies to promote the orderly development of oil reserves in such remote areas as the Persian Gulf. It would supplement Western Hemisphere reserves and protect against their early exhaustion. Moreover, U.S. control, through its nationals, of such a vital resource as oil added to the power and influence of the nation. Therefore, on December 11 Forrestal recommended that ownership of oil reserves outside the country by U.S. nationals be actively encouraged, and those holdings already in U.S. hands be protected. The War and Navy departments gave their official approval to the State Department's plans on December 20.[34]

The Near East Division's interest in providing aid to Saudi Arabia resulted from more than a concern over Saudi finances. Wallace Murray and other Middle East hands in the department were worried about the Palestine issue, and wanted to do something for Ibn Saud to offset whatever the president might do on Palestine. Although President Roosevelt made vague promises of future support for Zionist objectives, he was careful to keep the door open to the Arabs through the "full consultation" formula, which promised that no decisions altering the basic situation of Palestine would be reached without "full consultation" with both Arabs and Jews. In early 1944, resolutions were introduced into both houses of Congress calling for unrestricted Jewish immigration into Palestine and for the reconstitution of Palestine as a "free and democratic Jewish Commonwealth." To head off trouble, Assistant Secretary of State Breckinridge Long secretly met with Senators Tom Connally, Alben W. Barkley, Walter George, Arthur Vandenberg, and Robert LaFollette to point out the threat to U.S. interests in Saudi Arabian oil. According to Long, the senators were carried away with the importance of Saudi Arabian oil and strongly agreed that nothing should be allowed to interfere with continuing U.S. control of this rich resource. In addition, Secretary of War Henry L. Stimpson sent letters to the Senate Foreign Relations Committee and the House Committee on Foreign Affairs expressing the War Department's deep concern that the "war effort would be seriously prejudiced by the passage of the resolutions." After a second letter from Stimpson and executive session testimony by Army Chief of Staff George C. Marshall, the resolutions were withdrawn.[35]

The 1944 Democratic and Republican platforms supported the Zionist position. Moreover, just before the election, both President Roosevelt and the Republican candidate, New York Governor Thomas E. Dewey, endorsed the idea of a Jewish state in Palestine and pledged to work toward its realization. These actions led Murray to warn that the "mistaken Allied policy of overemphasizing Jewish aims" was turning the Arabs to the Soviets. If Ibn Saud adopted the policies of the Arab militants on the Palestine issue "our great economic stake in Saudi Arabia may become so compromised that our efforts through ordinary diplomatic channels to protect our interests there might be unavailing." Murray feared that the president's endorsement of a Jewish state would "have a very definite bearing upon our relations with Saudi Arabia and upon the future of the American oil concessions in that country, which is of incalculable value to the present and future strategic requirements of the United States." After the election, a resolution supporting the establishment of a "Jewish Commonwealth" in Palestine was reintroduced in the Senate, but the new Secretary of State, Edward R. Stettinius, Jr., was able to persuade the senators to shelve the resolution.[36]

In submitting its plans for aiding Saudi Arabia to President Roosevelt on December 22, the State Department argued that "an American national interest, basically strategic in character," existed in Saudi Arabia. Saudi Arabia was strategically located and its vast oil reserves should be safeguarded and developed in order to supplement Western Hemisphere oil reserves. In addition, the military urgently desired to construct certain facilities in Saudi Arabia to aid in the prosecution of the war. The king was experiencing budgetary deficits, and until oil revenues and the revival of normal trade provided sufficient income to meet his government's requirements, Saudi Arabia would need help from abroad. If such help was not provided by the United States, "undoubtedly it will be supplied by some other nation which might thus acquire a dominant position in that country inimical to the welfare of Saudi Arabia and to the national interest of the United States." Lend-lease would not be sufficient since the Foreign Economic Administration refused to continue aid beyond the fiscal year, and there was no assurance that lend-lease would be available on a long-term basis. Therefore, the president should request Congress to appropriate funds to meet Saudi Arabia's urgent needs. The president should also ask the Export-Import Bank to make a commitment in principle to development loans for Saudi Arabia. In the interim, the military should look into projects such as airfield and road construction and training missions that could provide short-term assistance to Saudi

Arabia. President Roosevelt approved the department's recommendations on January 8, 1945.[37]

Underlining the growing importance of Saudi Arabia, President Roosevelt met with King Ibn Saud of Saudi Arabia at the Great Bitter Lake in the Suez Canal on his way home from the Yalta Conference. That the president of the United States had gone out of his way to hold talks with the ruler of a hitherto obscure nation reflected, as Aaron Miller has pointed out, the United States' "growing awareness of its interests in Saudi Arabia—if not the entire Middle East." The main thrust of their five "very intensive hours" of conversation on February 14, 1945, was the Palestine issue. Roosevelt assured the king that he "would do nothing to assist the Jews against the Arabs and would make no move hostile to the Arab people." A month later, however, the president told his friend Rabbi Stephen Wise that he stood by his earlier pledge to support a Jewish homeland. The president's action led to widespread protests by the leading Arab governments, and Murray again warned that continued support of Zionist objectives could have far-reaching consequences for U.S. interests in the Arab world, and especially for the "immensely valuable oil concession in Saudi Arabia."[38]

By early 1945, the question was no longer whether the United States should provide aid to Saudi Arabia but how such aid could be provided. The War and Navy departments, while in favor of aiding Saudi Arabia, were reluctant to spend their own funds for this purpose. The War Department was interested in having an airfield at Dhahran, but felt that it could not justify spending $20 to $30 million for this purpose alone. Likewise the Navy, which was interested in the oil aspects of the problem, was hesitant to commit funds for oil in the ground in Saudi Arabia. Both departments felt that Congress would have to be approached for the necessary funds, though they cautioned that the matter would have to be handled carefully in order to avoid allegations that public money was to be used to benefit a private interest.[39]

The matter of aid to Saudi Arabia was turned over to an ad hoc subcommittee of the newly created State-War-Navy Coordinating Committee (SWNCC). On February 22, the subcommittee issued a report which noted that, while the military was interested in Saudi Arabia because of its location along the most direct air route to the Far East, "it is the oil of Saudi Arabia which makes that country of particular importance to the armed services." Oil revenues would eventually provide a solution to Saudi Arabia's financial problems, but Saudi Arabia would need financial assistance for the next few years to bridge the gap between its expected revenues and its needs. Interim, indirect assistance could be provided through the construction of military airfields and roads and the sending of a military mission. The War Department was now ready to extend such assistance provided that the State Department furnished assurances that such action was in the national interest. Long-range assistance would probably involve oil and could be in the form of loans to either Saudi Arabia or ARAMCO, government purchase of oil either above or below ground, or government assistance in the development of facilities needed to increase oil production. While not recommending any particular option, the subcommittee favored approaches that would involve the "active and direct participation by the private companies in the financial burdens involved." SWNCC quickly approved the subcommittee's recommendations.[40]

On March 8, accompanied by War and Navy department representatives, Dean Acheson, assistant secretary of state for Congressional Relations, presented the government's case to a select group of House leaders. Acheson told the congressmen that Saudi Arabia needed $50 million over the next five years to cover its budget deficits. If these deficits were not met the U.S. oil concession in Saudi Arabia would be endangered. Aid could be in the form of outright grants, secured or unsecured loans, or in return for guarantees regarding the concession and rights to a billion barrels of oil at a price reduction that would return the money to the government. Speaker of the House Sam Rayburn felt that the matter was important enough to justify grants. Carl Vinson, chairman of the House Naval Affairs Committee, as well as the other congressmen present were strongly in favor of a method that would provide the United States with a *quid pro quo* for any aid extended. Acheson noted that the State Department would prefer to avoid formal appropriations, since this would involve hearings, and suggested that general funds could be provided to the State Department, which could reach an informal understanding with the appropriate committees on how the money would be spent.[41]

Rayburn and Vinson were not representative of congressional views on oil. Vinson, in particular, was identified with advocating the establishment of naval petroleum reserves overseas. In contrast, spokesmen for the domestic oil industry, such as Senator Tom Connally and Senator E. H. Moore, opposed any federal involvement in oil matters. Domestic oil interests were still up in arms over the PRC pipeline plan and the Anglo-American Oil Agreement, and any affirmative action by the government in regard to Saudi Arabian oil was sure to draw fire from the oil industry and Congress. Emilio Collado, director of the Office for Finance and Development Policy, pointed out that Navy purchase of an oil reserve in Saudi Arabia would be attacked as government entry into the oil business, and if the United States gave or loaned money to the Saudi Arabian government, the oil industry would "recognize the same horse in a different blanket," and would charge that the government was spending public money to create a favorable environment for the development of low-cost competition.[42]

Moreover, Paul F. McGuire of the Division of Financial and Monetary Policy argued that government aid to ARAMCO was not necessary. As long as ARAMCO remained the "sole beneficiary of the huge potential profits from Arabian oil," it should "under the American private enterprise system . . . provide risk capital to the extent of its financial capacity, appealing, if necessary, to the open capital market." The department would be "treading on very dangerous ground" if it espoused a doctrine that the national interest justified use of government funds to support the operations of private U.S. companies abroad that competed with private domestic companies. If ARAMCO did not have the financial resources to enable it to function as an "adequate vehicle for the national interest now entrusted to it," it should be required to let other U.S. companies share in the concession. This approach would not only provide more money to develop the concession, but would provide additional marketing facilities and give more U.S. companies an interest in the development of Arabian oil. It would also "reduce the potential opposition to an integrated use of the world's petroleum reserves in the interest of security and conservation."[43]

Collado tried unsuccessfully to have the matter reconsidered, arguing that instead of asking Congress for funds directly, the department work for Export-Import Bank loans, additional advances by ARAMCO, and continuation of lend-lease through 1945. Collado, who left the department in late 1947 to take a position with Jersey, also felt that ARAMCO should be approached about bringing in additional companies "as the one company may not have sufficient markets to handle the concession." In addition, he argued that it was not necessary for Arabian oil to remain in the hands of a U.S. company in order to realize the strategic goal of reducing the drain on Western Hemisphere reserves. Indeed, this result might be obtained more quickly if Saudi oil were in British hands since this course would avoid dollar payment difficulties which threatened to limit the market for Arabian oil. The desired replacement of Western Hemisphere oil in European markets by Persian Gulf oil could then take place through British competition rather than affirmative action by the U.S. government, which would be opposed by most of the U.S. oil industry.[44]

These arguments convinced Assistant Secretary of State for Economic Affairs William L. Clayton that the national interest as well as the interests of the oil companies concerned would better be served if the oil companies themselves provided Ibn Saud with the required assistance. An international cotton trader who shared Hull's vision of an open world economy, Clayton urged that U.S. government aid be limited to the "usual diplomatic support accorded to all American business interests abroad, plus sustained efforts to work out a satisfactory international petroleum agreement within the framework of an international security organization." This course would avoid the opposition of the oil industry and the precedent of U.S. aid to finance the budgetary deficits of countries with which U.S. companies were doing business. If the oil concession could not be adequately protected by the private interest entitled to the profits from its exploitation, "doubt is cast upon the adequacy of the American private enterprise system in the international field."[45]

In the midst of the dispute between Clayton's economists and the Near East Division, the British informed the State Department that they had decided to reduce their 1945 subsidy to Ibn Saud to half the 1944 figure. Rather than risk losing Saudi Arabia, the State Department decided to go ahead with aid to Saudi Arabia while simultaneously exploring the possibility of "broadening the participation of the American oil industry in the potentially huge reserves represented by this concession." Although there were informal contacts between Navy Department officials and ARAMCO on the possibility of bringing other U.S. companies into ARAMCO, the matter was not pursued. On the question of aid to Saudi Arabia, the State Department, after discussing the matter with selected senators, decided to explore two plans—a loan through the Export-Import Bank, possibly in conjunction with a collateral contract between the Navy and ARAMCO; and a purchase of a military petroleum reserve. After being assured that prompt solution of the problem was essential to assure the stability and independence of Saudi Arabia, President Truman gave his approval to go ahead with the aid plans.[46]

Meanwhile, interim aid was running into difficulties. SWNCC plans for aiding Saudi Arabia called for interim aid to be furnished through the construction of a military airfield at Dhahran. Acting on the SWNCC recommendation, the Joint Chiefs of Staff (JCS) approved the construction of the airfield. After receiving British approval in April, the State Department began negotiations with Ibn Saud. JCS approval was on the basis of military necessity—the redeployment of troops to the Pacific theater required a landing and servicing station between Cairo and Karachi—and in June the War Department informed SWNCC that the course of the Pacific War and changes in redeployment plans had "resulted in a substantial diminution of the military necessity of this airfield." The war could easily be over before the project was completed, and expenditure of funds on the basis of military necessity might be of doubtful legal validity. Upset at the War Department's reevaluation of the project, the State Department warned that the effect of canceling the project would be to "strain to the limit Saudi Arabian confidence in the United States." Continued U.S. control of Saudi Arabia's oil was dependent on political stability in the country, and construction of the airfield would be a manifestation of U.S. interest aside from oil, would tend to strengthen the political integrity of the country, and would provide conditions under which the expansion of the oil concession could proceed, providing the revenue for long-term stability.[47]

President Truman approved continuation of work on the airfield on June 28, and in early August agreement was reached with Ibn Saud. In late August, however, the War Department again questioned the necessity of constructing the airfield. With the war almost over, the War Department felt that the State Department should assume full responsibility for the facility "entirely as an implementation of United States national interest." SWNCC again backed the State Department. Although the airfield was no longer a military necessity, the presence of the base would "create conditions in which American oil and other enterprises in that country can maintain their position and operate successfully." It was in the U.S. national interest that Saudi Arabia's huge oil reserves remain in U.S. hands so that they could be developed on a scale sufficient to lessen the drain on Western Hemisphere reserves. President Truman authorized completion of the airfield by the War Department at its own expense on September 28.[48]

Plans for long-term assistance were not going well. Plan I called for the Saudi Arabian government to assign to the U.S. government title to a billion barrels of oil underground. Royalty on this oil would be fifteen cents per barrel as opposed to the twenty-three cents per barrel paid by ARAMCO. In return for the reduced rate, the United States would make advance royalty payments up to a total of $50 million, to the extent necessary to balance the Saudi budget over the next five years. Royalties on oil produced while advances were outstanding would be applied to paying off advances. ARAMCO would agree to produce, transport, and refine this oil at cost. Under Plan II, the Export-Import Bank would loan money to the Saudi Arabian government to meet its deficits, with a 4 percent annual service charge on outstanding advances. Amortization would begin after ten years and would extend over a ten-year period. To ensure that the Saudi Arabian government would have sufficient dollars to service the loan, the plan called for the Navy to agree to buy a certain amount of oil from ARAMCO each year so that the company could pay its royalties in dollars. In return, ARAMCO was to agree to set aside a one billion barrel reserve for use by U.S. armed forces, to make this oil available to the military at a 25 percent discount, and to build a pipeline to the Mediterranean.[49]

There were problems with both plans. Although the War Department favored the reserve plan, State and Navy were uneasy about the possible reaction of the oil industry. The plan would also require legislation to amend the Naval Reserves Act, could create technical production difficulties, and could prove costly in terms of the interest on the public funds invested unless the oil was used soon. The key problem with the loan plan involved dollar exchange. ARAMCO had the option of paying royalties in either dollars or sterling, and with most of its probable markets in the sterling area was reluctant to guarantee royalties in dollars. Without dollars, however, the Saudi Arabian government would not be able to pay off its dollar loans. While the plan called for Navy purchases of ARAMCO oil to remedy this problem, the Navy was reluctant to commit itself in advance to buy any given amount of oil, especially in view of probable objections from the rest of the oil industry.[50]

Faced with these difficulties, the State Department decided to push for an Export-Import Bank loan to Saudi Arabia unconnected with oil. In mid-October, Emilio Collado presented a proposal to the head of the Export-Import Bank for a four-year, $25-million aid program. Collado explained that, although in the past his office had opposed such loans because of a concern that Saudi Arabia would lack sufficient dollars to pay interest and amortization, he had reevaluated the matter and decided that Saudi Arabia was a sound banking risk. Collado urged the bank to give the matter its immediate attention so that a proposal could be ready by December 1, and arrangements completed by the end of the year, when lend-lease ran out.[51]

A plan was worked out by mid-December, though technical difficulties prevented formal approval until early January 1946. The new plan called for the Export-Import Bank to establish a $25-million line of credit for Saudi Arabia, $5 million for development projects and $20 million to finance acquisition of U.S. products and services. The bank would maintain strict control of the money, and a service charge of 3 percent per annum would be assessed. Repayment would commence on October 31, 1951. In addition, the Saudi Arabian government would be required to agree "not to sell, encumber or in any manner dispose of to any third party its interest in any present or future oil concessions" while credits were outstanding, without the consent of the bank.[52]

The Saudi reaction was not favorable. Eddy had warned the department that the loan was more than a bank transaction: "It is a political commitment of long standing." Delays in developing the plan as well as the reduction of the amount from $50 million to $25 million and its duration from five to two and a half years had not pleased the Saudis. In March 1946, King Ibn Saud rejected the loan, protesting that the conditions attached to it were unacceptable and that he would not submit to bank control of funds. Specifically, the king objected to the "sweeping mortgage" of his present and future oil earnings, which he felt reflected unfavorably on his trustworthiness, and to the service charge which was in effect interest, which his religion prohibited. Rather than press the matter, Eddy recommended that the loan "be permitted to die quietly without obituary."[53]

Failure of the loan plan did not lead to financial disaster in Saudi Arabia, however, because oil revenues increased faster than anticipated. In 1945, ARAMCO paid the Saudi Arabian government $5 million in royalties, up from $2.5 million in 1944. Payments more than doubled in 1946, to $12.5 million, as production increased from just under 60,000 barrels per day (bpd) in 1945 to over 160,000 bpd in 1946. In this more relaxed atmosphere, a plan for a simple, straightforward Export-Import Bank loan of $10 million was developed, and in August 1946, the Saudis agreed to the new plan.[54]

By the end of the war, the main elements of a corporatist foreign oil policy were becoming evident. First, there was opposition to economic nationalism. Second, national security considerations were beginning to transform U.S. relations with the rest of the world, including the Middle East. In this regard, U.S. efforts to maintain the security and stability of the Middle East could be justified without being tied explicitly to oil. This shift was important because the structure of the U.S. oil industry meant that any affirmative action by the government to promote the development of Middle East oil ran up against the opposition of those segments of the oil industry whose interests would be adversely affected by the government's actions. The third element was a growing realization that the U.S. goals of security and access could be realized through the private operations of the major oil companies. Thus, rather than directly intervening in international oil matters, the United States should focus on maintaining an international environment in which U.S. oil companies could operate with security and profit.

Chapter 5

Public-Private Partnership

Changes in the patterns of oil use in the United States had a major impact on the nation's strategic position. Moreover, U.S. assumption of responsibility for the economic recovery of Western Europe increased the importance of Middle East oil. Between 1945 and 1948, U.S. foreign policy toward the Middle East underwent a dramatic transformation as the United States recognized and acted on its new strategic and economic interests. The solution to the problems facing U.S. foreign oil policy in the region was found in a series of private arrangements among the major oil companies which promised to accelerate the development of Middle East oil and thus reduce the drain on Western Hemisphere reserves. At the same time, the Truman Doctrine, with its call for the global containment of Communism, provided an ideological basis for maintaining the security and stability of the Middle East. U.S. policy with regard to Palestine further shaped the emerging public-private partnership and led to the U.S. government relying heavily on private interests to maintain U.S. influence in the region.

Supply, Demand, and Security

The United States in 1945 was acutely conscious of the vital role oil had played in World War II, both on the battle and home fronts. Between 1941 and 1945 U.S. oil production increased from 1.4 billion to 1.7 billion barrels. In the process the excess productive capacity that had existed in 1941 disappeared. By 1945, some U.S. fields were being drawn upon at slightly above the maximum efficient rate (MER). In addition, shortages of steel and labor, coupled with increased consumption, resulted in a decline in the rate of reproduction—the difference between marginal additions to proved reserves and gross annual consumption. During the war, the net annual increase in reserves averaged 0.4 billion barrels, as the growth in wartime demand almost equaled additions to reserves.[1]

The implications of the wartime experience were a matter of controversy among experts. Some took an optimistic view and argued that the decline in discoveries was a result of wartime conditions, in particular the lack of steel for drilling, and that once drilling activity resumed on a large scale the discovery rate would go up. Other experts were concerned by the duration of the decline in the rate of discovery and by the declining ratio of proved reserves to current consumption (see table 2). DeGolyer and McNaughton, the nation's leading oil consulting firm, advised the Navy in an April 1947 report that the United States would not be able to continue to find oil at a rate of 1 billion barrels per year, much less at the rate of 1.8 billion barrels needed to replace consumption. In a few years domestic production would begin to decline, with obvious consequences for the economy and for national security.[2]

The experience of the U.S. oil industry in the decade following World War II can be interpreted as supporting either the optimists or the pessimists. U.S. domestic oil production expanded from 1.7 billion barrels in 1945 to 1.9 billion barrels in 1950 and 2.4 billion barrels in 1955. Proved reserves also increased, from 20.8 billion barrels in 1945 to 25.2 billion barrels in 1950 and 30 billion barrels in 1955. On the other hand, the postwar increases in production and reserves were achieved only through greatly expanded drilling efforts. Not only were more wells drilled, but they were, on the average, deeper and more expensive than before the war, an indication that oil was becoming more difficult to find. Moreover, a large part of the increase in reserves was due to the development of already discovered fields and to improved recovery from existing fields rather than to the discovery of new fields. According to statistics first available in the 1960s, discovery of new fields declined sharply after World War II, especially in the largest size classes. In addition, consumption grew so rapidly that the ratio of proved reserves to current consumption fell from almost 14 to 1 in 1942 to around 11 to 1 by 1950.[3]

Both government and the oil industry had assumed that after the war

Table 2. The U.S. Oil Position, 1941–1950 *(in Thousands of Barrels)*

Year	Reserves	Consumption	Reserve / Consumption Ratio
1941	19,589,296	1,485,779	13.18
1942	20,082,793	1,449,908	13.85
1943	20,064,152	1,521,426	13.18
1944	20,453,231	1,671,263	12.23
1945	20,826,813	1,772,685	11.74
1946	20,873,566	1,792,786	11.64
1947	21,487,685	1,989,803	10.79
1948	23,280,444	2,113,678	11.01
1949	24,649,489	2,118,250	11.63
1950	25,268,398	2,375,057	10.63

Sources: Reserves: American Petroleum Institute, *Petroleum Facts and Figures,* Centennial Edition, 1959 (New York: American Petroleum Institute, 1959), p. 62. Consumption: DeGolyer & McNaughton, *Twentieth-Century Petroleum Statistics* (Dallas: DeGolyer & McNaughton, 1980), p. 100.

ended cutbacks in military oil consumption would leave an ample margin of capacity to meet postwar civilian needs. While military demand fell as expected, civilian consumption increased sharply as the booming postwar economy generated a ravenous demand for petroleum products. Between 1945 and 1955 oil increased its share of total U.S. energy consumption from 30.5 percent in 1945 to 37.2 percent in 1950 and 40.8 percent in 1955. Annual per capita consumption of oil rose from 13.4 barrels in 1945 to 15.6 barrels in 1950 and 18.7 barrels in 1955. Most of the increase came at the expense of coal, as the solid mineral's share of consumption fell from 50.7 percent in 1941 to 37.8 percent in 1950 and 29.3 percent in 1955.[4]

A key factor in the increase in oil consumption was the continuing transformation of the U.S. transportation system. Increasingly, automobiles, trucks, buses, airplanes, and diesel-powered locomotives replaced steam and electric-powered modes of transportation. The number of gasoline-powered motor vehicles rose from 30 million in 1945 to 58 million in 1955, and average annual consumption per vehicle increased from 641 to 790 gallons. The rise in automobile use was accompanied by the decline of public transportation. Ridership, which had swelled to 23.2 million per day in 1945 as a result of gasoline and tire rationing, dropped quickly and by 1955 was only 11.5 million. While the upsurge in motor fuel demand led the postwar expansion of oil consumption, the consumption of other oil products also increased. In spite of increasing competition from natural gas, oil continued to make inroads in the residential heating market. Even industries and utilities began to turn to oil as the coal industry became increasingly prone to labor-related disruptions. After an initial lag, kerosene shared in the increased demand for oil products as the advent of the jet engine opened a new market for this fuel. In addition, a new petrochemical industry began to emerge, further increasing the demand for crude oil.[5]

The result was a shortage in productive capacity and increases in the price of oil. Between the end of the war and December 1947, the price of 36° API midcontinent crude oil skyrocketed from $1.17 a barrel to $2.57 a barrel. Another sign of the short supply situation was the increase in oil imports. In 1948, total imports of oil increased from 160 million barrels to 186 million barrels, and the United States became a net importer of oil for the first time since the early 1920s. These manifestations of shortage had a great impact on oil policy. Between the end of the war and early 1948 some seventeen congressional committees inquired into various aspects of the oil situation and possible remedies. The executive branch was scarcely less active. In a draft letter to Congressman Dewey Short of the House Armed Services Committee on July 2, 1948, President Harry S. Truman noted that at least seven executive branch agencies and five special committees were keeping him "informed with respect to petroleum developments and prospects, alert to the petroleum problems confronting the nation, and vigilant in solution of those problems."[6]

Closely connected to concerns over the U.S. supply position was the fear of losing access to Middle East oil. In February 1947, the Joint Chiefs of Staff warned that in a future war the United States could run out of oil. Although the Middle East could produce much more oil if developed and offered the best return in terms of increased production per unit of input, the region was "very susceptible to enemy interference" and a "poor risk from the standpoint of accessibility of production in case of war." In view of these facts, the report's authors recommended the prompt development of a synthetic fuel production capacity utilizing domestic natural gas, coal, and shale reserves; the acquisition and maintenance of control of proved reserves in foreign countries, with special emphasis on reserves in the Western Hemisphere and the East Indies; and the conservation of domestic and more easily defensible U.S.-controlled foreign reserves through maximum peacetime importation of Middle East oil.[7]

Despite the JCS report's optimistic view that the development of synthetic fuels offered "the most promise as the safest and most prolific means to provide the required supply of petroleum and petroleum products to meet the demands of a future war," a large-scale program of synthetic fuel production was never undertaken. Under the authority of the Synthetic Liquid Fuels Act of 1944, the Bureau of Mines constructed and operated several experimental synthetic fuel plants in various parts of the country. For a variety of reasons, however, the synthetic fuels program was not able to generate the support necessary to make the transition from experimental to commercial operations. Synthetic fuel production required massive amounts of steel, much more than that required to find and develop natural petroleum, and steel was already in short supply during this period. In an emergency the availability of sufficient steel would be a crucial limiting factor. More importantly, synthetic fuels never caught on with the oil industry, in part because of costs, which were higher for synthetic fuels, though the margin of the differential was a matter of dispute. In addition, synthetic fuels, while a potential boon to the ailing coal industry, represented a source of competition to the oil industry.[8]

Reacting to the higher prices resulting from the postwar shortage, oil companies all over the world greatly expanded their investment in productive capacity. The result was a substantial increase in world productive capacity by the end of 1948. In addition, demand in the United States, which accounted for some 60 percent of world consumption, began to taper off during 1948 and declined slightly in 1949 because of a recession and an unusually mild winter. By the end of 1948, petroleum stocks were up 20 percent over the previous year, and the Texas Railroad Commission began cutting back production in Texas to bring supply into line with demand. Under the commission's order, production in Texas, which accounted for 45 percent of U.S. production, was reduced in stages from 2,710,000 bpd in December 1948 to 1,872,000 bpd in July 1949. Foreign production also fell, partly as a consequence of the decline in U.S. demand, and partly as a result of a slight drop in demand in the rest of the world.[9]

Imports, on the other hand, continued at a high level in 1949; Middle Eastern oil accounted for around one-sixth of the total. Although net imports constituted only 3 percent of domestic crude oil output, they took on increased significance since demand for oil was declining. In order to maintain prices, the Texas Railroad Commission further cut back allowable production. Although prices remained stable, restricting supply by reducing domestic production while maintaining import volume resulted in independent producers bearing the brunt of the burden of cutbacks. Independent producers reacted by calling for restrictions on imports.[10]

Given these circumstances, the U.S. oil industry vigorously opposed the government's synthetic fuels program. Domestic producers vociferously objected to the development of new competition at public expense. Companies with large sums of money invested in foreign reserves were under pressure from host governments to increase production, and did not want to lose the U.S. market. Moreover, oil imports were highly profitable, at least as long as domestic prorationing bodies maintained the high U.S. price. During 1949, the major oil companies canceled their own synthetic fuel experiments, and began attacking government plans for promoting synthetic fuel development. These efforts were ultimately successful as the Eisenhower administration terminated the nation's synthetic fuels program in 1954.[11]

Conservation of Western Hemisphere reserves through the development and commercialization of Middle East oil had been a U.S. policy goal since the middle of the war. As the United States gradually assumed responsibility for the security and stability of Western Europe, the need to develop Middle East oil became even more urgent. The goal was to increase Middle East production to the point at which it could supply all of the Eastern Hemisphere's requirements, thereby relieving the pressure on Western Hemisphere reserves. As Middle East production expanded, small amounts of Middle East crude oil also began making their way into the United States. To a point, imports of Middle East oil during peacetime would add to the nation's security. The problem was one of balance. If large amounts of low-cost Middle East oil were brought into the United States, many U.S. producers would be forced out of business and U.S. production would fall as exploration and development of new fields would be discouraged. Thus this course would "save" U.S. oil resources in the sense that they would not be used, but the result, paradoxically, might be detrimental to national security. From a military point of view, undeveloped oil reserves were useless, and dependence on Middle East oil imports could leave the United States in a situation in which, if access to Middle East oil were lost, the country would not have discovered sufficient oil reserves and developed sufficient productive capacity to meet emergency requirements. On the other hand, keeping oil production in the United States and the Western Hemisphere at high levels by shutting out Middle East oil imports would, at least for a few years, ensure the development needed to meet potential emergency requirements. But seeking to maintain domestic productive capacity by encouraging the consumption of domestic supplies of a nonrenewable resource, when the crux of the problem was to conserve domestic supplies, was clearly self-defeating.[12]

Several schemes were devised in an effort to overcome this dilemma. Strategist Bernard Brodie suggested importing large amounts of foreign oil in order to relieve the strain on domestic reserves, while promoting domestic exploration and the development of a synthetic fuels industry through a system of government subsidies. Noted oil geologist E. L. DeGolyer suggested a plan whereby the U.S. government would purchase domestic oil fields, to be held in reserve, and pay the owner(s) in oil purchased in the Middle East. This plan would allow the development of shut-in productive capacity since the purchased fields would be developed but not produced. Finally, Wallace Pratt, Jersey's chief geologist for many years before becoming the National Security Resources Board's oil expert, recommended that oil production in the United States and northern South America—the "minimum strategic area"—be restricted to a rate 20 percent below the MER of each field in order to develop a reserve capacity. Pratt would then restrict imports into this area. The resulting higher prices would, he argued, stimulate exploration and development of new fields as well as a synthetic fuels industry.[13]

None of the plans was ever tried. All shared the common "flaw" of requiring an active, interventionist role for the federal government. In addition, there were technical problems with both the DeGolyer and Pratt schemes since they required precise estimates of the MER of disparate oil fields. Estimating the MER of oil fields was not an exact science, a fatal flaw because small differences in calculations could mean thousands of dollars to the owners. Pratt's plan was also too obviously a scheme for raising prices by limiting production, and provoked an uproar when a copy was leaked to the press. Finally, all would cost large amounts of money to implement and enforce.[14]

Thus the United States was left with two alternatives—stimulating development of petroleum resources in the Western Hemisphere outside the United States or finding ways to maintain access to Middle East oil. (Reducing or at least slowing the growth of rapidly rising consumption was also an alternative but was not considered. The reason for this "nondecision" will be discussed in the conclusion.[15]) U.S. efforts to encourage Latin American oil development are dealt with in the following chapter. The rest of this chapter focuses on efforts to protect and promote U.S. interests in the Middle East.

The Great Oil Deals

During the war ARAMCO had discovered four major oil fields with proved reserves estimated in billions of barrels, and it was clear that great amounts of oil remained to be discovered. But to realize this potential would require an immense investment as well as markets for the increased production. ARAMCO's owners, Standard Oil of California (SOCAL) and the Texas Company, had already invested some $80 million in Saudi Arabia, and to expand their markets against the competition of the established international marketing companies would be expensive. Reaching markets west of Suez at competitive costs would require a 1,500-mile pipeline from the Persian Gulf oil fields to the Mediterranean, costing over $100 million. Seeking to force their oil into markets in Europe and elsewhere could lead to disruptive price wars, which would cut into profits even as sales expanded. Adding to the risks was the political instability of the Middle East and its vulnerability to Soviet takeover should the nascent cold war intensify. With King Ibn Saud urging immediate expansion to increase his revenues, SOCAL and Texas began to consider the idea of bringing in other companies to provide the capital and markets needed to expand production and stabilize the concession.[16]

Jersey and Socony-Vacuum were obvious prospects. Although both had interests all over the world, their position in the Middle East was limited to a joint 23.75 percent share of the Iraq Petroleum Company (IPC). In 1946, this share amounted to around 400 million barrels of reserves and 9,300 barrels per day (bpd) of production apiece. This relatively minor position stemmed primarily from the Red Line Agreement of 1928, which bound the owners of IPC not to acquire concessions independently of the others within the boundaries of the old Ottoman Empire. The Anglo-Iranian Oil Company (AIOC), the dominant member of IPC, had all the oil it needed in Iran and Kuwait, and seemed interested in other areas mainly to protect its position in these two countries. As a consequence, large areas of the Middle East had been left to other companies.[17]

Jersey's domestic and Venezuelan operations had previously been sufficient to supply most of its market outlets in both the Eastern and Western hemispheres, but the company's long-range supply and demand forecast at the end of the war indicated that Western Hemisphere demand alone would eventually absorb its U.S. and South American production. If Jersey were to maintain its position in Eastern Hemisphere markets, it needed to acquire a larger share of Middle Eastern production. Furthermore, as a March 1946 company strategy document pointed out, obtaining a substantial interest in Saudi Arabia would not only provide additional supplies which probably could be converted into cash and profits more rapidly than IPC oil, but would also have the advantage of "easing the pressure" that would otherwise come from ARAMCO as SOCAL and Texas sought markets for Saudi Arabian oil.[18]

In the spring of 1946, the ARAMCO partners and Jersey began informal discussions on the possibility of broadening ARAMCO's ownership. An initial problem arose over the Texas Company's insistence that Jersey pay $650 million for a one-third share in ARAMCO. Jersey refused to consider such a sum, and argued that it should be permitted to participate in ARAMCO for a nominal sum since its markets would so raise the company's total sales that SOCAL and Texas would be adequately compensated through higher profits. As talks continued, Texas reduced its demands considerably, and price never became a major issue.[19]

More serious opposition to the proposed sale emerged from within the ARAMCO organization. ARAMCO field management almost unanimously opposed the idea. Speaking for this group, Ronald C. Stoner, head of SOCAL's Production Division and a company director, argued that ARAMCO was too valuable to share with Jersey. Stoner pointed out that in Saudi Arabia ARAMCO possessed the world's largest and lowest cost reserves. This position made the company competitively strong vis-à-vis Jersey, whose foreign production was limited by joint agreements with Shell and AIOC. There was a danger that if Jersey got into ARAMCO it would try to retard the company's development in order to protect its extensive interests in other concessions. Stoner believed that a better way for ARAMCO to acquire more market outlets and capital would be through sales or exchange contracts with such medium-sized U.S. companies as Phillips, Atlantic, Standard of Indiana, and Sinclair, which were seeking additional sources of supply. Bringing in several smaller companies would also provide better political "protection" for ARAMCO.[20]

To counter Stoner's arguments, SOCAL head Henry D. Collier, whose background was in international marketing, had a profitability study prepared which showed that a one-third interest in ARAMCO with Jersey participating was worth more than a one-half interest without Jersey because of a doubling of sales. A Texas Company study reached a similar conclusion. Reserves in Saudi Arabia were greater than could be marketed over the lifetime of the concession, with or without Jersey as a partner, and the added volume would reduce unit operating costs. Cooperating with Jersey rather than competing with it would probably result in "better stabilization" of world markets, and reduce Jersey's, and possibly Socony's, desire to explore or develop other of their holdings in the Middle East and Far East. Bringing Jersey in would also justify early construction of the needed pipeline to the Mediterranean as well as reduce Texas's share from one-half to one-third in the cost of this and other projects. All these things would increase royalties to the king, lessen the need to make advances, and add to the stability of the concession. Equally important, the sale would eliminate Texas's "Arabian Risk" and provide the company with cash it could use elsewhere. In addition to materially reducing political, commercial, and economic risks, the deal would help assure, Texas believed, more support from the U.S. government for ARAMCO.[21]

If Jersey, and Socony which was also interested, were to participate in ARAMCO, they would have to do something about the Red Line Agreement. Jersey and Socony had been chipping at the Red Line restrictions for some time. In October 1945, they had declared that the restrictions against independent purchases of crude oil from within the Red Line area did not apply to purchases of products derived from crude produced within the Red Line. In a thinly veiled reference to the IPC Group Agreement, of which the Red Line was a part, Jersey President Eugene Holman had informed the visiting head of Compagnie Française des Pétroles (CFP) in January 1946 that "there had been a substantial change in the attitude of the American public and government toward restrictive agreements and, under current conditions, reaffirmation of the agreement seemed inadvisable." In July 1946, Jersey received word from its regular legal counsel in London that, in their opinion, the IPC Group Agreement had been terminated in 1940 when two of the owners of IPC— CFP and Calouste Sarkis Gulbenkian—had come under "enemy domination" as a result of the German occupation of France. If this opinion were upheld in court, the Red Line restrictions would no longer apply, and Jersey and Socony would be free to purchase an interest in ARAMCO while retaining their interest in IPC. Because of the importance of the matter, the head of Jersey's law department turned to several eminent law firms in London for their opinion on the matter. These firms were unanimous in their opinion that the Red Line Agreement was no longer binding on the signatory companies.[22]

Armed with these opinions, Jersey and Socony pressed on with their negotiations with ARAMCO. In late August, Orville Harden of Jersey and Harold F. Sheets of Socony met with Assistant Secretary of State William L. Clayton, Clayton's assistant George C. McGhee, and State Department Petroleum Adviser Charles Rayner. The oilmen informed the State Department of their lawyers' opinion that the IPC Group Agreement had been dissolved since 1940. Although they were happy to be rid of the Red Line restrictions, without the Group Agreement their right to take oil from IPC according to their ownership shares ceased. Unless a new agreement affirming their right to oil were negotiated, their interest in IPC would become merely a financial interest, an interest in earnings rather than oil. The oilmen maintained that such an outcome would be contrary to the national interest, and asked for State Department assistance in pushing for a new agreement that would reaffirm their right to IPC oil, but without the restrictive provisions of the old agreement. Rayner noted that the Justice Department would probably not allow any new restrictions, and promised that the State Department would support the companies in reaffirming their rights to IPC oil. He also urged them to try to work out the matter amicably with their foreign partners.[23]

In September, Harden and Sheets traveled to London to set out their case to their partners in IPC. They found the British more concerned about the politics of the matter than about legal technicalities. Sir William Fraser, the head of AIOC, felt that the maintenance of a united front was vitally important at a time when the dangers of government intervention were, in his opinion, so great. He viewed the group agreement as a defense against the dangers facing the oil industry, and argued that if a new agreement could not be worked out, it would be better to continue with the old agreement than to "rock the boat." Shell was more sympathetic to the U.S. position, but felt that it had to follow AIOC's lead. Shell was also concerned that the French and Gulbenkian might "drag the ugly matter into the public arena." Initially the French seemed willing to cooperate and to agree to a revised group agreement which would eliminate most of the earlier agreement's restrictive clauses. But as they became aware of the implications of such an action, they informed their U.S. partners that although they might agree to eliminating the restrictions against purchases of Red Line crude the other restrictions should be retained. Gulbenkian, whose fortune was based on his 5 percent share in IPC, refused to accept that the group agreement had been dissolved, and threatened to take the matter to court.[24]

To secure AIOC's support, Jersey and Socony began negotiations with Fraser to purchase large amounts of oil over a long period of time. This arrangement would provide AIOC with outlets for its vast holdings in Iran and Kuwait, and thus make any expansion of Saudi Arabian production less threatening. Meanwhile, Shell, which had been buying oil from Gulf since the middle of 1946, was negotiating a long-term supply arrangement with Gulf. Gulf would supply Shell with large amounts of oil over a ten-year period from its share in the Kuwait concession. Shell would then refine, transport, and market this oil in the Eastern Hemisphere and share the profits equally with Gulf. Shell's obligation to take oil from Gulf would be proportionately reduced if Gulf made inroads into Shell's Eastern Hemisphere markets. Thus Shell, too, need not depend solely on IPC for its supply needs.[25]

Additional assistance came from the State Department. During informal Anglo-American oil talks in London in late November, the State Department representatives pointed out that provisions of such agreements as the IPC Group Agreement and the Kuwait Oil Company agreement were contrary to the Anglo-American Oil Agreement and to the preservation of competition in the international oil trade. Therefore, the U.S. government was advising the U.S. members of IPC that the restrictive clauses in the

IPC Group Agreement should not be accepted when the agreement was renegotiated. The U.S. government would, however, support Jersey and Socony with respect to the preservation of IPC as a consortium based on shares of oil rather than shares of profits.[26]

Socony's chief counsel, C. V. Holton, advised his company in late October that the two companies' acquisition of an interest in ARAMCO should not in itself constitute a violation of the antitrust laws because no unreasonable restraint of U.S. commerce would be involved. Still, the arrangement would place practical control of crude oil reserves in the Eastern Hemisphere in the hands of seven companies, five of them U.S.-owned, all of which also had substantial holdings in the Western Hemisphere. Since such concentrated control could lend itself to restrictive arrangements which could affect the import and export trade of the United States, great caution would have to be exercised to stay within the law. Holton felt that the danger that the mere existence of potential control would be construed as an antitrust violation was slight, but added that he did not personally believe that "comparatively few companies for any great length of time are going to be permitted to control world oil resources without some sort of regulation."[27]

To head off trouble, Harden of Jersey explained his company's plans to the State Department on December 3. According to Harden, the deals were part of a long-range plan whereby Jersey would obtain all the oil for its European and other Eastern Hemisphere markets from the Middle East. Western Hemisphere oil would be retained exclusively for Western Hemisphere markets and thus conserved. The arrangement with AIOC was necessary since even with an interest in ARAMCO and its share in IPC Jersey needed additional oil to meet the demands of its market outlets. A week later, Gulf head J. Frank Drake also outlined his company's plans for the State Department. Drake explained that it would take Gulf years to build up market outlets for Kuwait oil on its own, and the sheik of Kuwait was demanding immediate and large increases in revenue. Thus the deal with Shell was necessary to retain the U.S. interest in the Kuwait concession. The State Department raised no objections to either of these deals.[28]

Holton and the general counsels of Jersey and the Texas Company discussed the ARAMCO negotiations with Attorney General Tom Clark on December 4. Texas Company President and General Counsel Harry T. Klein told Clark, a Texan with close ties to his state's business and political interests, that Jersey and Socony were being brought into ARAMCO because SOCAL and Texas did not have sufficient outlets for the quantity of oil that needed to be sold in order to satisfy the Saudi Arabian government and thus assure the stability of the concession. Clark expressed no objections to the planned transaction, and requested only that he be kept informed and that drafts of the contracts be sent to him when they were ready.[29]

With this groundwork completed, Jersey and Socony formally advised their IPC partners that they considered the IPC Group Agreement dissolved. Then on December 14, they initialed an agreement with SOCAL and Texas under which Jersey would acquire a 30 percent and Socony a 10 percent interest in ARAMCO. They also arranged to acquire proportionate interests in the Trans-Arabian Pipeline Company (TAPLINE), which had been organized by SOCAL and Texas to build a thirty-inch pipeline from the Persian Gulf oil fields to the Mediterranean. Less than a week later, Jersey signed an agreement in principle with AIOC to purchase, on a "cost-plus" basis, 133 million tons of crude oil over a twenty-year period, and to participate equally in the construction and use of a 300,000 bpd pipeline from the Persian Gulf to the Mediterranean. Socony later acquired 20 percent of Jersey's interest in this oil and the pipeline.[30]

AIOC and Shell agreed that the IPC Group Agreement was no longer valid. AIOC's apprehensions about competition from growing Arabian production had been lessened by the prospects of large sales to Jersey and Socony. Shell had seen to its needs with a long-term sales contract with Gulf. CFP and Gulbenkian, however, maintained that the group agreement was still valid, and threatened to take Jersey and Socony to court for their breach of the Red Line. CFP, as trustee of France's interest in Middle East oil, was especially upset at the prospect of being excluded from Saudi Arabia, fearing that if Jersey and Socony became deeply involved in Saudi Arabia they would neglect developing Iraq. At the IPC Group meeting on December 19, the CFP representative proposed that the IPC partners as a group share in any interest in ARAMCO that SOCAL and Texas were prepared to sell. Gulbenkian's representative supported this idea, but the British groups demurred from taking action. This avenue blocked, CFP threatened to "seek the protection of the courts" in order to preserve the group agreement. Jersey's and Socony's claim that U.S. antitrust laws prevented them from accepting such restrictions as the Red Line was "merely an excuse upon which to hang a contemplated and now effected breach of the Group Agreement." The French noted, however, that they would take no further action if the group agreement were changed to require members to share any interest in concessions or pipelines they obtained within the Red Line area with the other members. CFP also wanted assurances that IPC would be properly developed, including the completion of the two sixteen-inch pipelines under construction and the construction of a twenty-four-inch line from Iraq to the Mediterranean.[31]

The French government began putting pressure on Jersey and Socony operations in France, to the extent that the Socony representative in Paris warned that business would be adversely affected if CFP were not placated.

In addition, the French ambassador to the United States protested to the State Department that the action by Jersey and Socony would retard the development of the IPC concession and thus adversely affect the French economy by denying France substantial amounts of oil on a favorable basis in terms of foreign exchange. The French argued that the U.S. government was morally obligated to see that the terms of the IPC Group Agreement were enforced. The State Department maintained, however, that the problem was one between private parties, and that French interests were already adequately protected by existing plans to expand IPC's operations.[32]

Jersey and Socony received further assurance of State Department support on January 9. While urging the companies to try to accommodate legitimate French demands, department officials stressed that they did not want the companies to withdraw from the ARAMCO deal or to accept new restrictive clauses. Department officials also agreed with the companies that neither the British nor the French should be allowed into ARAMCO. According to company sources and to the Near Eastern Division of the State Department, King Ibn Saud insisted that the concession remain exclusively in U.S. hands. Moreover, the ARAMCO and Gulf deals provided a means for channeling increased revenues to the governments of Saudi Arabia and Kuwait, further helping to ensure the continuity of the concessions in U.S. hands. Finally, the deals served U.S. strategic interests, as defined by the State and Navy departments, by helping hasten the development of Middle East oil and its large-scale entry into world markets, thus reducing the drain on Western Hemisphere reserves. Without the deals, potentially destabilizing price wars could result if the companies with more production than market outlets—ARAMCO, Gulf, and AIOC— tried to force their way into world markets. Such a situation could increase the strain on Western Hemisphere reserves if the "established marketers"— Jersey, Socony, and Shell—continued to feed crude oil from their Western Hemisphere holdings into the European market rather than contract their distributive operations.[33]

The impacts of the deals on world oil markets were also a source of concern. As noted above, the three deals shared the common characteristic of transferring crude oil for marketing purposes from companies whose market outlets in the Eastern Hemisphere were not large enough for their potential production to companies whose crude supply was inadequate for their existing market outlets. Thus, in addition to reducing the drain on Western Hemisphere reserves, the deals had the cumulative effect of preserving the "competitive balance of power" among the major oil companies in world markets. Because of their scope and duration, and the existing control of the rest of Eastern Hemisphere production by the same companies, the deals also effectively deprived other companies of access to Middle East oil. Control of supply would allow these companies to maintain their control of markets and prices, especially in view of the tight web of interlocking arrangements among the major companies (see Appendix 1). Given the inescapable relationship of the positive to the negative aspects of the deals, the Petroleum Division recommended that the State Department neither approve nor disapprove of the proposed transactions. Instead, the department should "perceive no objection" to the deals, while asking the companies for assurances that the new arrangements would not involve allocation of markets and other practices "incompatible with United States trade policies and objectives."[34]

The difficulties the companies were having trying to accommodate the French also caused concern. In late February, Paul Nitze, deputy director of the Office of International Trade Policy, who had come to the department from the investment banking firm of Dillon, Read and Company, proposed that Socony could buy Jersey's share in IPC and then withdraw from the ARAMCO deal. This proposal would provide Socony with more oil from IPC and free Jersey to enter ARAMCO. Besides being a "simple, clear-cut solution" to the problem with the French, the plan would "simplify and perhaps arrest the trend toward multiplication of the interlocking arrangements between and among the oil companies engaged in the international oil trade." Nitze hoped that this indirect result would counter criticism of the deal from the Senate and from "outside" U.S. oil companies. As an alternative, Nitze suggested that Jersey sell its interests in IPC to a group composed of any U.S. oil companies desiring to participate. The Petroleum Division supported Nitze's idea and pointed out that "considering the repercussions of the deals, as well as a frank recognition of the possible consequences of the existing ownership base ... some other feasible pattern of ownership" of Middle East oil might be in the public interest and more consistent with U.S. foreign economic policy.[35]

Nitze presented his plan to Jersey and Socony on March 7, but the two companies were not interested. They had considered such an arrangement earlier in the year when it appeared that there might not be any other way around the Red Line. But as it became clear that the IPC Group Agreement was no longer valid under English law, they had dropped the idea, preferring, for commercial reasons, to spread their holdings. In addition, an analysis by Socony showed that the amount that it could afford to pay Jersey for its share in IPC was barely more than Jersey's annual profit from IPC. As for criticism of the deals, Jersey head Harden maintained that the deals did not exclude other companies from the Middle East. If other companies wanted in they could try to obtain a concession or buy into an existing one as Jersey had done.[36]

To deal with CFP, which along with Gulbenkian had started legal proceedings in London in February to block the deal, Jersey and Socony devised a plan which they felt would meet French requirements. The heart of the plan was to allow IPC members to obtain oil "in quantities in excess of basic proportions." Thus if the French wanted more oil than their interest in IPC entitled them to, they could obtain it at a special price halfway between cost and the world market price. In addition, Jersey and Socony agreed that IPC would build the two sixteen-inch and the twenty-four-inch pipelines to the Mediterranean that the French desired. Consequently IPC would be able to increase its production to 25 million tons per year, six times the 1947 level. In return, all restrictive provisions of the 1928 group agreement would be eliminated, allowing Jersey and Socony to acquire an interest in ARAMCO. Jersey and Socony felt that these arrangements were fair to CFP, and urged the State Department to help them convince the French to accept them.[37]

No agreement with the French or Gulbenkian was reached by March 12, 1947, the date set by SOCAL and Texas for the deal to be concluded. Jersey and Socony were forced to work out standstill agreements with the ARAMCO partners which held in abeyance their entry into ARAMCO and TAPLINE pending clarification of the legal issues. In the meantime, the two prospective buyers agreed to guarantee a bank loan of $102 million to ARAMCO, the amount they were to pay directly for their shares in the company, and to forgo over $300 million in dividends over a period of years. They also arranged to buy oil from ARAMCO in the interim and to guarantee their proportionate shares in a $125 million loan for TAPLINE. These agreements were signed on March 12, 1947.[38]

The French gave up in May, and the four major groups reached tentative agreement on the basis of the proposal put forward by Jersey and Socony in February. Gulbenkian, however, held out until November 3, 1948, one day before arguments on his suit and a countersuit by Jersey and Socony were scheduled to be heard. With his holdings confined to IPC, and with no refining or marketing organization of his own, Gulbenkian wanted compensation for agreeing to the ARAMCO deal and assured markets for his oil. The settlement gave him rights to a large amount of IPC production over and above his 5 percent interest for a fifteen-year period and guarantees that the other partners would market this oil and pay him in dollars. This matter settled, the restrictive provisions of the group agreement were eliminated, and in the so-called ARAMCO release Jersey and Socony were permitted to complete their deal with ARAMCO. The lawsuits were then withdrawn, and in December 1948, Jersey and Socony completed their purchase of ARAMCO stock.[39]

The private initiatives of the oil companies provided a solution to the problem of integrating Middle East oil into world markets and thus helped to secure U.S. interests in the region. At the same time, U.S. foreign policy was undergoing a momentous change that would result in a U.S. commitment to maintain the security and stability of the Middle East.

Toward the Truman Doctrine

After the crisis resulting from the scramble for oil concession rights in 1944, the United States returned to its previous policy of maintaining Iran as a buffer between the Soviet Union and the oil fields of the Persian Gulf. The State Department began to concentrate on ending the Anglo-Soviet occupation of Iran as soon as possible and on protecting U.S. interests by strengthening the Iranian security forces under the direction of U.S. advisers. Oil was not to be mentioned until all foreign troops had left. At the Yalta and Potsdam conferences the United States sought to secure assurances that the Soviet and British forces would be withdrawn from Iran as soon as possible. Secretary of State Edward R. Stettinius and British Foreign Secretary Anthony Eden reached an agreement at Malta in January 1945 to press the Soviets to agree to a mutual withdrawal of forces and to recognize the right of the Iranian government to refuse to negotiate oil concessions as long as the country was occupied by foreign troops. It was not until the September 1945 foreign ministers' meeting, however, that the Soviets agreed to withdraw their forces. At the meeting, Soviet Foreign Minister V. M. Molotov, in response to British queries about the date troops would be withdrawn, stated that the "Soviet Government attaches exceptional importance to the strict fulfillment of obligations undertaken." This statement was taken to mean that Soviet troops would be withdrawn by March 2, 1946, six months after the end of the war with Japan.[40]

As economic conditions in Iran deteriorated during 1945, Wallace Murray, who had taken over the ambassadorship in June in a move to bolster the U.S. presence in the strategically located country, reported that he regarded the situation with "grave concern." Political, economic, and social conditions were "deplorable," and the "present ruling class showed little evidence of either the will or the ability to improve them." Such conditions provided the British and the Soviets with an opportunity and an excuse to intervene. Murray felt that British objectives in Iran were "purely defensive." Although the present aims of the Soviets were "probably limited to the maintenance of a buffer zone in Iran as protection against attack from the south," the staunchly anti-Communist Murray suspected that they might include establishment of a "friendly" government in Tehran on the model of recent events in Eastern Europe. This arrangement would be harmful to U.S. interests because it would mean the exclusion of U.S. airlines from Iran, the loss of Iranian trade, and "most important of all it would mean extension of Soviet influence to the shores of the Persian Gulf creating a potential threat to our immensely rich oil holdings in Saudi Arabia, Bahrain, and Kuwait." The time had come for the United States "to take a positive stand against the continuance of present Soviet activities."[41]

The oil concession crisis of 1944 had also been a watershed for Soviet policy in Iran. Actual Soviet interference in Iranian affairs before the crisis had been limited. Faced with the possibility of being excluded from even minimal influence in northern Iran even as the influence of their only strong potential enemy grew, however, the Soviets apparently decided to secure their interests before the deadline for troop withdrawal interfered with their leverage. Beginning in the summer of 1945, incidents of antigovernment activity began to take place in Soviet-occupied Azerbaijan and Kurdistan. Encouraged but not created by the Soviets, the scattered outbreaks turned into a full-scale revolt against the government in Tehran by the fall. In mid-November, the Soviets prevented the Tehran government from sending troops into Azerbaijan to quell the revolt, claiming that to do so would threaten the security of their forces. The Iranians immediately called on the United States for help. The United States sent a note to the Soviet government on November 23, and suggested that all Allied troops should be withdrawn from Iran by January 1, 1946. The recommendation did not apply to the U.S. advisers to the Iranian military and gendarmerie.[42]

The crisis deepened in early December when the separatist forces in the north established the Autonomous Republic of Azerbaijan and the Kurdish People's Republic. Accurate information on the situation in northern Iran was difficult to obtain. Nonetheless, the State Department quickly accepted the Iranian interpretation of the uprisings as the product of Soviet aggression. The U.S. consul in Azerbaijan, whose reports emphasized the domestic roots of the revolt and the complexity of the situation, was removed. His replacement, an inexperienced twenty-seven-year old, was quick to place events in the north in their "proper perspective." The Soviets rebuffed U.S. efforts to obtain an early withdrawal of forces, pointing out that under the Tripartite Treaty they had a right to keep their troops in Iran until March 2, 1946. This initial effort failing, the United States encouraged Iran to present its case to the United Nations in order to put pressure on the Soviets to end their interference in Iran's internal affairs.[43]

The Iranian crisis coincided with, and contributed to, an important shift in U.S. policy toward the Soviet Union in the autumn of 1945. Soviet policy in Eastern Europe seemed to violate U.S. ideals if not self-interest, and revelations of Soviet espionage in Canada increased concern over the security of the U.S. atomic deterrent. Influenced by the State Department's interpretation of the situation, Truman viewed Soviet actions in Iran, along with events in Greece and Turkey, as part of "a giant pincers movement against the oil-rich areas of the Near East and the warm-water ports of the Mediterranean." The situation in Iran was an "outrage," the president complained in early January 1946, and he was "tired of babying the Soviets." The president's attitude marked a hardening of U.S. official views toward the Soviet Union. George Kennan's "long telegram" of

February 22 summed up these concerns over Soviet actions and stressed the necessity of countering what he viewed as the expansionist dynamic of Soviet policy.[44]

This shift in attitude influenced U.S. policy toward Iran. When the Soviets failed to withdraw their troops on March 2, reaction in the United States was sharp. Then on March 5, the new U.S. consul in Azerbaijan warned of ominous Soviet troop and armor movements toward Turkey as well as Tehran. Secretary of State James F. Byrnes, who had been stung by criticism of his handling of the Soviets, decided to take a tough public stand on Iran in order to silence his critics. To make sure that the purity of U.S. motives would not be questioned, Byrnes issued strict instructions to the U.S. Embassy in Iran that there should be no discussion of oil concessions until after the crisis. State Department officials worked closely with the Iranian ambassador to place Iran's complaint before the U.N. Security Council, while in Iran Ambassador Murray urged the new government of Qavam Saltaneh to stand firm.[45]

Meanwhile, Qavam, a veteran politician who had been prime minister in 1921 and 1922, had been engaged in negotiations with the Soviets. A journey to Moscow in mid-February produced no results, but on April 4, Iran and the Soviet Union reached agreement whereby the Soviets would withdraw their troops; a joint Soviet-Iranian oil company would be established, subject to the approval of the Majlis, to develop the oil of northern Iran; and Azerbaijan would be recognized to be an internal Iranian matter. The Soviets withdrew their forces, as promised, on May 6, ending this stage of the crisis.[46]

The events in Iran led to a reevaluation of U.S. policy toward the Middle East. In a lengthy report sent to President Truman on September 24, 1946, Special Counsel Clark M. Clifford argued that the United States should be ready to use force to guard its vital interests, warning that "continued access to oil in the Middle East is especially threatened by Soviet penetration into Iran." Then, in response to a series of questions from the State Department on U.S. interests in Iran, the Joint Chiefs of Staff reported on October 12 that it was "to the strategic interest of the United States to keep Soviet influence and Soviet armed forces as far as possible from oil resources in Iran, Iraq, and the Near and Middle East." As a source of oil, Iran was of "major strategic interest" to the United States, while its difficult terrain provided an important barrier between the Soviet Union and U.S.-controlled oil resources in Saudi Arabia. The Soviets did not have enough oil to fight a major war, and denial of the oil of the Middle East to the Soviets would force them to fight an "oil-starved war." On the other hand, loss of this oil would force the United States and its Allies to fight an "oil-starved war."[47]

Meanwhile, the new U.S. Ambassador to Iran, George V. Allen, had established close relations with the young shah. At forty-two the youngest ambassador in the foreign service, the North Carolina–born Allen utilized his weekly tennis match with the shah to cultivate the younger man's friendship. Allen assured both the shah and Qavam that the United States would support their nation's independence, and began pressing Qavam to move against the Iranian left. After repeated promises of arms and support, Iranian forces were sent into Azerbaijan in early December 1946, where they easily overthrew the leftist government. The Soviets, apparently fearful of endangering Majlis approval of their oil concession, refrained from taking any action beyond diplomatic protests.[48]

The strategic importance of Middle East oil influenced U.S. policy toward Greece and Turkey as well as Iran. Although circumstances differed greatly in the three countries, the United States came increasingly to interpret events in all three as being part of an overall Soviet design "to break through the Iran-Turkey-Greece barrier to the south and establish Soviet power in the eastern Mediterranean, the Middle East, and the Indian Ocean." In the wake of the Iranian crisis of early 1946, and renewed Soviet pressure on Turkey in August, the United States reformulated its policy toward the three countries of the northern tier. Together with the shift toward an active policy in Iran, the change in policy toward Greece and Turkey set the stage for the declaration of the Truman Doctrine on March 12, 1947. As Dean Acheson explained to a group of congressional leaders on February 27, Greece, Turkey, and Iran were all linked. If one fell, the others would go, and with them control of the eastern Mediterranean and the Middle East. Although Acheson had reference to the oil of the Middle East deleted from the president's address, concern over U.S. access to the region's chief resource played an important role in the "revolution" in U.S. foreign policy.[49]

The ouster of the leftist regime in Azerbaijan coupled with moves against the pro-Soviet Tudeh party in the rest of Iran removed a major source of Soviet influence. Nevertheless, a Soviet oil concession in northern Iran, as called for in the April 1946 agreement between Iran and the Soviet Union, could provide another avenue for possible Soviet political and economic penetration. Ambassador Allen warned in early 1947 that it was doubtful if the Soviets were "capable of owning economic concessions in foreign countries without inevitably violating the sovereignty of that country." Washington shared Allen's views. In an August 1947 policy statement, the State Department argued that it would be impossible for Iran to devise safeguards that could prevent the Soviets from using the concession as a "spearhead for the penetration of Iran." Therefore it would be "contrary to the interests of the United States for Iran to grant any kind of oil concession to the Soviet Union."[50]

The Iranians also wanted a way out of their agreement with the Soviet Union, but were concerned about the possible consequences of displeasing their powerful northern neighbor. As a solution, the Iranians suggested placing all their oil development, including the existing AIOC concession, under some sort of international supervision. The United States opposed this idea, pointing out that it could not support any solution that would interfere with valid concession rights. AIOC did not have a perfect record in Iran, Allen admitted to the shah, but "on balance the concession had been mutually beneficial." Allen also pointed to Mexico's loss of markets and revenue as an example of what would result from expropriation. Failing to obtain U.S. support for an "international" solution, the Iranians sought to get the United States to commit itself to defending Iran in the event that Soviet displeasure over Iranian refusal took aggressive forms.[51]

To help overcome Iran's fear of being left alone to face Soviet displeasure, the State Department instructed Allen to emphasize that the United States would support Iran's freedom to decide the oil concession issue without external pressure. The United States extended a $25 million credit to Iran in June to purchase surplus military equipment, and in early October renewed the contract of the U.S. Military Mission to the Iranian Army. As Soviet pressure on Iran to ratify the 1946 agreement increased, Allen delivered a speech in Tehran that strongly supported Iran's right to dispose of its resources as it saw fit, free from external threats and intimidation.[52]

These assurances of U.S. support turned the tide, and on October 22, the Majlis voted 102 - 2 to reject the agreement made by Qavam in April 1946. In addition to rejecting the Soviet oil concession, the law passed by the Majlis forbade the granting of oil concessions to foreigners or the formation of joint companies with foreigners to develop Iranian oil; established a government agency to collect information on the oil industry; and specifically instructed the government "to undertake such negotiations and measures as may be necessary to secure the national rights, in all cases where the rights of the people have been violated in respect to the natural wealth of the country including its underground resources, with special reference to the southern oil," a reference to AIOC's concession in southern Iran. The Soviets confined their protests to complaints over Iran's violation of the 1946 agreement and the "categoric discrimination against the U.S.S.R.," resulting from the rejection of the Soviet offer of a joint oil company for northern Iran at the same time that the British oil concession in the south was maintained. The Soviets, as well as the United States and Britain, seemed to have missed the larger significance of the Iranian action, for the law rejecting the Soviet concession also contained the seeds of future conflict with the British over the concession held by AIOC.[53]

The Iranian action in October 1947, coupled with aid to Greece and Turkey under the Truman Doctrine, consolidated the U.S. position along the northern tier of countries which protected U.S. interests in the eastern

Mediterranean and the Middle East. Just as this policy seemed crowned with success, a new problem arose which threatened to undo U.S. efforts to maintain secure access to the region's resources.

The Palestine Issue and Middle East Oil

The emerging public-private partnership in foreign oil policy underwent its first test with the Palestine issue. Reliance on private oil companies as the vehicles of the national interest in foreign oil allowed the United States to maintain a distinction between the actions of the U.S. government and the attitudes of U.S. oil companies. Thus U.S. policy on the Palestine issue appears in a different light if examined against the background of the public-private partnership on Middle East oil. Rather than a tragedy in which President Harry S. Truman and his political advisers acted contrary to the collective wisdom of the departments of State and Defense, as some scholars have contended, U.S. policy cautiously balanced sympathy for a Jewish homeland in Palestine with the nation's apparent economic and strategic interests.

Truman frequently expressed support for increasing Jewish immigration to Palestine and for the idea of a Jewish "national home" there, but he never committed the United States to any specific course of action in support of these goals. The British, not the United States, controlled immigration into Palestine. The United States, while supporting partition as the majority recommendation of a United Nations committee, carefully avoided being put into a position of having to impose partition with military force. Likewise, concerned over the possibility of Soviet penetration in the Middle East, the United States opposed U.N. enforcement of partition. Finally, U.S. recognition of Israel represented acceptance of the reality of Israel's existence rather than an attempt to foster the birth of a new nation. Significantly, the United States refrained from providing Israel with military support for fear of further alienating the Arab world and losing access to the oil of the Middle East. This outcome was not the result of a conscious policy. Rather it was the result of a process of adjustment between U.S. ideals and U.S. interests.

Saudi Arabia and its oil were the focus of U.S. concern over Arab reaction to the Palestine issue. In August 1945, State Department Middle East expert Loy W. Henderson warned that U.S. support of Zionism would have a "strongly adverse effect" on U.S. interests throughout the Near and Middle East. It would be "almost inevitable" that U.S. cultural, educational, and religious institutions in the region would suffer; that U.S. trade would be boycotted; and that U.S. economic interests "including our oil concessions in Saudi Arabia and in other Arab countries would be jeopardized." After Truman endorsed the Harrison report's recommendation that 100,000 immigration certificates be made available immediately for Jewish refugees in Europe to go to Palestine, Henderson elicited from the War Department an estimate that it would take 400,000 troops, of which the United States would have to furnish some 200,000 to 300,000, to keep order in Palestine if the restrictions on Jewish immigration were lifted.[54]

In late April 1946, an Anglo-American Commission of Inquiry, which had been established the previous fall, recommended the creation of a nonsectarian, multinational state in Palestine. Truman issued a statement on the report which welcomed the committee's endorsement of the immediate entry of 100,000 Jewish refugees, but stated that the rest of the report would require careful study. The Joint Chiefs of Staff (JCS) cautioned that implementation of the Anglo-American Commission report by force could prejudice U.S. and British interests in the Middle East. Use of U.S. troops would lead to "serious disturbances throughout the area," and the Soviet Union "might replace the United States and Britain in influence and power through the Middle East." The United States had a "vital security interest" in the Middle East. A stable Middle East, friendly to the Western powers, was a "buffer between Russia and the British Mediterranean life line." If the people of the Middle East turned to the Soviet Union, Turkey's position would become untenable, and the West would lose control of Middle East oil. The region's oil reserves were "probably the one large undeveloped reserve in a world which may come to the limits of its oil resources within this generation without having developed any substitute. A great part of our military strength, as well as our standard of living, is based on oil."[55]

In August 1946 the Jewish Agency proposed immediate partition and the "establishment of a viable Jewish state in an adequate area of Palestine." Shortly thereafter on the eve of Yom Kippur Truman issued a statement: "To such a solution our Government could give its support." The State Department immediately warned that U.S. support of partition would result in the destruction of U.S. interests and influence in the Arab world. This assessment seemed to be supported by reports from Saudi Arabia. In an early November meeting with department officials, James Terry Duce, ARAMCO's chief lobbyist, warned that Ibn Saud was reaching the point at which it would no longer be possible for him to maintain a distinction between the policies of the U.S. government and ARAMCO. If U.S. policies continued to ignore Arab wishes, the result could be the loss of U.S. control of Saudi oil.[56]

A key element in postwar plans to promote the development of Saudi Arabian oil was the Trans-Arabian Pipeline (TAPLINE). This line would greatly enhance Saudi Arabia's price advantage in European markets by shortening the distance from the oil fields to Europe by some 3,500 miles and avoiding Suez Canal tolls of $0.13 to $0.18 per barrel, and would reduce the need for oil tankers in a period of anticipated shortage. The markets TAPLINE would open for Saudi Arabian oil would enable

ARAMCO to increase production, and therefore royalties, and thus aid in securing the concession. Moreover, the pipeline would cross several countries in the region, and refinery facilities would probably be constructed at the terminus. In order to guarantee these countries the "reciprocal benefits" of petroleum development, the State Department urged ARAMCO to break the pattern of past contracts and grant the countries through which the pipeline passed transit fees in addition to the usual nominal security fees.[57]

After the State Department put pressure on the reluctant British Colonial Office to allow such generous treatment of the "natives," ARAMCO was able to complete a pipeline convention with Transjordan in August 1946, which provided for the payment of transit fees. Negotiations with Saudi Arabia also went smoothly, and were completed in July 1947. Syria, however, insisted on higher transit fees as well as a pipeline terminus on the Syrian coast, and did not agree to a convention until September after the personal intervention of Ibn Saud. The Syrian government's agreement, however, was contingent upon ratification by the Syrian Parliament.[58]

Meanwhile, TAPLINE was facing problems in the United States because of the Commerce Department's reluctance to issue needed export licenses for scarce steel pipe and tubing. Eager to begin work on the line before winter, ARAMCO's Duce met with State Department officials and sought to convince them that ARAMCO's interests were identical to U.S. strategic interests in the Middle East. Duce explained that ARAMCO had already advanced a steel company money for the special thirty-inch pipe before export controls were imposed. This pipe would not help increase U.S. production since U.S. fields were already producing at near capacity levels. On the other hand, world demand for oil was increasing at over 5 percent per year, and only the Middle East possessed the reserves necessary to meet this demand. Duce reminded his sympathetic listeners that development of Middle East production would help reduce the drain on Western Hemisphere reserves by displacing Caribbean oil from European markets. European requirements for oil were large and growing, and it was important to the success of the Marshall Plan that these requirements be met. In addition, the pipeline would provide benefits to the countries through which it passed and thus aid in improving the welfare and stability of the region.[59]

ARAMCO's arguments received a sympathetic hearing in the State and Defense departments. In early September, Under Secretary of State Robert A. Lovett, a former investment banker who had been assistant secretary of war during World War II, wrote the Commerce Department that it was the "considered opinion of the Department of State that the national interest would best be served by the immediate construction of this pipeline." Secretary of Defense James V. Forrestal, long an advocate of

Arabian oil development, also contacted the Commerce Department in support of the project.[60]

The potential explosiveness of the Middle East was an important factor in the State Department's support of TAPLINE. Consideration of the export license coincided with the publication of the United Nations Special Committee on Palestine (UNSCOP) report recommending partition. Weary of the responsibility of policing an increasingly violent territory, Britain turned the Palestine question over to the United Nations in February 1947. At the initiative of the United States a Special Committee on Palestine was established with broad powers to investigate all aspects of the problem. The UNSCOP report, signed in Geneva August 31, contained two plans. The majority plan, advocated by seven of the committee's eleven members, called for partition of Palestine into an Arab state, a Jewish state, and an internationalized city of Jerusalem, though all three were to be joined in an economic union. The Arab and Jewish states were to become independent following a two-year transition period during which Britain would continue to administer the country under the auspices of the United Nations. The United Nations itself would administer Jerusalem under the international trusteeship system. The minority plan, advocated by three committee members, proposed the creation of an independent federal state following a transition period not to exceed three years, during which it would be administered by an authority designated by the U.N. General Assembly.[61]

The State Department and the military feared the impact of partition on U.S. interests in the Middle East. On September 22, Loy W. Henderson sent Secretary of State George C. Marshall a long paper which warned that the U.S. position on the Palestine issue would have far-reaching effects on relations with the peoples of the Middle East and on Moslems everywhere. U.S. support for the establishment of a Jewish state in Palestine would undermine the West's relations with the Arabs at a time when the West needed Arab cooperation for bases in the Middle East and Arab resources for the reconstruction of Europe. Chief among these resources was oil. Finally, Henderson, who had begun his career as a Soviet specialist, warned that establishment of a Jewish state in Palestine could lead the Arabs to regard the United States as their enemy and to turn to the Soviets for help against the common enemy. Henderson informed Lovett on September 24 that the "entire Arab world" was "exercised" over the trend of U.S. policy in Palestine, and argued that a "beneficent economic policy" was one of the few means the United States had of offsetting, in some degree, the effects of its Palestine policy. TAPLINE had received widespread publicity in the Middle East, and a decision delaying or canceling it would be viewed by the Arab states as linked to the "unfriendly" U.S. policy on Palestine.[62]

TAPLINE won a key supporter when Max Ball, head of the Interior Department's Oil and Gas Division, came out in favor of export licenses. Ball, an independent oilman, had previously opposed granting the licenses because of the domestic oil and gas industry's need for steel. Ball reluctantly changed his mind when it became clear that sufficient tankers would not be available in the future to move Arabian oil to European markets. Failure to supply Europe, "would have disastrous political and economic repercussions and would at the very least retard world rehabilitation for many years." Since Europe could be supplied from the Western Hemisphere only by taking oil away from Western Hemisphere consumers, the United States should supply the necessary materials to increase Middle East oil production. Ball's conversion proved decisive. On September 26, the Commerce Department announced that it had approved issuance of a license authorizing TAPLINE to ship 20,000 tons of steel to Saudi Arabia for the fourth quarter of 1947. The announcement stressed that the construction of the pipeline would lower costs and thus greatly increase incentives for more rapid development of Middle East oil, "thereby permitting earlier relief of the present drain on Western Hemisphere reserves of petroleum."[63]

Domestic oil producers were not pleased with the decision. Russell Brown of the Independent Petroleum Association of America argued that the domestic oil and gas industry needed every pound of steel available in order to avoid future oil shortages. The long-range factors offered in support of the licenses were "far out-weighed by the immediate necessities of oil and gas consumers in the United States." Nebraska Senator Kenneth S. Wherry, chairman of the Senate Special Committee on Small Business wrote Secretary of Commerce Averell Harriman demanding to know how the export of such a large amount of steel could be justified in light of the steel shortage in the United States. Wherry also summoned Secretary of Defense Forrestal and Under Secretary of State Lovett before an executive session of his committee to explain their decision. Forrestal told the committee that "until there were indications of new fields of substantial magnitude in the Western Hemisphere . . . pipe for the Arabian pipeline should have precedence over pipe for similar projects in this country." Lovett dismissed Wherry's arguments concerning the "explosiveness" of the Middle East, and pointed out that nowhere in the Western Hemisphere was there the same possibility for "prompt and sizeable development" as there was in the Middle East.[64]

Partition and U.S. Security

Under tremendous political pressure to endorse partition, Truman overruled the State Department and decided to support the majority report. The official announcement of U.S. support for the majority recommendations, however, came only after Truman had assured the

State Department that the United States would provide no financial assistance nor would it assume any responsibility for maintaining order in Palestine, except through the United Nations.[65]

The president's decision to support the UNSCOP majority report went against the advice of the State Department, the military, and the Central Intelligence Agency. According to these agencies, U.S. support of partition would undermine the West's relations with the Arabs at a time when Arab cooperation for bases and Arab oil for the reconstruction of Europe were sorely needed. It would jeopardize U.S. strategic interests in the Middle East, and reduce U.S. influence in the region to that which could be maintained by force. The unrest resulting from partition, moreover, could enable the Soviet Union to replace the United States and Britain in power and influence throughout the region, with the result that access to the oil of Iran, Iraq, and Saudi Arabia would be "gravely prejudiced." Even if the Arab governments did not immediately cancel U.S.-owned oil concessions, partition would lead to such unrest and instability in the region that Soviet penetration would increase and eventually achieve the same result. According to the CIA, the Palestine issue was "capable of changing the development of the Arab world from one of evolution in cooperation with the West to one of revolution with the support of the USSR."[66]

Initial Saudi reaction to the U.S. decision to support the majority report was unfavorable. King Ibn Saud wrote President Truman that U.S. support of partition would "without doubt . . . lead to a deathblow to American interests in the Arab countries." Later reports from Saudi Arabia, however, made it clear that Ibn Saud was reluctant to take action against his main source of revenue. According to former Ambassador William A. Eddy, now an ARAMCO official, Ibn Saud had rejected a demand by the Iraqis that he cancel ARAMCO's concession. "You have no cooperation with Americans to sacrifice," he reportedly pointed out to his dynastic rivals. "When you have canceled the British oil concession in Iraq and have driven them from the administration of your railroads and ports, then you would have a better right to make suggestions to me, not before." These reports notwithstanding, the CIA feared that even though Ibn Saud knew where his best interests lay, he might not be able to prevent his people from attacking U.S. oil installations and pipelines.[67]

On November 29, 1947, the U.N. General Assembly voted to recommend the partition of Palestine. Widespread violence followed the partition vote. Arab groups attacked Jewish settlements in Palestine, and there were attacks against U.S. diplomatic missions in Syria and Iraq. The kings of Egypt and Transjordan warned that the Arab states were prepared to resist partition by force of arms. As a result of the unrest, TAPLINE construction was suspended and field personnel in Syria and Transjordan were evacuated. The Syrian legislature refused to ratify the TAPLINE convention, further delaying progress on the line. King Ibn Saud, however, assured the United States that although the two countries differed "enormously" on the question of Palestine, they had their own "mutual interests and friendship to safeguard." The king explained that because of his position in the Arab world he would have to make common cause with the other Arab states on Palestine, but he did not anticipate that any situation would arise that would draw him into conflict with the United States. Eddy informed the Defense Department that, although the Arabs were extremely displeased with the policy of the U.S. government, U.S. private interests had been spared as demands for the cancellation of educational and economic concessions had been rejected as "suicide and self-impoverishment." Eddy felt that as long as the United States left implementation of partition to the United Nations, Arab resentment would be displaced. But if the United States tried to enforce partition with arms, men, and money, U.S. concessions would be canceled, U.S. property expropriated, U.S. military bases lost, and, as a last resort, the Arabs would turn to the Soviet Union for help.[68]

These warnings intensified State and Defense efforts to try to reverse U.S. policy. Secretary of Defense Forrestal was convinced that domestic political considerations were dominating U.S. policy, and repeatedly urged President Truman to "lift the Jewish-Palestine issue out of politics and to be more aware of the security needs of the United States." His Wall Street background and personal ties with major oil company executives undoubtedly influenced his views on the Palestine issue, but Forrestal also genuinely believed that a pro-Zionist policy would be fatal to U.S. interests. Forrestal warned the cabinet in mid-January 1948 that without access to Middle East oil the Marshall Plan would not succeed and the United States would not be able to maintain the "tempo" of its peacetime economy, much less fight a major war.[69]

Key elements in the State Department also worked to change U.S. policy on Palestine. In mid-December 1947, a draft State Department paper emphasized the "impossibility of implementing partition," and recommended that the United States either fall back to support of a trusteeship or insist that no troops be sent to implement the U.N. decision. In a major paper in mid-January 1948, the Policy Planning Staff (PPS) warned that if the United States helped implement partition the result would be loss of military bases, oil concessions and pipelines, and trade as well as loss of access to British bases in the Middle East. The success of the Marshall Plan was dependent on increased oil production in the Middle East since other producing areas had no reserve capacity. But expanded development of Middle East oil had been thrown into doubt by the disorder growing out of the Palestine issue. On February 11, PPS Director George Kennan proposed that the United States reverse its support of partition, and back instead the establishment of a United Nations trusteeship or a federal state in Palestine.[70]

The National Security Council (NSC) staff pointed out that "unrestricted access" to the oil resources of the Middle East was essential in meeting the threat to U.S. security posed by the Soviet Union and its "aggressive program of Communist expansion." Middle East oil was essential to the economy of the United States and the economic recovery of Europe, and would figure prominently in the successful prosecution of any war with the Soviets. A "friendly or at least neutral attitude" on the part of the Arabs toward the United States was necessary to maintain access to Middle East oil and to the region's strategic areas. Thus the security of the eastern Mediterranean and the Middle East was vital to the security of the United States, and the Soviets could not be allowed to obtain a lodgment there. U.S. support of partition jeopardized these strategic objectives by alienating the Moslem world and providing a "vehicle for Soviet expansion" into the area. The majority of the NSC staff concluded that the United States should fulfill its "moral obligation" to support partition, but such support should stop short of the use of armed forces to impose the U.N. decision. The military members of the staff, however, recommended that the United States repudiate partition and urge the United Nations to establish an international trusteeship.[71]

The vast outpouring of opposition to partition no doubt impressed President Truman, but partition also had its articulate defenders. The president's special counsel, Clark Clifford, argued that the United States had to support the U.N. decision on partition because to do otherwise would be a "body blow" to the United Nations. The United Nations was a "God-given vehicle" for resisting the Soviets and maintaining the United States' "historic interests." According to Clifford, a politically astute lawyer from St. Louis, the United States had already "crossed the Rubicon" when it backed the majority plan before the General Assembly. The only course left was to take affirmative action to prevent chaos which could draw the Soviet Union into the matter. Fears that support of partition would cost the West access to Middle East oil were exaggerated. Without peace there would be no oil, and the only way to have peace was to back the United Nations. In addition, the Arabs could not afford to cut off the oil since they had to have oil royalties or "go broke." Adopting a pro-Soviet orientation would be "suicide" for the "Arab ruling classes," so this threat too was hollow. "Political and economic self-preservation" would compel the Arabs to continue to do business with the United States. "Their need of us is greater than our need of them."[72]

Meanwhile, the future of the Trans-Arabian Pipeline was becoming more and more uncertain. Faced with unrest in Transjordan and the refusal of the Syrian Parliament to ratify the pipeline convention, AR-

AMCO's owners had begun to consider abandoning the project. In early 1948, TAPLINE had once again faced the problem of obtaining export licenses for steel pipe and tubing. The steadily worsening situation in the Middle East coupled with the continuing steel shortage in the United States raised questions about the wisdom of supporting a project whose success depended on stability in an increasingly volatile region. The oil staff of the newly created National Security Resources Board (NSRB) pointed out that "from a national security resources point of view," construction of the pipeline was not justified. The United States was running a "grave risk" of becoming dependent on a transportation mode that could easily be disrupted in the event of war. Oil tankers cost more and used more steel to move an equivalent amount of oil, but were more difficult to interdict, and could be used elsewhere if the Middle East were lost.[73]

The State Department, in contrast, argued that abandonment of the TAPLINE project would seriously handicap the European Recovery Program. Tankers were in short supply, and it was "impractical" to try to construct sufficient tanker tonnage. Under the Marshall Plan, the United States had assumed responsibility for European recovery, and TAPLINE would make more oil available at a lower cost and with less steel than would alternative means. Increased utilization of Middle East oil would reduce the drain on Western Hemisphere reserves and free Caribbean oil for the U.S. market. Without TAPLINE, Europe would have to rely on the Western Hemisphere for its oil needs, and such reliance would not only increase the drain on Western Hemisphere reserves, but could also lead to shortages and rationing in the United States. The Interior Department's oil expert agreed with the State Department assessment, noting that "so far as oil is concerned, the success of the European Recovery Program will depend upon the things which advance or retard oil development in the Middle East."[74]

The military also supported the continuation of TAPLINE. A study by the Joint Chiefs of Staff in mid-March concluded that in view of the necessity of the continued availability of Middle East oil, its development should be continued "until the necessity for their stoppage is unmistakable." The JCS recommended that export licenses for the pipeline be granted, but also advised that because of the trend of developments in the Middle East, licenses should be periodically reexamined. After receiving the JCS study, Secretary of Defense Forrestal informed Senator Wherry that he believed that it was necessary to continue shipments of steel pipe for the pipeline. Wherry agreed to abide by the secretary of defense's decision, and this agreement cleared the way for the Commerce Department to issue the necessary licenses.[75]

While the battle over the TAPLINE was going on, the State Department attempted a reversal of U.S. policy at the United Nations. On March 19,

U.S. Ambassador to the United Nations Warren R. Austin asked the Security Council to suspend efforts to implement partition and to call a special session of the General Assembly to reconsider the Palestine problem. While this was being done, Austin proposed that Palestine be governed under a temporary trusteeship administered by the U.N. Trusteeship Council. It is not clear whether President Truman approved this drastic shift in policy, and if he did, whether he fully understood its implications. In any event, the resulting reaction to Austin's action greatly embarrassed Truman, and nothing ever came of the State Department's scheme. The main obstacle to the idea was that establishment of a trusteeship, like partition, would require the use of force. According to estimates by the Joint Chiefs, 50,000 to 100,000 troops would be needed, and the United States could not spare more than 15,000 without partial mobilization. In addition, sending troops to Palestine would "affect materially" U.S. ability to use its force elsewhere.[76]

As events in Palestine swept toward May 15, the day the British mandate was to end, the United States had to decide whether or not to recognize the state the Zionists planned to proclaim. At a tense, high-level meeting at the White House on May 12, Clark Clifford argued the case for recognition. Partition was a fact, Clifford pointed out, because the Jews were making it a fact. In these circumstances, the United States should get a jump on the Soviets by granting recognition first. Under Secretary of State Robert Lovett pointed out that the matter was still before the U.N. Security Council, and the United States should not preempt the United Nations. Secretary of State Marshall complained that Clifford's arguments were a "transparent dodge" to win votes. Marshall announced that if the president followed Clifford's advice, he would vote against the president in the next election. The secretary of state's stiff opposition gave Truman pause, and he did not announce his intention to recognize Israel at his press conference the next day, as Clifford had urged. Nevertheless, under tremendous political pressure to recognize the Jewish state promptly, Truman recognized Israel on May 15, only eleven minutes after the new state was proclaimed.[77]

Arab reaction to U.S. recognition of Israel was surprisingly mild. The Arab League informally advised the United States that as long as it made no further moves to assist the Jews, action against U.S. oil interests would be postponed. Though the Saudis were "profoundly shocked" by the U.S. action, Ibn Saud nevertheless assured ARAMCO officials that he drew a distinction between official U.S. policy and private U.S. interests. Even though he considered ARAMCO personnel to be like members of his own household, Ibn Saud was under great pressure from other Arab states to take action against the company. How long he could resist depended, to a large degree, on whether or not the United States actively supported the new state of Israel.[78]

Drawing on these and similar reports, the State Department devised a strategy for minimizing the threat to U.S. interests. The key was to avoid any action that could lead the United States to assume responsibility for the maintenance and security of Israel. In practice this policy meant dropping efforts in the United Nations proposing U.N. intervention in the fighting that had erupted in Palestine, and fighting off domestic pressures to end the arms embargo which the United States had imposed in late 1947. Although the Democratic party platform for 1948 called for revision of the embargo to permit arms sales to Israel, the State Department, with the support of the Defense Department, was able to resist efforts to end the embargo.[79]

To demonstrate the dangers to U.S. interests, the Petroleum Division (PED) prepared an analysis of the impact of the loss of Arab oil on the world petroleum situation. According to the analysis, the loss of Middle East oil, with the exception of Iranian production, would deprive the "free world" of approximately 6 percent of its total output. Since Middle East oil was distributed unevenly in world markets, a 6 percent overall reduction represented only 1.7 percent of U.S. needs, but 26 percent of the import needs of the Marshall Plan countries and 21.3 percent of the requirements of other Eastern Hemisphere countries. If the losses were spread proportionately to all markets, the shortfall would be reduced to 5.9 percent, assuming that the Iranian oil that normally went to Arab countries was diverted to other markets. Even on this basis, severe rationing would be required in Europe and mild rationing in the United States. Spreading the loss would require close cooperation of U.S. oil companies marketing abroad, and would entail an increase in U.S. oil exports. Time and inadequate steel supplies limited the degree to which the loss could be made up by increasing production elsewhere.[80]

These estimates were revised after the British pointed out that Iranian production would probably be lost in the event of an Arab oil embargo, either as a result of official sympathetic action or of Communist actions in the oil fields. Iranian production accounted for nearly 5 percent of total world production, and when this additional loss was factored in, the European deficit increased to 37 percent, and that for the rest of the Eastern Hemisphere to 43 percent. Spread proportionately among all markets this represented an 11 percent shortfall. Such a curtailment of consumption could not be absorbed in the United States without formal rationing and price controls. Loss of Middle East production would thus force the United States to choose between substantial curtailment of domestic consumption in order to make oil available for Europe, and "serious difficulties . . . if not failure" of the Marshall Plan.[81]

After the establishment of Israel, the State Department convinced ARAMCO to postpone further requests for export licenses for TAPLINE until the situation in the Middle East improved. Construction continued at a reduced pace during the summer of 1948 as the company sought to maintain its organization. Efforts by the company in August to resume shipments of steel were unsuccessful as doubts within the Commerce Department and the military about sending so much pipe abroad prevented action on the company's requests. In February 1949, however, the granting of export licenses was resumed after the pipeline company signed a secret agreement with the U.S. Navy giving the U.S. military an option to utilize 5 percent of the line's capacity at cost in an emergency. The Syrian Parliament finally ratified the pipeline agreement in May 1949. Construction was completed in September 1950, and in November 1950, the first barrel of oil from the Persian Gulf arrived at the line's Mediterranean terminus.[82]

Although the solution to the major problems facing U.S. foreign oil policy was found in private initiatives, the government provided invaluable assistance in making a corporatist solution possible: It provided the companies with diplomatic support and reinterpreted the Open Door to allow whatever cooperative arrangements the companies deemed necessary. Moreover, maintaining an environment in which private interests could freely operate involved the United States deeply in the politics of the Middle East, with consequences for U.S. foreign policy as well as for oil policy.

Though the United States was able to maintain Iran as a buffer against Soviet influence and to minimize the damage to its interests from the Palestine problem, these successes stopped short of making Middle East oil totally secure. The oil fields of the Middle East still lay on the Soviet Union's "doorstep," and were thought to be impossible to defend from determined Soviet attack. The tenuous nature of Western control of Middle East oil was reflected in U.S. military plans for the region. Early in 1948, the Joint Chiefs of Staff and the National Security Council began work on plans to deny the region's oil facilities and fields to the Soviets in the event of war. The plans called for advance preparations to destroy aboveground facilities and to plug the wells with concrete blocks. The latter measure would make the wells unusable without the waste of reserves that would result from the use of fire or explosives to destroy the wells.[83] In addition to steps to deny Middle East oil to the Soviets, the United States began to look elsewhere for oil to meet its growing strategic and economic requirements.

Chapter 6

Containing Economic Nationalism

The doubtful security of Middle East oil coupled with growing U.S. requirements for oil led the United States to take a renewed interest in Latin American oil development. In Venezuela, the United States continued to support the cooperative arrangements between the oil companies and the Venezuelan government worked out in 1943. Such a solution proved impossible in Mexico because of divisions among U.S. oil companies and within the U.S. government and continued Mexican opposition to foreign control of their oil industry.

The Problem of Economic Nationalism

Early in November 1947, Secretary of Defense James Forrestal warned Secretary of State George C. Marshall that the "estimated petroleum requirements for the military and civilian efforts in the event of a major war are such as to exceed the production now obtainable in areas which we can consider securely within our reach." The United States was becoming a net importer of oil, and even though there was plenty of oil in the Middle East, it would be difficult to maintain access to this oil in the event of hostilities. In contrast, Latin American oil would be "securely available" to the United States in wartime. The oil potential of Latin America was great, but economic nationalism had resulted in restrictive laws and other conditions "inimical to the freedom of operation of capital . . . essential to rapid development."[1]

As Forrestal noted, U.S. efforts to promote oil development in Latin America ran up against the oil policies of the region's nations. With the notable exception of Venezuela, most Latin American countries were reluctant to turn oil development over to foreign companies. The United States, in contrast, believed that "the most efficient method for achieving rapid expansion of petroleum resources" was through private industry. In the context of the oil industry "private" meant U.S. or British companies since there were no private companies in Latin America with the necessary resources and skills. The United States opposed national oil companies, which existed in many Latin American countries because of the inadequacy of private resources. State-owned companies were "objectionable," an interdepartmental committee explained in 1947, because they were inefficient and would receive special privileges, and because they would restrict business and investment opportunities for U.S. nationals.[2]

Opposition to state-owned enterprises was part of a general ideological orientation that went beyond oil policy. Beginning with the assumption that political freedom and economic freedom were inseparable, U.S. policymakers believed that "the maintenance of the system of private enterprise and private property on a sufficient scale throughout the world" was necessary to give the U.S. system a "congenial atmosphere in which to function." Private companies would not invest unless they were given a reasonable opportunity to earn profits commensurate with the risks involved. Thus the problem, as defined by U.S. policymakers, was to establish a basis whereby private U.S. oil companies could freely operate in Latin America.[3]

The United States and Venezuelan Nationalism

The key tenet of U.S. oil policy toward Latin America was to take no actions that could in any way jeopardize U.S. control of Venezuelan oil. Venezuela allowed foreign oil companies wide latitude in developing the country's oil resources, and Venezuelan oil played an important role in "oiling" the Allied victory in World War II. The 1943 reform of Venezuela's oil laws led to a tremendous expansion of the industry, as production grew from 178 million barrels in 1943 to 323 million barrels in 1945 and to 490 million barrels in 1948. Reserves in 1947 were estimated at around 8 billion barrels. Less than 3 percent of Venezuelan production was consumed locally, while 30 percent went to the United States, 30 percent to Western Europe, and some 25 percent to other Western Hemisphere markets. Although providing Venezuela with greatly increased revenues— from $110 million in 1943 to almost $415 million in 1947—the 1943 reforms left control of the nation's oil industry completely in the hands of foreign oil companies. In 1947, Jersey's subsidiary, Creole Petroleum Company, accounted for 52.5 percent of Venezuelan production, Royal Dutch/Shell, 27.7 percent, and Mene Grande (a subsidiary of the Gulf Oil Corporation), 14.6 percent. The remainder was divided among several other, mostly U.S.-owned, companies, the most important of which were the Texas Company and Socony-Vacuum.[4]

In October 1945, a coalition of junior military officers and reformist politicians from the left-of-center Acción Democrática (AD) party overthrew the conservative, pro-U.S. government of President Isaías Medina Angarita. The officers acted largely because of dissatisfaction with low pay, limited opportunities for promotion, and poor military standards. Their civilian allies, in contrast, belatedly joined the disaffected officers when an informal agreement over the presidential succession collapsed

Source: U.S. Department of State, Office of Public Affairs, "Venezuela: Oil Transforms a Nation," February 1953, p. 6.

and dashed their hopes of ever attaining power through the electoral process. The successful revolutionaries set up a seven-man junta on October 19, composed of four members of the AD, two officers, and an independent civilian politician. Control of all governmental matters except those related to defense and national security were placed in civilian hands, and the junta in its initial statements promised a new constitution and free national elections.[5]

The prominent role of the AD in the new government raised fears that the oil industry would be nationalized. The AD had built its reputation on criticism of the role of the foreign oil companies in Venezuela, and the party's chief constituency was the oil workers. These fears increased when Juan Pablo Pérez Alfonso, the leading critic of the 1943 oil law, was appointed minister of development. To everyone's surprise, AD followed a relatively cautious policy toward the oil industry. AD leader Rómulo Betancourt later explained that this caution was due to the great importance of oil to the Venezuelan economy. By 1945, oil income accounted for over 90 percent of Venezuela's export earnings, over half of government revenues, and approximately 30 percent of gross domestic product. Given this dependence on oil, and the lack of an indigenous technical capacity to run the industry, Betancourt and his comrades felt that it would be "suicidal" for Venezuela to seek to emulate Mexico's bold example.[6]

The thirty-seven-year old Betancourt had led a student revolt against the Gómez regime in 1928 and had spent several years in exile before returning to Venezuela in 1941, when he helped form Acción Democrática. Seasoned by this experience, Betancourt took immediate steps to assure the oil companies and the U.S. government that the new government posed no threat to their basic interests. Betancourt met with the heads of Creole, Mene Grande, and Shell on October 23 and assured them of the junta's intention to honor the 1943 oil law and valid concession contracts. The companies were warned, however, that they would be expected to recognize and bargain with the oil workers' unions. Pérez Alfonso repeated these assurances in a signed statement to the oil industry publication, the *Oil and Gas Journal*, which explained that the Junta was not interested in nationalizing the oil industry, but rather in ensuring that the 1943 code and existing tax laws were honestly enforced. The U.S. Embassy was given similar assurances, and on October 24, U.S. Ambassador Frank P. Corrigan reported to the department his belief that U.S. interests were not adversely affected by the change in government. Corrigan, a New Deal physician whom Roosevelt had appointed in 1939, opposed efforts by the oil companies and the British government to tie recognition to a formal pledge that the basic principles of the 1943 oil law would be respected. On October 30, the United States recognized the junta as the legal government of Venezuela.[7]

While accepting the basic framework established by the 1943 oil code, the new government was determined to enforce the law vigorously. On the last day of 1945, the junta decreed an extraordinary excess profits tax on oil industry earnings, charging that the Medina regime had manipulated income tax figures to justify its claim that it had received 50 percent of oil company profits in 1944. The tax cost the companies some $27 million, and was designed to equalize the companies' profits and the government's oil revenues for 1945. Although upset by the government's action, the three major oil companies, Creole, Shell, and Mene Grande, decided to avoid a confrontation. Creole head Arthur Proudfit, who had been one of the advocates of the 1943 reforms, later explained that the company did not oppose the tax because there was no legal ground to contest it. Technically, the tax was not a violation of the 1943 code, which supposedly had intended to equalize Venezuela's oil income and the companies' profits. Proudfit himself had admitted in November 1945 that "no one could object if the junta or its successor, seeking more revenue, should increase the income tax rate and surtaxes." Moreover,

Creole, Shell, and Mene Grande could afford to pay the extra tax. Profits as well as production had expanded greatly since 1943. In 1945, for example, Creole earned $94 million even after it had paid an additional $18 million because of the new tax.[8]

The initial U.S. reaction was one of "shocked surprise." Chargé Allan Dawson warned that there was "every indication" that the tax was "only the beginning of a campaign to milk the companies of everything possible." Creole's attitude, however, coupled with assurances from Betancourt that the tax was a one-time measure, calmed State Department concerns. The smaller oil companies, which together produced only 5 percent of Venezuela's production, were not so "profit-fat" as the large companies, and one of the small companies, Pantepec Oil, tried to get the State Department to protest the new tax. Although Pantepec engaged former Assistant Secretary of War John J. McCloy to present its case, the department refused to interfere in a domestic Venezuelan matter, and advised Pantepec not to protest the tax.[9]

In the spring, the junta pressured the oil companies to accept new labor contracts that granted oil workers pay raises ranging from 35 to 50 percent, two weeks' paid vacation, and sick leave. Anxious to avoid any disruption in production and profits, Creole and Shell agreed to meet all the union demands except those relating to management prerogatives. The smaller companies, less able to afford a generous settlement, wanted to force a "showdown" with the government over the issue. Abandoned by Creole and Shell, they eventually gave in after Betancourt and Labor Minister Raúl Leoni threatened to impose a settlement. On June 14, 1946, a collective bargaining contract was signed. The State Department took its signals from Creole and supported the settlement. When Secretary of War Robert Patterson complained that the unions' demands would "force the companies out of business or effectively curtail their control of production," the State Department replied that the terms of the labor agreement were neither unreasonable nor impossible. U.S. interests would not be served by encouraging the companies to resist the reasonable demands of labor.[10]

Encouraged by these successes, the Venezuelan government began to explore other means to increase its revenues from oil. In mid-1946, Pérez Alfonso informed the companies that the government would begin taking part of its royalties in oil so it could take advantage of the high level of world demand. (Because of the postwar world oil shortage, it was possible to obtain premium prices by making spot sales to crude-short independents.) The major companies did not like the idea of such a large amount of oil—around two tankers a day—escaping their control, and quickly made arrangements to buy most of the royalty oil back from the government at premium prices. Nevertheless, the Venezuelan government continued to take a portion of its royalties in oil for sale to the highest bidder.[11]

In the fall of 1946, the Venezuelan government began considering enacting an excess profits tax. Because of the rapid increase in world oil prices, Venezuela was receiving somewhat less than the 50 percent of oil company profits envisaged in the 1943 code. Dawson reported in late September that there was "no sound economic reason" to oppose an excess profits tax "in view of the relatively low income taxes now prevailing and the huge profits which are being made by the monied interests in this boom period in Venezuela." Creole's earnings for 1946 would be "in the neighborhood of $100 million, and those of the Shell group proportionately large." The companies were not happy about the tax, but worked closely with the government to delay implementation. Finally, on November 12, 1948, Venezuela enacted a new income tax law, which included a 50 percent rate on any sum by which a company's net profits exceeded the government's share of the company's earnings. With the passage of the new tax, the fifty-fifty sharing arrangement promised in the 1943 code became a reality.[12]

The AD government was not content to work for greater benefits for Venezuela within the framework established by the 1943 petroleum code. The party's strategy called for the piecemeal introduction of reforms that would lay the groundwork for an alternate system of petroleum development, one in which ultimate control would be in Venezuelan rather than foreign hands. Taking royalties in kind was part of this strategy. Another key step in the strategy was a policy of refusing to lease new concessions. Part of the rationale behind this policy was that the government was having difficulty in efficiently utilizing the oil income it was already receiving because of shortages of machinery and materials. In addition, the AD was interested in securing Venezuela's petroleum future through conservation. To aid in this endeavor, Pérez Alfonso turned for technical assistance to the Texas Railroad Commission, the leading oil conservation organization in the world. In 1948, he arranged for twelve of the commission's engineers to visit Venezuela, and for Venezuelan engineers to go to the United States for further training. Most of the companies operating in Venezuela had acquired sufficient concession rights to keep them busy for several years, so the no-concession policy posed no immediate threat. Long-term prospects were more uncertain, however, as it was not clear whether the government would, at some point, resume the granting of concessions to private companies, or if it intended to restrict future petroleum development to government agencies.[13]

In March 1948, the Venezuelan government appointed a special commission to study the creation of a national oil agency to guide the nation's oil policies. The new agency, it was envisaged, would include a national oil company in order to encourage governmental development of new oil lands under mixed company arrangements with private interests. In this way AD hoped to nurture the development of an indigenous technical and managerial capacity that could gradually provide Venezuela with the ability necessary to control its petroleum destiny. The United States did not welcome these plans, and tried to persuade Venezuela of the "advantages of the present system of private enterprise operations with any regulatory safeguards that might reasonably be required." In July 1948, the U.S. Embassy in Venezuela was instructed to "express both surprise and concern" over the government's plans to set up a national oil company.[14]

As during the Medina years, the United States accepted reforms in the relationship between the oil companies and the Venezuelan government because they did not threaten continued U.S. control of Venezuela's strategically important oil. Following the lead of the major oil companies, and in particular, Creole, the State Department sought accommodation with Acción Democrática as a means of channeling AD's economic nationalism into paths consistent with U.S. interests. In addition, the State Department viewed AD as a guarantor of a stable and pro-Western Venezuela. According to a January 1948 CIA analysis, AD's overwhelming victory in the December 1947 presidential and congressional elections not only placed "the stamp of unmistakable popular approval" on AD's policies but also assured the continuation in power of "an active non-Communist party which can be counted on to obstruct the development of a strong Communist movement." For its part AD was careful to assure the U.S. government and the oil companies of the moderation of its intentions and to support the United States in the cold war and on regional issues.[15]

As the cold war intensified, however, the United States became concerned about AD's capability to take the steps necessary to ensure the security of oil installations against sabotage and enemy attack. In addition, the U.S. Army, perhaps suspicious because of Betancourt's involvement with the Communist Party of Costa Rica in the early 1930s and AD's cooperation with Venezuelan Communists in 1945, maintained an active distrust of AD. Army Intelligence dissented from the January 1948 CIA report. The army argued that AD's nationalism represented "a potential danger to United States interests in Venezuelan oil" and that AD's pro-U.S. stance represented expediency. Moreover, growing internal opposition to AD's policies during 1948 cast doubt on its ability to ensure stability. As the elected government of Rómulo Gallegos came under increasing pressure from the military in the fall of 1948, U.S. Ambassador Walter Donnelly, a career diplomat with long experience in Latin America, followed a policy of strict nonintervention. On November 24, 1948, four days after the United States extended recognition to the military junta that had overthrown the constitutional government of Peru, the Venezuelan military overthrew the Gallegos government.[16]

According to recent scholarship, the coup of November 24, 1948, had little to do with oil. Although AD had won the support of over 70 percent of Venezuelans in the free elections of December 1947, its plans to transform Venezuelan society had alienated many powerful groups. The party's advocacy of labor rights cost it the support of business groups, while its enactment of an agrarian reform law in October 1948 threatened the power of the landed elite by calling for land redistribution. In addition, the AD education program, which emphasized secular values, impinged on the traditional prerogatives of the Catholic Church. Most importantly, AD lost the support of the armed forces. Although AD was careful to meet military concerns by substantially increasing military spending, the military leadership, for a variety of reasons, sided with the landowners, businessmen, and clergy who decried the government's policies. The major oil companies seem not to have been involved in the coup, though it is possible that some of the smaller companies may have provided financial support. After a period of soul-searching, the United States recognized the new government in January 1949.[17]

Although the coup itself was relatively bloodless, it inaugurated ten years of increasingly brutal military dictatorship. Military rule proved to be a bonanza for the oil companies. In early 1949, the military dissolved the oil workers' unions, and offered no protests as the companies cut their production and laid off large numbers of employees. The new government retained the taxes enacted by the AD governments, but revisions of royalty pricing arrangements and other tax adjustments in favor of the companies cost Venezuela an estimated $13.5 million dollars in revenue losses between 1949 and 1954. Plans for government development of new areas were scrapped. Oil company profits slumped in 1949 as Venezuelan production was cut back by over 20 percent between January and May 1949 because of the world oil surplus. They quickly recovered, however, and were soon back to high levels.[18]

Not surprisingly, the oil companies quickly warmed to the new regime. Creole pressed the State Department to recognize the new regime largely out of fear of disorder and concern that Shell might profit if British recognition preceded U.S. action. By the end of January 1949, Creole officials were reporting "greatly improved conditions" and less danger of "subversive activities" under the new regime. In late November 1949, Creole head Arthur Proudfit urged the State Department not to pressure the Venezuelan government to hold elections since fair elections would be won by AD and this outcome would be a "disaster." The State Department, which had also made its peace with the new regime, assured Proudfit that the United States was not unfavorably disposed toward the present Venezuelan government, and had no intention of pressuring Venezuela to hold elections.[19]

Mexican Oil and U.S. Security

Corporatist policies continued to prove effective in maintaining the U.S. position in Venezuela. The U.S. government and the major U.S. oil companies were able to retain good relations with changing Venezuelan governments. In Mexico, however, there were no U.S. oil companies to serve as instruments of U.S. foreign oil policy. U.S. efforts during World War II to reverse the nationalization of the Mexican oil industry had failed because of Mexican resolve and Harold Ickes's interference with State Department plans. Nevertheless, the State Department had managed by the end of the war to gain presidential sanction for its policy of withholding aid to Petróleos Mexicanos (PEMEX) until U.S. oil companies were permitted to operate freely in Mexico. Ickes's departure from the government in early 1946, and the dismantling of the Petroleum Administration for War, removed the State Department's major rival in foreign oil policy, and cleared the way for a new attempt to influence the direction of Mexico's oil policy. Although Mexican production in 1945 was only 43.5 million barrels and proved reserves a mere 800 million barrels, many oilmen believed that given sufficient investment Mexico could once again rival Venezuela as a major producer and exporter.[20]

The victory of Miguel Alemán Valdés in the 1946 Mexican presidential elections gave the United States new hope that Mexico would finally change its policies on oil. Alemán's plans for the oil industry, contained in a report by several PEMEX engineers, called for large increases in production and reserves. Such expansion would require massive investment, and Alemán's advisers recommended that private companies be utilized to undertake two-thirds of the necessary exploration and development. The report made it clear that private participation would be under the control of PEMEX and subject to strict terms. Alemán also moved to increase the efficiency of PEMEX management by appointing Antonio Bermúdez, a wealthy industrialist and dynamic administrator, to head the government oil agency. Bermúdez was reported to favor private participation in Mexican oil development.[21]

On August 19, 1946, the State Department called together seven leading oilmen to discuss "American Capital in the Mexican Petroleum Industry." Under Secretary of State for Economic Affairs William L. Clayton explained that the United States had an interest in seeing that its southern neighbor's petroleum potential was developed and utilized to aid the Mexican economy as well as to contribute to hemispheric defense. This U.S. interest in the solution of Mexico's petroleum problem was "partly that of a friendly neighbor, offering advice based on experience and knowledge, and partly that of a country holding an important place in world oil affairs with a direct responsibility for the efficient and orderly development of world petroleum resources."[22]

While the United States recognized the "sovereign right" of Mexico to determine its own oil policy, it felt that Mexico should consider the "wisdom of a change in its attitude towards inviting foreign capital back to Mexico to assist in developing its petroleum resources." PEMEX was "incapable" of developing Mexico's oil either efficiently or profitably, and the United States felt that by nationalization of its oil industry Mexico was not carrying out the spirit of Articles 2 and 6 of the Economic Charter of the Americas signed at Chapultepec in March 1945.[23]

To remedy this situation, Clayton outlined a new plan that provided for private participation in Mexican oil development through the device of an operating and management contract, thus avoiding the historically sensitive question of subsoil rights. Under the plan, the Mexican government would set aside prospecting zones from which the companies could select claims in 10,000 hectare blocks to explore under four-year investigation contracts. If oil were found, the companies would receive fifty-year exploitation contracts which would assure the Mexican government of 35 percent of the net profits, or, alternatively, 12 to 15 percent of the oil produced. The foreign contractor would be exempt from taxes, import and export duties, and have "exclusive control" of any refineries and pipelines built, as well as "absolute control" of the administration of the enterprise. The proposal concluded with a flat statement that the U.S. government would not consider making any loans for the commercial development of the Mexican petroleum industry. The oilmen were quite pleased with the State Department paper, and asked only that the section on profits be amended to include explicitly the right of the contractor to remit profits to its home office. This change was made, and the paper sent to the U.S. Embassy in Mexico on August 27 to serve as the basis of the U.S. negotiating position.

The oilmen at the August 19 meeting all represented large, integrated firms. Jersey had been the largest U.S. company in the Mexican oil industry before expropriation, and had received the lion's share of compensation. Moreover, Jersey was vitally concerned about the effects U.S. policy toward Mexican oil might have on its rich holdings in Venezuela, which constituted the backbone of its supply system. Jersey's unwavering position on Mexican oil was that before U.S. companies could come back to Mexico the Mexicans would have to revise their petroleum laws and labor code, and would have to allow the companies to have full managerial control of their operations and the opportunity to make a reasonable profit. Jersey warned that the participation of small, independent oil companies in Mexico before Mexico made the necessary changes in its oil laws would have "disastrous effects" on the development of a sound oil policy. Such participation, in Jersey's opinion, would be illegal under Mexican law and could only lead to trouble.[24]

George Messersmith, U.S. ambassador to Mexico during the war, had agreed with Jersey on this point, and had constantly cautioned against letting small companies into Mexico before Mexico's general oil policy was revised. Spruille Braden, assistant secretary of state for American Republic Affairs, an independent-minded man whose family controlled extensive copper holdings in Chile, disagreed. Braden felt that it was all right to warn the Mexicans against dealing with small-time operators or "coyotes," but the department should not take the position that the door was closed to all independent companies and negotiations with the Mexicans reserved for the major companies. There were several "substantially large and financially sound" independent oil companies that were looking for reserves abroad, and Braden felt that they should be given equal assistance and encouragement to enter Mexico. Arrangements by the independents, Braden argued, could constitute a "first step" in the reentry of U.S. companies, and in some cases might even have the approval or tacit consent of the major oil companies. Braden soon left the department, however. Messersmith's successor, Walter Thurston, a career diplomat who had served in various Latin American posts, continued the policies begun under Messersmith.[25]

The Mexican government, while aware of the need for financial assistance, could not, for political reasons, permit foreign oil companies, large or small, to operate in Mexico except under the control of PEMEX. President Alemán explained to Ambassador Thurston in late January 1947 that PEMEX had made plans for an extensive exploration and development program, and intended to select suitable U.S. drilling contractors and independent oil companies to undertake much of the work under contract. General participation of foreign oil companies in the Mexican petroleum industry, however, was precluded by Mexican law. Alemán nonetheless took pains to assure Thurston that in an emergency Mexico's oil resources would be instantly at the disposal of the United States. Thurston concluded that Mexico seemed determined to "go it alone" for the time being, and that it would not be advisable to press the Mexicans about oil policy.[26]

In accordance with these plans, PEMEX negotiated drilling contracts with several U.S. companies, and participation contracts with Cities Service and one or two other independent oil companies. The drilling contracts called for the contractors to receive costs plus 10 percent of the profits, with payment in 6 percent Mexican government bonds convertible into oil if any of the wells were productive. The participation contracts called for the contractor to finance development operations by PEMEX, with reimbursement to be in government bonds and marketing rights to 50 percent of any oil produced.[27] Before any of the contracts could be completed, however, they fell prey to conflicts within the Mexican government over oil policy. Mexican Minister of Finance Ramón Beteta refused to authorize issuance of the bonds necessary to finance the contracts, ostensibly to avoid straining Mexico's creditor position. The actual reason may have been that Beteta and the Mexican ambassador to the United States, Antonio Espinosa de los Monteros, had developed a plan whereby they would establish a syndicate to finance development projects in Mexico. This plan caught the eye of Paul Shields, a U.S. businessman with extensive experience in Mexico, who had been authorized by Secretary of Defense Forrestal to explore means of increasing Mexican oil production. The Navy's interest in Mexican oil convinced Beteta and Espinosa that the U.S. government would be willing to provide funds for oil development, and in September they requested a $100 million loan to finance their proposed syndicate.[28]

The Mexican loan request brought into sharp focus the dilemma faced by the United States in dealing with Mexico's nationalized oil industry. Forrestal, who was becoming increasingly concerned about the threat to U.S. access to Middle East oil posed by the Palestine problem, saw Mexican oil as a possible answer to the military's need for a secure source of oil. What he wanted to know, Forrestal told a high-level meeting of State and Defense officials in late September, was "what do we have to do to get oil?" Shields argued that the only way the United States could be assured of getting oil from Mexico was to loan Mexico money to develop its oil. Ambassador Thurston, who had been summoned to Washington to participate in the discussions, argued that any loans given to PEMEX would be "throwing money down a rat hole." Moreover, a loan to PEMEX would be seen as condoning expropriation and would encourage other Latin American countries to nationalize their oil industries. The United States should first try to get Mexico to modify its oil laws to permit private companies to participate in oil development. Only if this course failed should the United States try to work out some basis for assisting Mexico in developing its oil under its present nationalized ownership. Convinced by these arguments, Forrestal agreed not to consider making any loans or advances to Mexico until it had been determined that a "reasonable and mutually acceptable basis for the re-entry of foreign oil companies into Mexico does not exist."[29]

To regain the initiative on Mexican oil policy Thurston was directed to begin discussions immediately with the Mexican government on the formulation of an oil program that would permit foreign oil companies to operate in Mexico on a "competitive and non-discriminatory basis." While the United States was "completely open-minded" as to the basis for the reentry of foreign oil companies, Thurston was to point out that any proposal that did not offer the companies "reasonable compensation" for the risks involved would fail. In addition, if the Mexicans raised the question of the creation of a joint military petroleum reserve in Mexico as proposed by President Roosevelt during the war, Thurston was to make it clear that the United States would no longer consider this approach to the problem.[30]

When Thurston presented the new policy to President Alemán in early December, Alemán informed him that PEMEX was negotiating not only with drilling contractors, but also with several independent U.S. oil companies, which would be allowed to undertake "wildcat" operations in areas designated by PEMEX, and to take a share of any oil found. According to Alemán, such arrangements would not necessitate any changes in Mexico's petroleum laws, which prohibited foreign participation in exploration and exploitation, since the companies would be operating on the basis of a contract with PEMEX. Legally, the companies would be agents of PEMEX so no violation of the law was involved.[31]

In early February 1948 the Mexican government presented the United States with a proposal setting forth the terms under which PEMEX would make contracts with foreign oil companies. PEMEX proposed to designate certain areas for exploration and development and allocate them to various oil companies under contracts. The oil companies would supply all the capital, equipment, and personnel, and conduct the operations, though PEMEX reserved to itself the right to supervise all operations and budgets. If no oil were discovered, the company would take the loss. If oil were found, the property would be turned over to PEMEX, and the company would be reimbursed for its expenses out of 80 percent of the net value of the oil produced. After its expenditures were recovered, the company would receive from 10 to 15 percent of the net value of the oil produced for a specified number of years as a reward for its investment and the risks taken.[32]

The proposal received a mixed reception from the U.S. oil industry. In accord with its longstanding policy on Mexican oil, Jersey rejected the whole setup as illegal and stated that the company could not tolerate the loss of managerial control or accept the provisions regarding financial returns.[33] Unlike Jersey, which had solved its supply problems by buying into ARAMCO, many independent U.S. oil companies needed new sources of oil. Moreover, most independents had no existing foreign investments that would be harmed by a deal with Mexico. Thus many independents felt that the problems of legality and managerial control stressed by Jersey could be resolved through negotiation. There were doubts, however. As one independent oilman pointed out, the provisions reserving supervision and control of operations to PEMEX were no more far-reaching than those contained in the U.S. Federal Leasing Law governing operations on public lands. The problem was that the oil companies did not have confidence in the Mexican government to use these powers with discretion. Oil operations were spread over many years, and whatever the attitude of the present Mexican government, there was no assurance that a future administration would not reverse policy. Several U.S. companies began negotiations with PEMEX on the basis of the Mexican proposal, and in April, Cities Service and PEMEX initialed a draft contract under which

Cities Service would provide PEMEX funds for oil development in return for a commission of thirty-five cents per barrel on 50 percent of any oil found for twenty-five years. The State Department was not pleased with this arrangement because it provided PEMEX with working capital to carry on its own operations, and thus reduced Mexico's incentive to bring in other U.S. companies.[34]

The domestic oil situation and concern over the security of the Middle East also led Congress to take an interest in Mexican oil matters. In January 1948, Senator Walter George of Georgia had strongly backed efforts by Southeastern Oil Company, an independent marketing company, to convince the Export-Import Bank to discount certificates of indebtedness issued by PEMEX so that Southeastern could undertake drilling operations in Mexico. The State Department opposed Southeastern's plans, and the bank turned the company down. In addition, the State Department's policy of supporting export licenses for the Trans-Arabian Pipeline project in the Middle East at a time of a steel shortage clashed with congressional interest in Mexican oil. In March, J. Edward Jones, a combative independent oilman from Louisiana charged before the Senate Special Committee to Study the Problems of American Small Business that the large oil companies were working with the steel companies to prevent him from getting the steel he needed to fulfill a drilling contract with PEMEX. Jones claimed that the State Department was working against him, and had discouraged the Navy from buying Mexican oil until Mexico made its peace with the major companies.[35]

Jones's charges resulted in a letter from the committee chairman, Senator Kenneth S. Wherry, demanding that the State Department clarify its policy on Mexican oil development. In particular, Wherry wanted to know the department's attitude toward the drilling contracts PEMEX was offering, and if the department would favor exporting oil well supplies to carry on the exploration and development program envisaged in the PEMEX contracts. The department's reply claimed that it was favorably disposed to the objectives of the PEMEX plan, but noted that it believed that the achievement of these objectives would depend upon the "establishment and maintenance" of conditions under which U.S. oil companies could participate in the oil industry of Mexico on a basis "mutually satisfactory to the companies and the Mexican government."[36]

Congressional interest in Mexican oil development did not escape the attention of PEMEX head Bermúdez, who in mid-June invited the members of the House Interstate and Foreign Commerce Committee to visit Mexico and survey PEMEX's operations. The committee, which was engaged in a study of the world oil situation, eagerly accepted the invitation, and arrived in Mexico in mid-August. During their two-week stay, the committee members visited the oil fields and other PEMEX projects, attended a special session of the Mexican Congress, and were given a private conference with President Alemán. Committee Chairman Charles A. Wolverton of New Jersey was interested in promoting Mexican oil development as a reliable alternative to Middle East oil, and came away impressed with such social aspects of PEMEX's operations as the construction of schools, hospitals, and low-cost housing.[37]

Bermúdez proposed to the congressmen that the U.S. government loan PEMEX $500 million to finance an ambitious development plan. Wolverton favored the idea, feeling that in return the United States could insist that private U.S. companies be allowed to participate in exploration and development. Ambassador Thurston, however, warned that any loan to PEMEX would be seen as a "reward for expropriation" and could set off a "chain reaction" in the rest of Latin America. Mervin Bohan of Thurston's staff cautioned that the adverse effects of a loan in the rest of Latin America could result in a net decrease in Western Hemisphere oil production. Wolverton was unimpressed by these arguments, and argued that Mexico was a much better risk for U.S. aid than Europe.[38]

The probability that Wolverton's committee would recommend a loan to PEMEX worried the State Department. Most department analysts believed that PEMEX could not be counted on to do more than develop enough oil to meet Mexico's growing internal needs, plus a small surplus for export. What the United States wanted was oil in vast quantities, so that there would be a large surplus available for export. In contrast, PEMEX, either because of inadequate resources and lack of technical and managerial skills or because of a different view of Mexico's needs, could not be relied on to furnish the oil the United States wanted. The "greatest danger" to the U.S. interests in building up a reserve capacity in the Western Hemisphere, Bohan told a gathering of State Department officials in early November, was that PEMEX would bring in just enough oil to satisfy Mexico's needs, and so lose interest in readmitting private capital. Moreover, the State Department feared that direct loans to PEMEX without a change in Mexico's policy toward foreign participation in oil development would "place the seal of United States government approval on expropriation," and were bound to have adverse effects in the rest of Latin America. In particular, no one wanted to take any action in Mexico that risked "losing" Venezuela.[39]

After his brief flirtation with the Shields loan scheme in the fall of 1947, Secretary of Defense Forrestal became a staunch supporter of the State Department's position on loans to Mexico. In the aftermath of the Wolverton committee's visit to Mexico, Forrestal assigned Assistant Secretary of the Navy M. E. Andrews to investigate the matter and make recommendations. Talks with Texas, Jersey, and Gulf convinced the Texas-born Andrews that only the combined efforts of the major U.S. oil companies and the investment of large sums over a ten-year period could achieve the degree of development of Mexican oil that the United States desired. A loan to PEMEX would only delay the beginning of this necessary program, since as long as there was hope for a loan the Mexicans would offer no encouragement to private companies. Forrestal approved Andrews's recommendation that the Defense Department oppose any loan to PEMEX.[40]

Holding the Line in Mexico

The Wolverton committee's report, issued on December 30, 1948, recommended U.S. government assistance to PEMEX, subject to an understanding that U.S. oil companies be allowed to participate in petroleum exploration and development activities on reasonable terms. The committee printed a new set of "bases" negotiated with Bermúdez in August: they provided that private oil companies, acting as PEMEX agents, could explore for oil on sites of their own choosing and at their own risk. If oil was found, PEMEX would take over the operation, though there was the possibility that the company might be allowed to stay on as a PEMEX agent. Fifty percent of gross production would be devoted to reimbursing the company for its expenses, and another 13.5 percent of gross production would go to the company as compensation for risk. The company would have the option to buy any remaining oil after Mexican requirements were met. According to calculations by an oil company representative interviewed by the committee, 13.5 percent of gross production was roughly equivalent to 35 percent of net income. The oilman pointed out that the actual split of net income in Venezuela was 55-45 in favor of the government rather than 50-50. Consequently, the committee concluded that the oil companies and PEMEX were not so far apart as to preclude a compromise.[41]

The committee argued that it was in the U.S. national interest to help Mexico develop its oil. Mexico was in imminent danger of becoming a net importer of oil since its domestic consumption was steadily increasing while production was gradually declining. In monetary terms, Mexico was already a net importer vis-à-vis the United States, as the value of the specialty products it imported from the United States exceeded the far larger quantity of crude oil it exported to its northern neighbor. This situation exacerbated Mexico's already severe balance of payments problem. Development of Mexican oil could reverse this situation by providing Mexico with the foreign exchange needed to finance essential imports for continuing its economic development. "Preservation of a sound economy in the nation abutting our southern border" by aiding Mexican oil development was "absolutely imperative" from the perspective of national security. Mexican oil, if developed, could go a long way toward meeting projected U.S. emergency needs, and it was more defensible than Venezuelan oil, not to speak of Middle East oil.[42]

Robert Crosser of Ohio replaced Wolverton as the committee's chairman after the Democrats regained control of the House in the 1948 elections. Crosser had been in Mexico at the time the committee's report was issued, and the State Department harbored hopes that he would oppose an oil loan. Crosser, however, was an advocate of nineteenth-century economist Henry George's ideas concerning the "single tax," and was convinced that this concept offered a solution to the Mexican oil problem. Crosser proposed that the United States lend PEMEX funds for oil exploration and development. Once oil was found and developed, PEMEX should contract out the operation of its properties to private companies on the basis of competitive bidding, subject to the requirement that the ground rent of the land involved was to be paid to the Mexican government. The purpose of this plan was to remove the justification for high profits by removing the risk to the companies. The State Department tried to discourage Crosser by arguing that exploration and development were activities for which private enterprise was best suited, and by pointing out that it was highly unlikely that the Mexican government would allow private companies back into domestic marketing and distribution. Crosser would not be dissuaded, and invited Bermúdez to Washington to discuss the possibility of a loan.[43]

Before coming to Washington, Bermúdez made a deal with a U.S. oil company in order to demonstrate his readiness to permit foreign participation in Mexican oil development. In early March, PEMEX signed a contract with a consortium of independent U.S. oil companies organized as Compañia Independiente México-Americana (CIMA). The contract ran for twenty-five years, and provided that the company would be reimbursed for all expenses, including dry holes, out of 50 percent of the proceeds of gross production. As payment for risk, the companies would receive 15 percent of production in upland areas, 16.25 percent for lagoon areas, and 18.25 percent for offshore sites. A sales contract signed concurrently with the exploration contract meant that compensation would actually be in oil rather than dollars. PEMEX would officially take over the wells after oil was found, but the companies would be allowed to stay on to provide technical assistance.[44]

Bermúdez presented the CIMA contract to the State Department in mid-March 1949, as evidence that there was no prejudice in Mexico against U.S. private capital. In return, Bermúdez asked for $203 million in aid, $60 million of which was for production equipment, and the remainder was earmarked for pipelines, refineries, and distribution equipment. The State Department, which had received White House authorization to handle all negotiations with Mexico, was cool to the PEMEX request. Oil loans were contrary to long-established policy, and the department was concerned about the repercussions in other oil-producing countries. In addition, the oil shortage of 1947–1948 had become a surplus, and the discovery of new fields in Canada promised to relieve military concern over oil reserves. Upping the stakes, Assistant Secretary of State for Economic Affairs Willard Thorp informed Bermúdez that uncompensated claims by U.S. citizens against the Mexican government would have to be settled before loans could be considered, since they could become a "source of opposition to consideration of the PEMEX proposal."[45]

The State Department's reluctance to aid PEMEX was shared by the departments of Defense and the Interior. Since early in the year, the International Petroleum Policy Committee, an interdepartmental group under the Executive Committee on Economic Foreign Policy, had been working on a policy statement on an oil loan to Mexico. Most of the committee's members opposed any loan to Mexico, and argued that increasing Venezuela's production was a more promising means of meeting strategic requirements than trying to stimulate Mexican development. Increases in Venezuelan production would be available for export whereas increases in Mexican production would be domestically consumed. In addition, the U.S. chargé in Venezuela was concerned that the arrangements under which private companies reentered Mexico might be more favorable to Mexico than comparable arrangements in Venezuela. Finally, the military representative insisted that the conditions attached to a loan would have to be explicit enough to "make it amply clear to all Latin American countries that nationalization in Mexico has failed."[46]

To meet these concerns, the committee devised a strategy whereby loans for transportation and refinery facilities would be considered only if the Mexican government established conditions under which private companies would be permitted to operate in Mexico on a sound legal basis. Unconditional loans and loans for exploration and development would not be considered. President Truman, however, wanted to help Mexico. On March 3, 1947, Truman had made the first official visit by a U.S. president to Mexico City in an effort to strengthen the ties between the two countries which had developed during the war. Alemán returned the visit in late April 1947 as U.S.-Mexican relations entered what Howard F. Cline has termed an "era of good feeling." In addition, U.S. investment in Mexico began to grow again in the 1940s as Alemán adopted an "open door" policy toward foreign capital. Finally, Truman was concerned that unless Mexico's oil problems were solved soon, economic conditions in Mexico could seriously deteriorate and lead to instability. Therefore, on May 23, 1949, he instructed the State Department to proceed "as expeditiously as possible to develop a solution on the basis of a conditional loan for distribution and refinery equipment."[47]

Despite the president's endorsement, there was little enthusiasm for loaning Mexico money for oil development. At a meeting of the heads of all concerned agencies in mid-June, the State Department assured the group that the strategy devised by the Petroleum Policy Committee "would constitute a reversal of nationalization." Nevertheless, Secretary of the Interior Julius A. Krug argued that Mexico should open all areas of its economy to U.S. companies before the United States loaned it any money. The assistant secretaries of treasury and commerce argued that assistance to PEMEX would encourage other countries to nationalize U.S. companies, and the military representatives once again stressed their concern that nothing be done that would jeopardize U.S. interests in Venezuela since the loss of Venezuela would more than offset any possible benefits to be gained from helping Mexico develop its oil. The vice-chairman of the Export-Import Bank argued that before any loans were even considered, oil should actually be produced on properties granted to U.S. companies. These objections notwithstanding, no one was willing to go against the president's wishes and openly oppose conditional loans.[48]

Since the Export-Import Bank would be the source of any loans, the State Department worked closely with bank officials to devise a procedure for implementing the president's wishes. Bank Chairman Herbert E. Gaston felt that the Export-Import Bank should not consider any kind of loan for petroleum development. In his opinion, petroleum was a proper area for private capital, and Mexico could obtain the capital it needed simply by changing its laws. Nevertheless, the State Department was able to get Gaston to agree to a step-by-step procedure that required Mexico to take and to continue "appropriate action to provide an adequate legal basis for private companies to undertake exploration, development and production activities." Loans would be dealt with on a piecemeal basis, each subsequent loan dependent on some further action by the Mexican government in favor of private development.[49]

Before informing the Mexicans of the U.S. position, the State Department added one further condition. In 1935, the Sabalo Transportation Company, a U.S.-owned company incorporated in Mexico, had obtained rights to 10,000 hectares in the Poza Rica oil field. Because of disagreements with the company, the Mexican government itself took over these rights, and in 1942 Sabalo was awarded $897,000 by the Cooke-Zevada Commission. After accepting all but the last installment of this amount, Sabalo's owners revived their claim with the assistance of the powerful law firm of Sullivan and Cromwell. As a result of Sullivan and Cromwell's lobbying efforts, the State Department dispatched two notes supporting Sabalo to the Mexican government during 1948, which were given directly to President Alemán by Ambassador Thurston. In early 1949, the Mexican Supreme Court ruled that Sabalo's rights had not been expropriated in 1938 and were still valid. The State Department therefore decided that during loan negotiations the United States would try to "extract" from Mexico assurances that Sabalo would be permitted to resume operations in Mexico under its original contract. Since the Sabalo claim included almost 40 percent of the rich Poza Rica field, Mexico's principal source of oil, and since Sabalo's owners had made arrangements with Jersey for the operation of their properties, what the department was proposing as a condition for a loan was no less than the return of Jersey to a major position in the Mexican oil industry.[50]

On July 6, 1949, the State Department presented the Mexican government with an Aide Mémoire enumerating the conditions under which the United States would be willing to aid Mexican oil development. Applications for financial assistance for the construction of refining, transportation, and other distribution facilities would be considered "provided that Mexico will take steps to assure an increase in oil production in Mexico through increased participation by private companies, including foreign companies." The "prerequisite" to such increased participation by private companies was "appropriate action" by the Mexican government to provide an accepted legal basis for foreign companies to undertake exploration, development, and production activities. If Mexico took such action, the United States would consider a loan application for a "relatively small project." Other loans would be contingent upon further positive action by the Mexican government. One of the ways by which the Mexican government could "indicate its attitude towards private capital in the field of petroleum exploration, development and production" was the "speedy settlement" of claims against it by U.S. companies, "particularly one large claim."[51]

The State Department was well aware that Mexico would not like the strict terms of the Aide Mémoire, but the opinion in Washington was that the time had come to make clear the U.S. position on the role private enterprise and capital should play in Mexican oil development. If the Mexican government wanted U.S. public funds to build refining and distribution facilities, it would have to follow the procedure laid out in the Aide Mémoire. Indeed, the head of the Office of Mexican Affairs told Thurston that many people would be pleased if the Mexicans turned it down. The department did not intend to "budge an inch" from the position set forth in the note. Even if the State Department were willing to compromise, it "would take an act of Congress" to get the Export-Import Bank to change its position. There would be nothing further done about oil loans until several U.S. oil companies were operating in Mexico on satisfactory terms.[52]

To no one's surprise, the Mexican government found the whole tone of the Aide Mémoire insulting, especially the step-by-step procedure with its requirement that the Mexican government meet certain standards in order to receive aid. It was interpreted as unconcealed mistrust of Mexico's good faith. Moreover, the inclusion of an "unmistakeable reference" to the Sabalo case was "especially distasteful." The Mexican government felt that the case was extraneous to its application for a loan, especially since talks were already underway with representatives of the company.

Most of all, the Mexicans resented the U.S. demand that in order to get a loan, Mexico would have to change its oil policy to suit the United States. Oil policy and "any other matter of an exclusively internal nature," the Mexican government informed the State Department on July 18, would be determined solely by Mexico.[53]

Congressman Crosser was also displeased with the department's position, and warned the State Department that he would work to reverse its policy. In addition, Wolverton pointed out that the United States had spent millions of dollars to build pipelines and develop oil thousands of miles from its shores, while overlooking oil on its "doorstep," and argued that loaning Mexico money for oil development would actually encourage U.S. companies to go to Mexico to assist in oil development. Wolverton criticized the State Department for holding up a loan, and pointed out that Assistant Secretary of State for Inter-American Affairs Edward G. Miller had handled the case for Sullivan and Cromwell before joining the department.[54]

The Mexican reaction to the Aide Mémoire coupled with congressional criticism and the personal intervention of President Truman led to a change in U.S. strategy. In late August, President Truman informed Secretary of State Dean Acheson that he was "exceedingly anxious" that President Alemán not be put into an embarrassing position over the loan issue. "Neither do I want the representatives of those American oil interests who caused our oil industry down there to be expropriated, to be put back into a position to embarrass us." Mexico's oil should be developed "for the benefit of the Mexicans themselves." If it could be done by U.S. companies with mutual benefits and profits, he would support that. But Truman was adamant that what he termed "the old Doheny interests and people who robbed the Mexicans before" not be permitted "to do it again and bring about the same sort of situation with which we are now faced."[55]

The president's actions got things moving again. In mid-September, Acheson, who had succeeded George Marshall in January 1949, instructed Thurston to seek a resumption of negotiations for a loan for refining and transportation equipment. In several meetings with Alemán and Bermúdez, Thurston gained their assurances that PEMEX was prepared to negotiate other contracts on the basis of the CIMA formula. In addition, the Mexicans assured him that CIMA-type contracts were "legally invulnerable," and that any loans would be used only for refinery or distribution facilities.[56]

These developments became the basis of a plan to reconcile President Truman's desires with the conditions outlined in the Aide Mémoire. The State Department proposed to accept the Mexican government's assurances in lieu of a government-to-government agreement on greater private participation in Mexican oil development. Then the department would issue a public statement noting the Mexican government's intention to facilitate increased participation of private companies in Mexican oil development. These actions would open the way for consideration of a small loan for a refinery or transportation project, with additional loans dependent on subsequent developments. The question of U.S. claims against Mexico, moreover, would be dissociated from the loan issue. In this way, the department hoped to provide Mexico with a face-saving way of meeting the conditions outlined in the July note. The change in strategy did not represent a retreat from the basic tenets of U.S. policy on Mexican oil development, however, since U.S. government aid would be provided only if the Mexicans opened the door to U.S. oil companies.[57]

Export-Import Bank Chairman Herbert E. Gaston undercut this strategy by refusing to agree to any departure from the Aide Mémoire. Gaston stuck to his firm belief that a substantial increase in Mexican oil production could only be accomplished by private companies. In his opinion, the recent statements and actions by the Mexican government failed to offer evidence of any intention to open fully Mexican oil development to U.S. companies. Gaston also feared that the State Department's plan would encourage other Latin American countries "to restrict or eliminate the present or potential operations of United States private oil companies in their countries."[58]

Meanwhile, PEMEX's strategy of relying on exploration and drilling contracts with independents was meeting with success. The first CIMA wildcat well discovered a major field, and the geology of the discovery suggested that huge reserves lay untapped. In addition, PEMEX's drilling contractor, Wiegand of San Antonio, brought in five producing wells in the first six wells drilled. These discoveries enabled PEMEX to obtain private financing for its most urgent refinery and pipeline projects, and demonstrated that PEMEX could develop Mexico's oil without turning to the major oil companies. Jersey's representative in Mexico, Walter Sollenberger, drew the obvious conclusion: "The return of the major United States oil companies . . . was never less likely," he told a member of the U.S. Embassy staff in October 1949. Mexico intended to develop its oil itself, and Sollenberger believed "she can do it." Jersey was "wasting its money" keeping him in Mexico.[59]

These developments, coupled with continued warnings from the embassies in Venezuela and Colombia that any loan to PEMEX would prejudice U.S. oil interests in those countries, convinced the State Department that little was to be gained by a loan. Accordingly, Acheson recommended to President Truman, in late January 1950, that the United States stay with the position taken in the Aide Mémoire. U.S. government financing of PEMEX "would be interpreted in other Latin American countries as approval in principle of state operation of the oil industry," especially since Mexican officials had said that a loan would "consecrate"

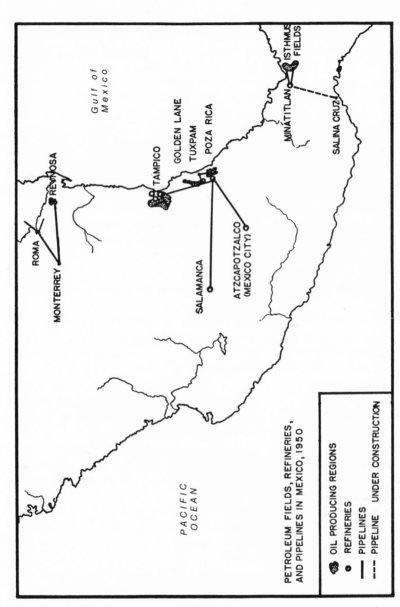

PETROLEUM FIELDS, REFINERIES,
AND PIPELINES IN MEXICO, 1950

🛢 OIL PRODUCING REGIONS
● REFINERIES
—— PIPELINES
--- PIPELINE UNDER CONSTRUCTION

Gulf of
Mexico

PACIFIC
OCEAN

REYNOSA
ROMA
MONTERREY
TAMPICO
GOLDEN LANE
TUXPAM
POZA RICA
SALAMANCA
ATZCAPOTZALCO
(MEXICO CITY)
MINATITLAN
ISTHMUS
FIELDS
SALINA CRUZ

Source: J. Richard Powell, *The Mexican Petroleum Industry, 1938–1950* (Berkeley and Los Angeles: University of California Press, 1956), p. 54. Used with the permission of the University of California Press.

the principle of expropriation. Thus loaning PEMEX money would "weaken the position of American investments abroad."[60]

The State Department had not counted on what N. Stephen Kane has termed Truman's "populist inclinations." The president rejected the State Department's recommendation, and directed his administrative assistant, Charles S. Murphy, to take charge of the matter. "This needs careful consideration. Standard of New Jersey and the Texas Company and also the Gulf have a finger in what has happened. I want a loan granted to Mexico for refinery and pipeline development. I want private arrangements made with our wildcat drillers for the proper extension of drilling. Something is slowing the program. Get me all the facts. Watch the successors of Teapot Dome and see if we can't help Mexico and the Mexican people."[61]

Acting on the president's instructions, Murphy and George M. Elsey, another administrative assistant, met with State Department officials over the next few months in an effort to find a basis for reviving loan talks. The State Department remained adamantly opposed to "any loan to Mexico for any phase of its petroleum industry," and inundated Truman with objections. The petroleum industry was facing an "over-supply situation." Furthermore, Mexico had failed to enter into any additional contracts with U.S. companies. A loan would not solve Mexico's distribution and refining problems nor would it bring about expanded exploration and development. In addition, the Export-Import Bank was considering other, more important loans to Mexico, and since Mexico's debt was near the limit of its servicing capacity, an oil loan would prejudice the other projects. Mexico's desire for a combination loan and credit was contrary to bank policy and would needlessly immobilize the bank's capital. Finally, a loan to PEMEX "would encourage a nationalistic approach to the problems of discovering and developing Latin American oil resources."[62]

Although Truman continued to insist on an *oil* loan, neither the State Department nor the Export-Import Bank relaxed their opposition. The Export-Import Bank's Gaston was particularly adamant, arguing that such a loan would be contrary to the bank's charter, which prohibited the bank from competing with private capital, would also block further private development in Mexico, and would encourage other countries to turn to the bank rather than to private sources for funds. Worst of all, an oil loan to Mexico would appear to condone expropriation and thus increase the likelihood of expropriation in countries like Venezuela. Although the White House staff felt that Gaston's arguments against a loan were inconclusive, Truman, increasingly preoccupied with other, more urgent matters, decided not to press the issue. At a meeting at the White House in late June with representatives of the State Department and the bank, the president "simply emphasized that he wanted to make his views known to the bank," and left it up to the bank to study the matter.[63]

A resolution to the issue was finally reached in August 1950. Apparently disturbed over the impact of the loan question on cabinet politics, Mexican President Alemán decided to withdraw Mexico's request for an oil loan. Instead, Mexico requested a general line of credit of $150 million to finance the renovation of its railway system and some irrigation projects. The State Department supported this plan because it avoided the oil loan issue, helped insure Mexican cooperation in the Korean War effort, and helped improve general U.S.-Mexican relations. Even though the plan in effect allowed Mexico to shift its limited funds to oil development, Gaston raised no objections, and on the last day of August, the Export-Import Bank approved a $150 million line of credit to finance projects in Mexico involving agriculture, transportation, communications, and electric power.[64]

Chapter 7

Securing the Middle East

Shorn of all embellishments, which have been designed for tactical purposes only, the relation of state and industry in respect of foreign operations is extremently simple: the oil people are dealing as agents for a principal who has elected to pay them commission on a generous scale.

—Paul Frankel, *Essentials of Petroleum*

U.S. foreign oil policy looked to the private initiatives of the major oil companies to protect and promote the U.S. national interest in Middle East oil. This reliance on private interests significantly shaped the U.S. response to the postwar dollar shortage and to producing country desires for increased revenues. In these cases, the position of the major oil companies as vehicles of the national interest in foreign oil enabled them to obtain public support while maintaining their autonomy and their ability to profit from the oligopolistic structure of the international oil industry.[1]

Oil and the Marshall Plan

By the late 1940s, the international economic order was in chaos. The earlier system of international trade, multilateral exchange, and free convertibility of currencies had largely disappeared under the impact of depression and world war. In its place stood what was virtually two worlds of international finance. One world consisted of the United States and other countries whose currencies were fully convertible into dollars, known as dollar or hard currency countries. Their hard currency reserves came largely from a favorable balance of trade with the United States. These countries enjoyed free convertibility of currencies and free movement of funds among one another. The other world was made up of national economies short of dollar exchange. Within this world the main group was the sterling area, composed of the United Kingdom and a number of other countries that had worked out with the British government certain arrangements regarding their foreign payments. Most of the members of the sterling area had some sort of connection with Great Britain, either as colonies, protectorates, or members of the British Commonwealth. The remaining soft currency countries were mainly in Western Europe and South America, and many had a tradition of trade in sterling. Despite the lack of convertibility into hard currencies, the amount of world trade in sterling was large and growing.[2]

Despite more than $9 billion in U.S. economic assistance between 1945 and 1947, the economy of Western Europe had not recovered from the ravages of war. Widespread physical destruction, worn-out and obsolete capital equipment, and exhausted, undernourished, and disorganized labor forces interacted to prevent economic recovery. In addition, the war had disrupted the patterns of trade within Europe and between Europe and the rest of the world. Western Europe was no longer able to obtain food from its traditional sources in Eastern Europe, and had lost many of its markets in the underdeveloped world to the United States. A substantial part of the invisible earnings on shipping, banking, insurance, and foreign investment, which before the war had financed Europe's trade deficit, was now gone. With imports of food, coal, and industrial equipment from the United States growing to meet relief and reconstruction needs, Europe was facing a massive dollar shortage.[3]

During the early postwar period, U.S. government aid channeled through the United Nations Relief and Rehabilitation Administration and the drawing down of gold and dollar reserves enabled Europe to obtain needed supplies from the United States. By early 1947, however, the financing of U.S. exports was becoming more difficult as foreign reserves neared depletion and U.N. relief efforts came to an end. The State-War-Navy Coordinating Committee warned in early 1947 that the "world will not be able to continue to buy U.S. exports at the 1946–1947 rate beyond another 12–18 months." Without new means of financing, a sharp decline in U.S. exports was inevitable, and this decline, in turn, could have a serious impact on the U.S. domestic economy. Moreover, with the Greek civil war focusing attention on the political consequences of economic dislocation, U.S. policymakers feared that without massive dollar aid the European economy would sink to a level so low that most of continental Europe would fall into the hands of the Communists. Communist parties were already gaining strength in France and Italy, and economic problems were undermining Britain's position as a world power. Out of these concerns came the plan for European reconstruction announced by Secretary of State George C. Marshall at Harvard University on June 5, 1947. As Fred Block has pointed out, the "genius" of the Marshall Plan was that "it simultaneously attacked all of the forces that were moving Western Europe away from multilateralism"; by strengthening the European economy, it aimed at undercutting the appeal of the left and the pull from the Soviet Union as well as the trend toward state capitalism in foreign trade.[4]

Before the war, Western Europe had depended on coal for over 90 percent of its energy requirements. Wartime destruction, dislocation, and overuse sharply curtailed the productive capacity of Western Europe and Britain. British coal production was down 20 percent from 1938 while coal output in the western part of Germany was only 40 percent of the prewar figure. Moreover, Western Europe lost access to the oil of Eastern Europe as the Soviets extended their control over this area after the war. Under such circumstances, the possibility of obtaining oil supplies from overseas seemed to offer a sure and almost immediate solution to the impending energy crisis. Although oil accounted for a little less than 10 percent of Europe's total energy supply in 1947, it was the only source of fuel for aviation and road haulage as well as an increasingly important source of fuel for inland and overseas shipping and railway transport. Oil was more efficient than coal, and politically more reliable, as many coal miners belonged to Communist-led unions. On the other hand, approximately half of Western Europe's oil in 1947 was supplied by U.S.-owned oil companies and thus required payment in dollars.[5]

Oil was the largest single item in the dollar budget of most of the Marshall Plan countries. Moreover, the sharp rise in the price of oil after World War II had been one of the most important factors in the deterioration of Europe's current dollar accounts. Savings of dollars on oil imports had the greatest potential of any single project for reducing Europe's dollar shortage. Oil bought from U.S. companies led to dollar claims against the foreign exchange supply of the purchasing country. U.S. companies needed dollars not only to remit profits and to amortize invested capital, but also to pay dollar salaries and buy dollar supplies. Even oil supplied for other currencies by non-U.S. companies entailed substantial dollar costs because of the dominant position of the United States in petroleum equipment, technology, and shipping. In Venezuela, for example, Shell had to pay royalties and taxes in dollars as well as buy all needed local currency in dollars. In Iran, the British government was obliged to convert a large portion of Anglo-Iranian's royalty payments into dollars to satisfy Iran's requirements for imports from the United States. Moreover, much of the oil equipment needed for maintenance and expansion of oil facilities was available only from the United States, and the majority of oil tankers were under either U.S. or Panamanian registry and so required dollars to lease. In short, if Europe were to meet its energy needs with imported oil, it would need dollars to pay the bill.[6]

For these reasons, oil became one of the key commodities in the European Recovery Program (ERP). More than 10 percent of the total aid extended under ERP was spent on oil, more than for any other single commodity. Between April 1948 and December 1951, 56 percent of the oil supplied to the Marshall Plan countries by U.S. companies was financed by the Economic Cooperation Administration (ECA) and its successor, the Mutual Security Agency (MSA). This aid not only helped provide Europe with the energy it needed for recovery; it also served to maintain markets for U.S. oil companies at a time when their potential customers would otherwise have been unable to obtain the necessary dollars. This aid was especially important to the U.S. oil companies operating in the Middle East since most of their potential markets were in Western Europe. As ECA Administrator Paul Hoffman, who had reluctantly left his position as head of the Studebaker Corporation to run the ERP, explained in early 1950, the Marshall Plan countries had remained "good customers of the [U.S.] petroleum industry" because of dollars furnished by ECA.[7]

Unlike most commodities supplied under the Marshall Plan, oil did not come primarily from the United States, but from U.S. companies operating abroad. At the time the Marshall Plan was being developed there was a worldwide shortage of petroleum. In order to ensure that helping Europe meet its energy needs would not impose an undue drain on U.S. resources or impair the "fulfillment of vital needs of the people of the United States," Section 112 of the Economic Cooperation Act of 1948 stipulated that the "procurement of petroleum and petroleum products under this title shall to the maximum extent practicable be made from petroleum sources outside the United States." Following these directives, ECA financed the major portion of its petroleum transactions from "offshore sources." The world's oil resources outside the United States were almost entirely owned by five U.S. and two British companies. Consequently, most of the oil financed by ECA was supplied by the five U.S. companies and their affiliates: 48.8 percent from Standard Oil of New Jersey; 14 percent from the California-Texas Oil Company (Caltex), the jointly owned marketing organization of Standard Oil of California (SOCAL) and the Texas Company; and 9.2 percent from the Socony-Vacuum Oil Company.[8]

The ECA legislation, while requiring ECA "to facilitate and maximize the use of private channels of trade," prohibited the financing of commodities at "prices higher than the market price prevailing in the United States at the time of purchase, adjusted for differences in cost of transportation to destination, quality, and terms of payment." The vertical integration of the major oil companies meant that the "sales" ECA financed were in fact transfers among the companies' respective corporate affiliates. Thus even though the prices charged on such transactions were lower than comparable U.S. prices, ECA could not assume that the prices reflected the operation of market forces. This was especially the case with oil from the Middle East, and throughout the program ECA engaged in a running battle with the oil companies over the prices charged for Middle East oil.[9]

Despite the apparent leverage possessed by ECA by virtue of its financing of a significant amount of their sales, the major oil companies were able to maintain the administrative pricing system they had designed to avoid price competition. Though the price of crude oil f.o.b. Persian Gulf (36° API) fell from $2.22 per barrel (36° API) in the spring of 1948 to $1.75 per barrel by the summer of 1949, this reduction did not affect the overall profits of the parent companies since they were the buyers of this oil as well as the sellers. Rather the main impact was to redistribute profits within the integrated organizations. ECA pressure undoubtedly played a role in the price reductions, but they were due mainly to decisions by the major companies to utilize more fully the large quantities of Persian Gulf crude oil that were becoming available. Lower crude oil prices served as a means to this end by making it profitable for the majors' refining affiliates to utilize Middle East rather than Western Hemisphere crude oil. This interest of the companies in lower prices for Middle East crude oil coincided with ECA's interest in reducing the cost of oil imports to Europe. In the case of refined products, the interests of the companies and ECA diverged, and no reductions occurred, leading ECA in the spring of 1951 to suspend financing of product shipments. In late June 1952, ECA's successor agency, the Mutual Security Agency (MSA), decided to stop financing purchases of Middle East crude oil. On August 22, 1952, the Justice Department filed suit against Caltex and its owners, SOCAL and the Texas Company, to recover $66 million in alleged overcharges.[10]

Lower oil prices saved dollars. Another way to save dollars was to expand refinery capacity in Europe, thus allowing the participating countries to shift their oil imports from more expensive refined products to less expensive crude oil. The Marshall Plan countries wanted ECA to provide dollars for the expansion of European-owned refinery capacity. Initially, such a program would require large dollar expenditures for specialized petroleum equipment, engineering fees and services, and other items available only from the United States. The Europeans argued, however, that the dollars spent on refinery capacity would ultimately result in large dollar savings, and thus a reduced burden on U.S. taxpayers.[11]

Lack of refinery capacity was also the key bottleneck holding back the replacement of dollar oil by sterling oil. British oil companies planned to double their production from 1.1 million barrels per day (bpd) in 1947 to 2.2 million bpd by 1953. Since world demand outside the United States was expected to increase from 2.5 million bpd to 3.6 million bpd in the same period, it appeared that the British planned to supply the entire increase in demand. The world shortage of dollars created considerable pressure for such a development to take place, but the main obstacle to such a development was the shortage of refinery capacity in the Eastern Hemisphere. As long as this was the case, Western Europe would be forced to rely to a certain extent on dollar refined products if not dollar crude oil. In light of this situation, the refinery expansion program of the Marshall Plan countries was of key importance to the future positions of sterling and dollar oil in world markets.[12]

The expansion of the British oil industry would not have been such a threat if U.S.-controlled foreign oil production could have been assured sufficient markets in the United States. State Department oil experts calculated that if net imports of U.S.-controlled foreign oil into the United States increased from the 1948 level of around 200,000 bpd to around 600,000 bpd, the planned British expansion should pose no threat to U.S. interests. While attracted by the profits from selling cheap Middle East oil at U.S. prices, the major companies had no intention of causing a collapse in the domestic prorationing system by forcing too much foreign oil into the U.S. market. The key impediment to increased imports, however, was the opposition domestic oil producers were likely to organize. A foretaste of this opposition came in 1949 as domestic producers reacted to continuing high levels of imports at a time when U.S. demand was declining. There were three major congressional investigations to determine whether oil imports were injuring independent domestic producers. These investigations attracted a great deal of attention and served both as reassurance to the independents and a threat to the majors that if imports were not kept at a reasonable level legislative action would be forthcoming.[13]

Given the "problematic net import needs of the United States," Walter Levy, ECA's chief oil expert, and the oil people in ECA opposed aiding the expansion of the European oil industry, arguing that to do so would harm U.S. oil companies operating overseas. Born and educated in Germany, Levy had worked as an oil journalist in London from 1933 to 1941 when he came to the United States. During the war, he had worked for the OSS. He had just accepted a position with Socony-Vacuum when ECA head Hoffman asked him to advise ECA on oil matters. Levy rejected the initial oil equipment program proposed by the oil committee of the Organization for European Economic Cooperation (OEEC) as unrealistic, and asked the committee to adjust the individual country estimates so that a "reasonable and coordinated" expansion program could be initiated. Levy also informed the OEEC that in view of the incomplete state of the expansion program ECA would not finance any part of the program at that time. In order to implement this policy, Levy had a qualification placed on ECA procurement authorizations under the machinery code to the effect that no authorizations could be used for petroleum equipment.[14]

On the other hand, some ECA officials, especially the economists in the Controller's Office, felt that the general interest of European recovery took precedence over the special interests of oil companies, and that ECA should support an activity that promised such great dollar savings. At an interdepartmental meeting in early March 1949, Levy warned that the loss of markets for U.S.-owned foreign oil could jeopardize the security of the concessions held by U.S. companies in Latin America and the Middle East. These warnings brought the military and the State Department into the controversy within ECA on the side of the oil people, as both agencies strongly opposed any actions that would adversely affect the position of U.S. companies in foreign oil.[15]

There was a "vital United States interest in American-owned foreign oil," the State Department asserted. U.S. security was dependent on the availability of oil from foreign sources in an emergency since U.S. domestic productive capacity was insufficient to meet prospective emergency demands. Oil supplies strategically located outside the United States would be militarily advantageous in the event of a war, and U.S. oil operations abroad added to U.S. prestige and influence. Moreover, oil investments accounted for over 70 percent of the net foreign investment of U.S. nationals and played an important role in economic development. Thus it was important to protect U.S. holdings in foreign oil from loss. Displacement of dollar oil by sterling oil in world markets could cause reductions in the foreign operations of U.S. oil companies. Forced shut-in of U.S.-controlled production in foreign countries while British production in the same countries or the same area was being rapidly expanded would place the U.S. companies in an untenable position, and they could lose their concessions.[16]

The State Department also argued that it was important that U.S. companies, rather than British companies, control the bulk of foreign oil reserves since U.S. companies would be more likely to cooperate with the U.S. government, particularly in providing the types of information needed for pre-emergency planning. The military tended to agree with this assessment, though the head of the Armed Services Petroleum Board admitted that "logistically speaking, it should not matter which friendly power developed Middle East oil production, since, if the area could be held and oil used during hostilities, ownership as between Allies would be unimportant." The military believed that the real strategic problem created by the dollar shortage was the danger that U.S. companies in the Middle East, in order to find outlets for their production, would force so much of their low-cost oil into Western Hemisphere markets as to harm Western Hemisphere producers and lower Western Hemisphere productive capacity.[17]

In late March 1949, the OEEC submitted a scaled-down refinery expansion program to ECA. ECA, however, decided that it would finance refinery expansion only to the point required to replace imports of dollar products. Further expansion, ECA pointed out, would replace imports of sterling products and would mean that these products would be seeking new markets outside Europe in competition with dollar products. Beyond this minimum program, ECA would approve only those projects that were "demonstrably in line with normal market growth." Projects whose economic vitality depended on currency discrimination or other "undesirable trade restrictions" would not be approved. In practice, therefore, ECA would not finance projects that threatened to compete with U.S. companies. Since almost anything that went beyond the minimum program would do this, ECA financed few refinery projects. In all, ECA provided only $24 million to increase European refinery capacity, a small amount compared to the $1.2 billion provided to finance purchases of oil from U.S. companies.[18]

ECA's refusal to finance European refinery expansion beyond a limited degree did not stop expansion. The Marshall Plan countries merely shifted their free dollars to refinery construction, and ECA could do little to prevent this. Drastic action, such as withholding funds, might have had disastrous consequences for the overall recovery program. Moreover, cutting funds could be counterproductive as it could force the countries affected to impose even more drastic limitations on dollar imports, and to give greater emphasis to expanding nondollar oil projects. Rather than trying to stop the expansion of the European oil industry, the United States sought instead to remove the British currency restrictions, which, in combination with the dollar shortage, made the expansion program a threat to the position of U.S. companies involved in foreign oil.[19]

The Sterling-Dollar Oil Controversy

Despite ECA aid, it appeared by late 1948 that the world dollar shortage would continue for the foreseeable future, and would worsen after the end of dollar aid under the Marshall Plan. British Treasury regulations prohibited U.S. companies from selling for sterling outside the sterling area, since under the Anglo-American Financial Agreement of 1945 Britain would be obligated to convert the sterling so earned into dollars at the companies' request. (U.S. companies were allowed to sell for sterling inside the sterling area.) In contrast, British companies could sell for sterling outside the sterling area and use the sterling freely wherever it was acceptable. With dollars in short supply, these regulations put U.S. companies in a potentially unfavorable position.[20]

By the spring of 1949, some countries had begun to make special arrangements with British oil companies to supply their oil needs. In April, Egypt and Argentina made bilateral deals with Britain to supply all their oil needs against payment in sterling. In both cases, U.S. companies lost their shares of the country's market. Then Britain established greater control over trade with other soft-currency countries and ordered British bankers to refuse to transfer funds in payment for U.S. oil from sterling balances in London of countries outside the sterling area. As a result, the Scandinavian countries, which were so short of dollars as to have to pay for oil with sterling, were no longer able to draw on their sterling balances in London to pay for imports from U.S. companies. This move also cost U.S. companies valuable markets.[21]

The major U.S. oil companies argued that the British actions were not necessary to save dollars and that it was possible to work out arrangements under which they could sell oil for sterling outside the sterling area without causing a significant dollar drain on the United Kingdom. The key difference between U.S. and British companies in regard to dollar needs, the U.S. companies maintained, was the net amount of dollars returned to the United States for U.S. taxes and dividends. To the extent that profits were reinvested abroad, Britain's overall balance of payments would not be affected. The companies conceded that in the long run dollar profits would enter into Britain's overall balance of payments picture, but argued that they were small in relation to the value of the oil supplied and to the overall balance of payments problem. If they were given the opportunity to sell for sterling outside the sterling area, they claimed that they could reduce their dollar costs to almost the same level as non-U.S. companies, and lower than the dollar costs involved in increasing British production to the point of displacing U.S. oil from its foreign markets.[22]

The State Department agreed with the companies' argument and moved to back the oil companies by having the sterling-dollar oil issue included in the American-British-Canadian (ABC) financial talks scheduled for September 1949. The Treasury Department, and at first ECA, opposed this course of action. They argued that no commodity should receive special treatment, and that the U.S. government should work to end all trade discrimination. Until then, however, the oil industry should not receive special privileges. Treasury Department economists especially opposed oil company proposals to sell for sterling outside the sterling area, arguing that to do so would increase Britain's dollar drain, and that the accumulation of inconvertible sterling by U.S. oil companies would delay the coming of convertibility. Secretary of the Treasury John W. Snyder overruled these objections, however, and it was decided that oil matters would be discussed at the ABC talks.[23]

Meanwhile, the U.S. recession, which began in late 1948 and continued through most of 1949, had led to a worsening of the position of the pound sterling, as U.S. imports from Britain and the sterling area declined. The consequent drop in sterling area dollar earnings precipitated a major crisis. By the second quarter of 1949, Britain's deficit with the dollar area reached $660 million, twice the amount of Marshall Plan dollar aid. In July, the Commonwealth Finance Ministers Conference agreed to a series of common efforts to reduce dollar expenditures by 25 percent from 1948 levels. The British government also decided at this time to seek more dollar aid from the United States. The sharpening of the sterling crisis in August and September led to the devaluation of the pound in mid-September.[24]

The U.S. Treasury had vigorously campaigned for devaluation in the hope that it would lead to an increase in British exports and ultimately to a removal of restrictions on international trade. As far as oil was concerned, however, devaluation made matters worse as British oil companies raised their sterling prices to maintain price equality with dollar oil. Moreover, with dollar oil now more expensive, the incentive to reduce imports of dollar oil increased. Thus the British also wanted to talk about oil at the ABC talks, but for reasons altogether different from those of the United States. British companies had 4 million tons of surplus oil available, and the British government wanted to use this oil to replace an equivalent amount of dollar oil then coming into the sterling area. The British calculated that this substitution would save them $40 million out of a net dollar drain on the petroleum account for fiscal year 1950 of $710 million.[25]

By the time discussions with the British began in mid-September, there were three oil company proposals on how to solve the sterling-dollar problem. All the company proposals claimed that the dollar costs of British and U.S. companies were very similar, and that consequently discrimination against U.S. companies to save dollars was unnecessary. The Caltex plan called for the company to change over to a large extent to a sterling basis by reducing the dollar element in its costs from approximately 84 percent in 1948 to around 20 percent by 1952. Caltex argued that this conversion would enable it to accept sterling payments without causing a drain on Britain's dollar reserves. Jersey also proposed to reduce its dollar costs and offered to convert only half of the sterling earned outside the sterling area to dollars. Socony-Vacuum's plan was more complicated. It called for both British and U.S. oil companies to require purchasing countries to pay the dollar costs of any oil they purchased in dollars, with the balance in sterling. Socony also felt that it could reduce its dollar costs to almost the same level as British companies.[26]

There was no U.S. position that all agencies agreed on until late October, when ECA's position paper was accepted as government policy. ECA's agreement was crucial since it controlled the expansion of British oil facilities as well as the funds for British purchases of dollar oil. ECA proposed that the United States urge the British government to negotiate agreements with the U.S. oil companies that would allow the companies to sell for sterling outside the sterling area. The U.S. government would not be directly involved in these negotiations, but would urge the companies to take all possible steps to "sterlingnize" their operations so as to reduce the potential dollar drain on the British Treasury. As for the British refinery expansion program, ECA reiterated its position that it would help finance only those facilities that could economically be employed in the absence of currency restrictions and other undesirable trade practices.[27]

There was opposition to this policy within the Treasury Department.

George A. Eddy, the chief Treasury official involved in the talks, argued that U.S. oil companies should offer better terms for their oil by accepting local currencies or by lowering their prices. Lower oil prices would encourage consumption and thus expand the markets for Middle East oil. Eddy felt that the oil companies had resisted this course of action because of the "cartel-monopoly nature" of the industry and because they believed that the U.S. government would help them out of their difficulties by forcing the British to grant them the general right to sell for sterling. It was a "ruinous reversal of proper values" to seek to maintain the relatively small royalty payments to Saudi Arabia at the "risk of British insolvency and serious economic disorder."[28]

Discussions between the British government and the oil companies went nowhere. The British maintained that any plan allowing U.S. oil companies to sell for sterling and convert the profits into dollars would increase the dollar drain on the United Kingdom's balance of payments to an unacceptable level. Therefore, the British pressed ahead with their plans to reduce the sterling area's demand for dollar oil to the extent of the expected surplus of sterling oil. In late November, the British informed the State Department and the oil companies that they intended to begin substituting surplus sterling oil for dollar oil in January 1950. The British planned to require the affiliates of U.S. companies to purchase the oil they needed in the sterling area from British companies to the extent that the British companies had surplus oil. A U.S. company, therefore, could bring its own oil into the sterling area only insofar as the volume required to meet its needs was more than what British companies could supply.[29]

The proposed British regulations, together with the other restrictions arising out of the dollar shortage meant the loss of outlets for large quantities of U.S.-owned foreign oil. Saudi Arabian production fell from 530,000 barrels per day (bpd) in February 1949 to less than 462,000 bpd by the end of the year. This drop represented a loss of some $25 million dollars to the Saudi Arabian government, around one-quarter of its oil income. To make matters worse, production in neighboring countries—largely controlled by British companies—continued to increase during 1949, as the British expanded their production in order to save dollars and to take over markets in dollar-short countries.[30] According to Jersey, its foreign producing affiliates lost outlets for 16 million barrels of crude oil and products during 1949 because of the dollar shortage and sterling restrictions. This loss represented 6 percent of Jersey sales outside the United States. By the spring of 1950, the State Department was estimating that sterling oil was displacing dollar oil at the rate of 135,000 bpd, equivalent to about 9 percent of the total overseas production of U.S. companies.[31]

The oil companies protested vigorously, and the State Department condemned the British action as "arbitrary and unwarranted" in view of the fact that discussions about the problem were still in progress. On January 3, 1950, ECA informed the British that not only would it finance no further expansion of oil facilities, but that if the British continued to use their free dollars to finance an expansion of their oil facilities which depended on currency discrimination and "other undesirable trade practices" to be economically viable, ECA would reduce the United Kingdom's allocation of ECA aid by a corresponding amount. These U.S. protests caused the British to delay the implementation of restrictions to February 15, 1950.[32]

The State Department was concerned about the possible political impact of the decline in Saudi revenues. In March 1950, the United States pointed out to Britain its concern that British restrictions on the sale of dollar oil had seriously affected production and government revenues in Saudi Arabia. "The basic British and American interest in the Middle East is indivisible." Saudi Arabia was strategically located at the "land, sea and air crossroads of Eurasia, south of the heart of Russia," and housed at Dhahran the most important U.S. military base in the area. In addition, Saudi Arabia had "consistently exerted a stabilizing and pro-Western influence in the area. During the Palestine War Saudi Arabia provided an important moderating influence on the other Arab States." Oil revenues played an important role in promoting continued political stability in Saudi Arabia. If unchecked, the cutbacks in oil production, and hence government revenues, could "place in jeopardy the stability of Saudi Arabia," and thus work to the detriment of both U.S. and British economic, political, and strategic interests in the Middle East.[33]

Discussions through the winter and spring of 1950 failed to find a basis for a comprehensive settlement of the dispute.[34] During the spring and summer, however, a series of separate agreements between the British and the oil companies, coupled with an upturn in economic activity and oil consumption in the United States and Europe, led to the end of the sterling-dollar controversy. By allowing U.S. companies to sell certain amounts of their oil for sterling in exchange for reducing their dollar costs, the various agreements reduced the handicaps placed on U.S. companies by the dollar shortage. In addition, Jersey, Texas, and Socony recovered a part of the Argentine market lost earlier to the British on the basis of a $125 million Export-Import Bank credit to Argentina. The Argentine government, which had been having trouble working out the details of its agreement with Britain, agreed to allow the U.S. companies to import enough crude oil to maintain their refinery operations. The companies were able to cushion further the impact of the dollar shortage by increasing their imports into the United States. Finally, the Korean War quickly took care of any "surplus" oil and enabled U.S. companies, both at home and abroad, to resume the upward trend of their production.[35]

Although unable to insulate U.S. oil companies completely from the effects of the dollar shortage, Marshall Plan aid and diplomatic support helped U.S. firms maintain sufficient markets to protect their positions in the Middle East. Government assistance was also instrumental in helping ARAMCO meet Saudi Arabia's demands for higher revenues. In this case, the chosen solution had the effect of transferring the cost of higher payments from ARAMCO to the U.S. Treasury.[36]

Fifty-Fifty for Saudi Arabia

In late February 1949, John Paul Getty's Pacific Western Oil Corporation signed a concession agreement with Saudi Arabia covering the Saudi share of the Neutral Zone between Saudi Arabia and Kuwait. Pacific Western paid a lump sum of $9.5 million, and guaranteed Saudi Arabia a minimum royalty of $1 million a year for the first three years whether or not oil was found. Royalty was set at $0.55 per barrel, and Saudi Arabia was given a 25 percent share in the company's profits. Interestingly, Pacific Western's concession contract stipulated that it would pay income tax to Saudi Arabia if it could claim these taxes as an exemption on its U.S. taxes. Getty had been preceded by the American Independent Oil Company (Aminol), a consortium of eight medium-sized U.S. oil companies headed by Ralph K. Davies, which obtained the concession to Kuwait's share of the Neutral Zone in June 1948. Aminol made a down payment of $7.5 million in cash, provided the ruler of Kuwait with a $1 million yacht, guaranteed Kuwait a minimum annual royalty of $625,000, and promised the small country 15 percent of the company's profits.[37]

The Pacific Western contract, by far the most generous in the Middle East, demonstrated that the oil companies could afford to pay more and still stay in business. Until 1948, ARAMCO had paid Saudi Arabia a royalty of four gold shillings per ton, approximately $0.22 per barrel, which was about the same, or slightly higher, than the royalty received by other governments in the Middle East. In 1948, ARAMCO settled a longstanding controversy with Saudi Arabia over the value of the gold shilling by agreeing to an increase in value which raised the effective royalty to around $0.33 per barrel. In the summer of 1949, the Saudi Arabian government began to press ARAMCO for higher royalties and a share of profits.[38]

The Saudi demands worried company officials. In a meeting with Assistant Secretary of State for Near Eastern Affairs George C. McGhee and other State Department officials, ARAMCO Vice-president James Terry Duce claimed that higher royalties would make ARAMCO non-competitive with its rival Persian Gulf producers by burdening the company with high fixed costs. Duce further explained that Saudi Arabia was aware that money paid by the company to the Saudi Arabian government in income taxes would be deductible from the company's U.S. taxes. A U.S. Treasury Department expert who had visited Saudi Arabia in 1948 to aid the kingdom with its finances had reviewed U.S. law in this regard for the Saudis. An income tax would not increase the company's costs—only the flow of taxes would change. ARAMCO's concession contract, however, contained a clause exempting the company from Saudi income taxes. ARAMCO was willing to work with the Saudis on shaping the tax structure to their mutual benefit, but the company was concerned that if it voluntarily waived its exemption, the U.S. Treasury would not allow it to deduct the Saudi tax from its U.S. taxes. McGhee, a Rhodes Scholar who had made a fortune in oil and son-in-law of noted geologist Everett Lee DeGolyer, sympathized with the company's problems, but pointed out that ARAMCO had an important public relations angle to consider in regard to U.S. taxes. Nevertheless, a week later, McGhee wrote Treasury explaining ARAMCO's problems.[39]

Treasury's reply was not encouraging. Assistant Secretary John J. Graham explained that the provision in U.S. income tax law regarding double taxation required that the foreign income tax for which credit was sought be of general application and not be discriminatory or a special duty. It was not designed to permit taxpayers operating abroad to obtain substitution of income taxes for items of cost, such as royalties or other payments to foreign governments that were customary in a particular industry. It was also not intended to allow taxpayers to negotiate voluntary payment of foreign income taxes that other taxpayers were not required to pay. The law was intended to prevent the burden of double taxation, and "should not be used for the purpose of shifting business costs or other contractual obligations to the Federal Government."[40]

By the spring of 1950, the Saudi Arabian government was feeling the effects of a decline in revenue because of cutbacks in ARAMCO's production caused by a drop in demand and British restrictions on dollar oil. Moreover, because of the increase in the ARAMCO off-take price in 1948, ARAMCO's profits had skyrocketed (see the discussion in Appendix 1). In 1949 the company's U.S. taxes exceeded royalties to the Saudi Arabian government. ARAMCO's payments to the Saudi Arabian government had increased from $4.3 million in 1945 to $39.1 million in 1949, while the company's profits had gone from $2.8 million in 1944 to $115,062,120 in 1949. ARAMCO paid U.S. income taxes of $43 million in 1949. As ARAMCO head Fred Davies later testified, the Saudis "weren't a darn bit happy about that." The Saudi finance minister informed the company in May that he knew what the company's profits were and how much the company paid in U.S. taxes. In late June, the finance minister demanded that ARAMCO increase its spending on projects such as roads, wells, and housing to $10 million a year; pay royalties on oil used in refinery operations; cover the costs of construction of a pier at Dammam and a railroad from Dammam to Abquaiq; and allow Saudi Arabia to defer payments on funds advanced by the company until 1951.[41]

Company representatives immediately reported these demands to the U.S. ambassador in Saudi Arabia, J. Rives Childs. According to ARAMCO officials, the company had reached the limit of the concessions it could make to Saudi Arabia. The other Persian Gulf producers operated on a much lower royalty basis so ARAMCO was already at a competitive disadvantage. It simply could not afford to pay more and remain competitive. Moreover, if it gave in to these demands, new ones would immediately be forthcoming. Childs tried to impress the company's concerns on the Saudi Arabian government, but to no avail. In August, the Saudis, upset at the lack of response from the company, demanded renegotiation of the concession contract in order to provide for an income tax on the company's profits.[42]

Meanwhile, the State Department, increasingly concerned about the threat of "communist aggression" in the wake of the Korean invasion, had invited a group of major oil company leaders to discuss the stability of their oil concessions in the Persian Gulf area. The department's policy paper for the meeting noted that "the advancement of U.S. policy objectives is tied into the stability of oil concessions." The revenues the oil companies provided helped achieve such policy goals as "economic and political stability, increased standards of living and the development of a Western orientation and democratic processes." Moreover, Middle East oil was essential to European security and prosperity as well as representing a major portion of U.S. overseas investment. The outstanding problem affecting the stability of concessions in the Middle East was the desire of the region's governments for greater financial returns from their concessions. It would be in the interests of world consumers, as well as of the companies, if the region's governments could be dissuaded from "new rounds of royalty increases." It would be impossible as well as unwise to try to "hold the line" totally against the desires of the region's governments for more revenue. Some retreat was inevitable. The goal should be to "make the retreat as beneficial and orderly as possible for all concerned."[43]

The State Department argued that the way to counter demands for higher royalties would be for the established companies to relinquish those parts of their concession areas that they did not plan to develop in the foreseeable future. This option would allow additional companies, more than likely U.S. companies, to obtain concessions in the Middle East. Not only would it provide additional sources of income for the region's governments and thus take some of the pressure off the established companies, it would also help eliminate the "serious criticism by both friendly and unfriendly states" of the control of the Middle East's resources by the major companies. On the other hand, the Middle East countries appeared to be moving in the direction of Venezuela in demanding 50 percent of the profits earned from the development of their resources. There would be pressures on the companies to accept income, corporate, and other taxes from which the companies in the region were currently exempt. The possibility of tax credits from the U.S. Treasury might make it seem desirable from the oil company's point of view to retreat, in the face of demands for increased income, in this direction. The U.S. government and public would come out the "loser," if this happened. Such a solution was also not in the best interests of the companies. "Failure of oil companies to pay U.S. taxes while benefiting from U.S. protection and support has already been subject to Senate investigation which damaged the reputation of the oil industry."[44]

As far as the oil companies were concerned, however, the relinquishment option had been tried, with disastrous consequences. At the State Department's urging, and in exchange for extending its concession to offshore areas, ARAMCO had relinquished its rights to the Saudi share of the Neutral Zone. The result was the entry of Getty's Pacific Western Corporation, which paid higher royalties. Independent oil consultant Everett Lee DeGolyer explained at the September 11 meeting that unless the companies were prepared to relinquish a share of world markets, they could not be expected to give up concession areas voluntarily. He had no doubt but that "Mr. Getty would be able to sell his oil."[45]

It soon became apparent that ARAMCO's owners preferred to "retreat in the direction of the income tax." At a meeting with Assistant Secretary McGhee and other department officials in early November, the high-level ARAMCO representatives present pointed out that the situation in Saudi Arabia could become serious if the Saudi demands for increased revenues were not met. To meet these demands, ARAMCO was considering agreeing to an income tax. The State Department officials pointed out that in effect this would amount to a subsidy of ARAMCO's position in Saudi Arabia by U.S. taxpayers. McGhee urged them to try to meet Saudi demands for additional benefits within the terms of the concession contract through such measures as promising increased production, providing more services, and accelerating the company's relinquishment program. The oilmen, however, argued that Saudi Arabia was justified in objecting to a situation in which the U.S. government received from the company's operations benefits that were the same as or greater than those received by Saudi Arabia. Only a major financial change in the contract such as profit sharing, an income tax, and/or increased royalties would satisfy the Saudi Arabian government. In their opinion, immediate and substantial increases in payments to the Saudi Arabian government were necessary to retain the concession. In the face of these arguments, which Ambassador Childs, whose experience in the Middle East dated from the 1920s and who was present, did not contest, McGhee assured the company representatives that the State Department would give the company full backing in its efforts to meet the Saudi demands. McGhee later testified that the department, "through the National Security Council, made known its views on the overall political situation, and in the Council the U.S. policy was put together which led the Treasury Department to making the tax credit concession." Although recent research has revealed that the NSC probably did not formally consider the matter, McGhee's recollection indicates that it was the subject of high-level concern.[46]

Just after this meeting, the Saudi Arabian government announced the establishment of an income tax of 20 percent on foreign companies operating in Saudi Arabia. The heads of ARAMCO's parent companies came to the State Department on November 13 to review the company's options in responding to this breach of their concession contract. R. G. Follis, vice-chairman of SOCAL, argued that the tax presented the company with the choice of either "virtually tearing up their concession or standing on it." To give in to "such arbitrary and unilateral action" would "create a precedent which struck at the heart of all company contracts throughout the world." Follis admitted that ARAMCO could not resist making some changes in the concession, but argued that rather than submit to unilateral action, the company should counter with an offer of its own so that the final result would be the product of mutual agreement rather than unilateral action. Brewster Jennings, president of Socony, agreed that ARAMCO could do little to resist Saudi Arabia's demand for higher revenues. The key, in his opinion, was whether protesting the tax would strengthen ARAMCO's case for a tax credit. The State Department, by this time thoroughly convinced of the need for ARAMCO to make greatly increased payments and impressed with the uselessness of trying to divert Saudi demands into other channels, assured the oilmen that it would "endeavor to set as favorable a stage for these discussions as possible."[47]

Negotiations between ARAMCO and the Saudi Arabian government began on November 28. The Saudis demanded that in addition to an equal share in profits they should continue to receive their basic royalty payments and other taxes. Moreover, they wanted ARAMCO to guarantee that any settlement would provide Saudi Arabia with a substantial per-unit increase over what it was then receiving. The ARAMCO negotiators took the position that royalties and other taxes would have to count as a credit against Saudi Arabia's equal share in net profits, and that the company would only agree to fifty-fifty profit sharing after Saudi Arabia withdrew its demands. After a month of difficult negotiations, which included several consultations with King Ibn Saud, agreement was finally reached on December 30. The Saudi Arabian government accepted the profit-sharing principle put forward by ARAMCO and officially withdrew its outstanding demands. The agreement was based on a royal tax decree issued on December 27, which provided that on companies "engaged in the production of petroleum or other hydrocarbons in the Kingdom of Saudi Arabia, there shall be imposed . . . an income tax of fifty per cent (50%) of the net operating income." Existing royalties and taxes were

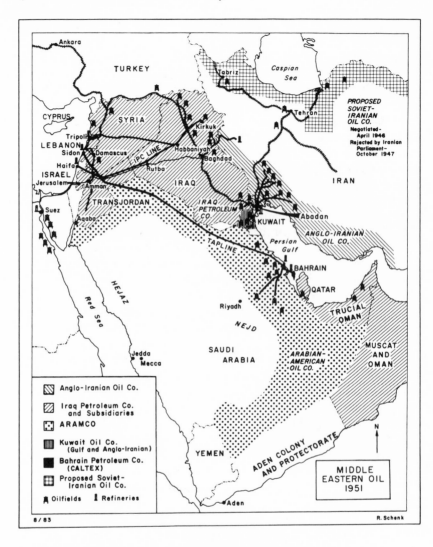

Source: William Roger Louis, *The British Empire in the Middle East, 1945–1951: Arab Nationalism, the United States, and Postwar Imperialism* (Oxford: Clarendon Press, 1984), p. 690. Used with the permission of the Clarendon Press.

counted as credits against the new income tax. Since the issue of a U.S. tax credit had not been resolved, ARAMCO insisted on, and received, a definition of net income that allowed it to deduct its U.S. taxes from its operating income before Saudi Arabia's 50 percent share was calculated. The December 30 agreement included a waiver by ARAMCO of its exemption from an income tax in exchange for a guarantee that its total taxes for any one year would not exceed 50 percent of its net income, as defined in the December 27 decree.[48]

The effect of the fifty-fifty agreement on the "flow of taxes" was immediate and dramatic. In 1950, ARAMCO paid the U.S. government $50 million in income tax, while the Saudis received $66 million in royalties and other payments from the company. In 1951, claiming a credit for taxes paid in Saudi Arabia, ARAMCO paid only $6 million in U.S. income taxes. Saudi Arabia, in contrast, received almost $110 million from the company. In 1952, the net tax paid by ARAMCO to the United States was less than $1 million, and in each of the years after 1952 the credit completely offset U.S. income tax. The Bureau of Internal Revenue ruled in favor of the tax credit in May 1955, after auditing ARAMCO's tax return for 1950.[49]

The State Department felt that the principle of fifty-fifty sharing of profits between oil companies and the governments of producing countries represented a basis on which Western interests in Middle East oil could be defended and stabilized. In the case of U.S. oil companies, the United States was willing to underwrite oil company generosity by providing a tax credit for income taxes paid to host governments. After the December 1950 agreement in Saudi Arabia, the United States urged the other oil companies operating in the region to follow ARAMCO's example.[50] Iran's oil, however, was controlled by the British-owned Anglo-Iranian Oil Company (AIOC). AIOC's failure to grant similar financial benefits to Iran led to the nationalization of the Iranian oil industry in March 1951 and a crisis that threatened the Western position in the Middle East.

Chapter 8

The Limits of Corporatist Foreign Policy

*The question arises, therefore, whether in a situation where a vital
power position of the United States is at stake, it can afford to
apply fully the normal and traditional laws of sovereign self-
determination to the control of underdeveloped countries over the
oil in their soil.*

—Walter Levy, 1952

The U.S. response to the nationalization of the Iranian oil industry
highlights the main elements of U.S. foreign oil policy—opposition to
economic nationalism, an activist role in maintaining the stability and
Western orientation of the Middle East, and public support for and
nonintervention in the private operations of the major oil companies. The
Iranian nationalization and the ensuing dispute between Britain and Iran
threatened the stability and security of the Middle East and the fabric of
the world oil industry. U.S. involvement in the overthrow of Iranian
nationalist leader Mohammed Mossadegh and the subsequent inclusion
of U.S. oil companies in a multinational consortium to run Iran's oil
industry answered the question posed by Levy in ways that had a lasting
impact on U.S. foreign policy.[1]

The Iranian Dilemma

The major problem facing the United States in the Middle East as the
1950s began was how to maintain and defend Western interests within
the region in the context of limited military resources and declining British
influence. The possibility of Soviet military action against the Middle East,
while always a danger, was not an immediate concern. Rather the major
threat to Western interests lay in instability within the countries of the
region and the growing anti-Western, particularly anti-British, orientation
of Middle Eastern nationalism. U.S. policymakers feared that these
problems could provide an opening for the expansion of Soviet influence
into the Middle East.[2]

These fears seemed to be materializing in Iran as a dispute between the

Iranian government and the British-owned Anglo-Iranian Oil Company (AIOC) threatened the stability and Western orientation of the region. Relations between Iran and AIOC had never been smooth. Although negotiations to revise AIOC's concession agreement resulted in the conclusion of a so-called supplemental agreement in July 1949, it fell far short of Iran's initial demands. The Majlis (Parliament) balked at ratifying the agreement, and in June 1950, referred it to a special oil committee headed by Dr. Mohammed Mossadegh, leader of the National Front, a coalition of various nationalist groups. At the end of November, the committee unanimously rejected the agreement. In his report to the Majlis, Mossadegh declared that nationalization of the oil industry was the only way Iran could secure its rights as a sovereign nation.[3]

Fearing that AIOC's unwillingness to offer more generous terms was creating political instability in Iran, the United States urged the British to reach a settlement. The British refused to bend until the fifty-fifty agreement between ARAMCO and Saudi Arabia at the end of 1950 undercut AIOC's arguments against increased payments. In early 1951 AIOC offered similar terms to Iran, but by this time it was too late. Iranian opinion had begun to swing sharply toward Mossadegh's position. The British government further fueled the fires of Iranian nationalism by announcing that it would help AIOC resist any attempts at nationalization. Iranian Prime Minister Ali Razmara, a pro-Western general, tried to halt the movement toward nationalization, warning that Iran did not possess the technical or managerial capacity to run the industry on its own and that compensation would be so heavy as to make national ownership unprofitable and cost Iran valuable foreign exchange. His efforts failed to convince the Majlis and cost him his life. On March 7, he was assassinated by a religious fanatic. The next day, the oil committee, which had already given preliminary approval to nationalization, voted unanimously to recommend nationalization to the Majlis. The Iranian Majlis (Parliament) voted to nationalize AIOC on March 15, and five days later the Senate unanimously approved the principle of nationalization. In April, a nine-point implementation law was passed. The laws went into effect on May 1, after receiving the shah's approval.[4]

The Iranian nationalization presented the United States with a dilemma. The United States had little sympathy for the British position; President Harry S. Truman told newspaper columnist Arthur Krock that Britain had dealt "ineptly and disastrously" with the whole matter. Truman also volunteered that the head of Anglo-Iranian looked like a "typical nineteenth-century colonial exploiter." Secretary of State Dean Acheson, a staunch Anglophile whose bushy mustache gave him the appearance of a nineteenth-century British colonial officer, later wrote that "never had so few lost so much so stupidly and so fast." Moreover, the United States feared that British efforts to reverse nationalization could lead to the loss of Iran. A shutdown of oil operations and a boycott of Iranian oil could easily lead to economic collapse and a Communist coup. Armed intervention by the British in the oil areas of southern Iran could lead to similar action by the Soviets in the north, possibly at Iran's request. In mid-1949, the National Security Council had warned that if Iran should come under Soviet domination, the independence of the entire region would be directly threatened and important U.S. security interests jeopardized. Soviet control of Iran would not only provide them with Iran's rich oil resources, but would enable them to threaten Western control of the rest of the region's oil. According to a 1951 Central Intelligence Agency (CIA) report, loss of Iran's oil would retard European recovery and impose severe financial hardships on Great Britain. Loss of all of Middle East oil would make the Western European rearmament program "impossible of accomplishment" and would force "profound changes" in Europe's economic structure.[5]

On the other hand, political ties and economic interests inclined the United States to support the British. British friendship was the cornerstone of U.S. foreign policy, especially in the Middle East, which was still an area of British responsibility. The United States was also well aware of the importance of Iranian oil to the British economy. AIOC's holdings in Iran were Britain's largest overseas investment. Estimates of the company's value ranged from £81 million (undepreciated book value) to £500 million (replacement cost). More important, the company's operation in Iran provided Britain with over $400 million per annum in foreign exchange. Loss of this foreign exchange threatened Britain with serious balance of payments difficulties. In addition to losing these earnings, Britain would be forced to develop alternative sources of oil, particularly refined products, and the large capital outlays required would exacerbate the trade and payments effects of the loss of Iranian oil. In addition, the United States could not ignore the impact of a successful nationalization on its own oil interests in the Middle East and Latin America. Any settlement in Iran that provided greater benefits than those received by other countries, the State Department warned, would lead to demands for equal treatment, "thus creating a pattern of 'most favored nation' demands." Nationalization was an unlikely prospect in most of the Middle East because of lack of technical capacity, but the Iranian nationalization still had the potential to harm the world oil industry.[6]

Assistant Secretary of State George C. McGhee, who was visiting Pakistan, was sent to Iran in late March to survey the situation. On the way back to the United States, McGhee stopped in London for talks with British officials, and tried to impress on them the seriousness of the situation in Iran. In mid-April, U.S. and British officials held informal discussions in Washington. The British believed they could sidestep the nationalization issue by offering Iran 50 percent of AIOC's profits and agreeing to phase in Iranian control over the remaining life of the concession. U.S. officials disagreed, pointing out that nationalization had overwhelming popular support, and warned that an unyielding attitude on the part of the British could create opportunities for the Communists. On the other hand, the shah and "responsible government officials" would try to reach a reasonable settlement if the British cooperated. Acheson warned British Ambassador Sir Oliver Franks on April 27 that, if Britain took a hard line against the Iranian nationalist movement, "one risks its immediately being captured by the U.S.S.R." In Acheson's opinion, the question of ownership was "irrelevant" so long as AIOC was allowed to manage the company and received 50 percent of the profits. The British were in no mood to compromise, and insisted that the essential point was not the right of a sovereign nation to nationalize commercial enterprises, but the Iranian waiver of this right in the 1933 agreement.[7]

In late April, the British began moving troops to bases in the Middle East and dispatched additional warships to the Persian Gulf. These moves prompted Acheson on May 17 to warn London that the United States would support the use of military force only on invitation of the Iranian government, or in the event of Soviet military intervention or a Communist *coup d'état*, or to rescue British nationals in danger of attack.[8]

While taking a firm line with the British on the use of force, the United States also took a strong stand against nationalization. After talks with officials of the leading U.S. oil companies, the State Department issued a statement of the U.S. position on May 18. Even though the United States sympathized with Iran's desire for increased control and benefits, it strongly opposed "any unilateral cancellation of clear contractual relationships." Efficient operation of the oil industry required not only technical knowledge and capital, but also transportation and marketing facilities. Elimination of AIOC would deprive Iran of these essentials, and the U.S. companies which possessed these capabilities had "indicated . . . that they would not in the face of unilateral action by Iran against the British company be willing to undertake operations in that country." Although Iran protested the statement as "interference" in its internal affairs, the United States "reaffirmed its stand against unilateral cancellation of contractual relationships and actions of a confiscatory nature."[9]

Meanwhile, matters were coming to a head. On May 26, the British had submitted the dispute to the International Court of Justice (ICJ) in The Hague. AIOC Vice-Chairman Basil Jackson traveled to Iran in mid-June but failed to reach an agreement with Iran despite the efforts of U.S. Ambassador Henry F. Grady. On June 22, the British requested an interim injunction from the ICJ which was granted on July 5. The ruling called on Iran to permit operations on a prenationalization basis until a settlement was reached. Iran rejected the court's competence to rule on what it considered to be a domestic matter, and began demanding that tanker

captains sign receipts acknowledging Iranian ownership of the oil they
were taking on. The captains refused, bringing exports to a halt. In the
wake of the ICJ's decision, Iran gave British employees of AIOC one week
to decide whether to work for the newly created National Iranian Oil
Company (NIOC) or to leave the country. In response, AIOC threatened
to close down all its operations. The Iranians countered with an antisabotage bill, which provided stiff penalties for interfering with oil
operations. In protest, the company withdrew its manager from the
Abadan refinery complex and warned that other personnel would soon
leave. Thereupon, the Majlis passed the antisabotage law, and the British
moved their warships into striking position.[10]

With the situation poised on the brink of war, President Truman
approved a National Security Council paper that called for the United
States to "bring its influence to bear in an effort to effect an early settlement
of the oil controversy between Iran and the United Kingdom, making
clear both our recognition of the rights of sovereign states to control their
natural resources and the importance we attach to international contractual
relationships." Truman proposed sending experienced statesman Averell
Harriman to Iran to find a basis for resuming negotiations. The Iranians
accepted immediately. The British were not enthusiastic about the idea.
Foreign Secretary Herbert Morrison complained to Acheson on July 7
that U.S. "indulgence" was encouraging the Iranians. What Britain wanted
from the United States was not an offer to mediate, "but a firm and
categorical statement that it is up to Persia to accept and follow the
recommendation of the International Court of Justice." Nevertheless, the
British agreed to let Harriman try to get talks started again.[11]

The Harriman Mission

Harriman arrived in Iran on July 15. Accompaning him was oil expert
Walter Levy and interpreter Vernon Walters, a career Army officer who
linguistic abilities would take him on many similar "silent missions" in
the future. There, they came face-to-face with Mossadegh, who had
become prime minister at the end of April. Born into the Iranian aristocracy
around 1880, educated in France and Switzerland, Mossadegh had become
the public and symbolic leader of Iranian nationalism during the Pahlavi
dictatorship. The core of Mossadegh's support lay among intellectuals
and young professionals, but by 1951 he had won widespread public
support. Harriman and Levy consistently made it plain that the Iranian
government "could not expect to obtain a financial return greater than
that of other countries under comparable conditions," and that the
operation in Iran had to be run on an efficient basis. Harriman and Levy
insisted that efficient operation could only be accomplished through a
"foreign-owned operating company with freedom in day to day management." Whatever the form of the final settlement, the oil companies

"would have to obtain the equivalent of 50 percent of the net receipts." Loss of oil revenues, Harriman warned Mossadegh, would lead to economic collapse and a Communist takeover. Harriman also met with Shah Mohammad Reza Pahlavi and discussed the possibility of replacing Mossadegh. The shah and his advisers explained, however, that this option was politically impossible since the country was solidly behind the prime minister on the oil issue. No one but Mossadegh could obtain popular approval for a settlement with the British, and even Mossadegh could not make a deal that did not meet the requirements of the nationalization law.[12]

Working with Mossadegh's cabinet and the Majlis oil committee Harriman devised a formula for resuming negotiations. The Iranians agreed that if the British would accept the principle of nationalization, they would agree to negotiate the manner in which nationalization would be carried out. Harriman urged the British to accept the Iranian offer, pointing out that Iran had considerably retreated from its previous positions. To press his argument in person, Harriman flew to London on July 27. The British were reluctant to resume negotiations, claiming concern over Iranian "interference" with the oil operations still in progress and over the safety of British citizens in Iran. Suspecting that London was laying the groundwork for military intervention, Acheson reminded the British that the United States would support only evacuation operations limited to covering the withdrawal of British citizens in immediate jeopardy from mob violence. This warning seemed to halt the drift toward intervention, and Harriman was able to convince the cabinet to send a negotiating team to Iran. Nevertheless, the U.S. ambassador to Britain warned that the government was still under pressure to send troops to Iran. Ostensibly it would do so to protect British lives, but the decision to send troops could easily evolve into a decision to occupy Abadan if not the whole southern oil region.[13]

The British delegation, headed by Lord Privy Seal Sir Richard Stokes, a wealthy businessman and prominent Labour party politician, arrived in Iran on August 4. Both sides requested Harriman to stay on in order to facilitate negotiations. After preliminary talks, Stokes presented an eight-point proposal on August 13. All AIOC assets in Iran would be transferred to the NIOC, with compensation for the main facilities to be included in the operating costs of the industry. Actual control of the industry would be in the hands of a "purchasing organization," which would provide an "operating organization" to manage the industry for the NIOC as well as an outlet for Iranian oil. The purchasing organization would contract for most of Iran's production under a twenty-five-year contract and at prices that would result in a fifty-fifty split of profits between the NIOC and the purchasing organization. Iran would be represented on the board of the purchasing organization, and the operating organization would employ non-Iranian staff only to the extent necessary to ensure efficiency. The small Kermanshah Petroleum Company, owned by AIOC, would be turned completely over to Iran "on favorable terms" in order to meet Iran's internal needs.[14]

Harriman vigorously lobbied for the Stokes plan, but Mossadegh objected to the size and power of the purchasing organization, its monopoly over exports, and its control of the operating organization, and charged that the plan would reestablish AIOC in a new form. Mossadegh also objected to the fact that compensation was not included in the British share of the fifty-fifty division of profits. Moreover, the plan would leave control of Iran's major industry in foreign hands, and thus assure the British of a dominant role in Iranian politics as well as economic life. The Iranians presented a counterproposal which called for NIOC operation of the industry utilizing the present AIOC personnel. NIOC would make sales contracts with AIOC's former customers on a national basis, while AIOC would receive only the share that Britain consumed. The fifty-fifty division of profits would only apply to the British share of exports, and compensation for the nationalized properties would be included in the British share.[15]

Harriman flatly rejected the counterproposal as "contrary to well-known commercial methods of the international supply and distribution of oil." For Iran to sell its oil it would have to make arrangements with the large companies that possessed the necessary transportation, distribution, and marketing facilities. AIOC was the only company in the position to market large quantities of Iranian oil so Iran would have to deal with it. Moreover, Iran could not expect to receive higher prices for its oil than other producing countries since the oil companies could buy or produce oil elsewhere. Finally, the Iranian proposal on compensation was "untenable." The United States held that "seizure by any government of foreign-owned properties without paying prompt, adequate, and effective compensation or working out new arrangements mutually satisfactory to former owner and government, is not nationalization, but confiscation." Since Iran had no funds to pay adequate and prompt compensation, it was obliged to work out an arrangement satisfactory to AIOC.[16]

Talks were suspended on August 22, and the British delegation returned to London the following day. Mossadegh received a vote of confidence from the Majlis for his handling of the negotiations. The British reaction to the end of talks was to withdraw all British personnel from the oil fields on August 27. Harriman was able to persuade the British not to send in troops. In talks with the British cabinet on his way back to the United States, he advised them to let the situation "simmer." Worsening economic conditions would weaken Mossadegh, only if they appeared to be the result of Mossadegh's policies. Overt external pressure, on the other hand, would rally the country behind the prime minister. Harriman recommended that the British work through the shah, and that above all they avoid provocative actions.[17]

The British ignored Harriman's advice. After the Iranian Senate authorized Mossadegh to expel the remaining AIOC staff in Abadan if negotiations were not resumed, the British government announced that negotiations were "no longer in suspense but broken off." Under pressure from the Conservatives over its "weak" policy in Iran and facing elections in October, the Labour government opted for a show of force. On September 8, the British dispatched four destroyers to the Persian Gulf to join the ten warships already stationed there. This action was followed by the suspension of the financial and trading privileges previously accorded Iran by the Bank of England, which denied Iran virtually all of its dollar exchange. Export licenses for scarce commodities destined for Iran were revoked and cargoes en route were halted. AIOC announced that it would institute legal proceedings against any and all purchasers of "stolen" Iranian oil. After the British rejected his offer to resume negotiations, Mossadegh announced that all remaining British employees of AIOC would be required to leave Iran by October 4. On September 27, Iranian troops occupied the Abadan refinery complex. Although a British strike force was prepared to seize Abadan island, the certainty of an adverse reaction from Washington gave London pause. Thus, instead of defending Abadan by force, as pledged earlier, the British called on the United Nations to order Iran to stop taking provocative actions. Iran, however, requested and received a delay in the Security Council's consideration of the case to allow Mossadegh to travel to New York to speak for Iran. The delay undercut the purpose of the British request, and on the morning of October 4 the remaining AIOC staff boarded the British cruiser *Mauritius* and, with the band playing, left Abadan.[18]

The Petroleum Administration for Defense

U.S. policy was not completely neutral. In the spring and summer of 1951, the United States, through the Petroleum Administration for Defense (PAD), had worked out a plan under which U.S. oil companies cooperated with AIOC in replacing Iranian oil. The cutoff of Iranian exports at the end of June had presented major problems for the world supply situation. Iranian production at the time was around 660,000 barrels per day (bpd), more than one-third of total Middle East production and around 7 percent of total world production. Almost all of Iran's production was exported, the bulk as refined products. The refinery complex at Abadan was the largest in the world, and supplied more than one-fourth of all refined products outside the Western Hemisphere. Loss of Iranian crude could be easily made up by increasing production in Kuwait, Saudi Arabia, and elsewhere in the Middle East, but replacement of Iranian products posed a difficult problem, especially for areas east of Suez. There was just enough refinery capacity in the non-Communist world to fill the gap left by the Iranian shutdown, but the logistical problems of getting the oil to where it was needed seemed to require the cooperation of the major oil companies.[19]

The world oil industry was prepared, having been alerted to the possible consequences of loss of Iranian supplies by a general strike in the Abadan area in April. Because of the impact of the strike, and because there were signs that the controversy in Iran could intensify, PAD brought together major oil company representatives in Washington at the end of April to discuss the steps necessary to ensure alternate sources of supply should Iranian supplies be shut off. In view of the cooperation among the oil companies that appeared necessary to redistribute the loss of Iranian oil without disruption, PAD decided that Section 708 of the Defense Production Act of 1950 would have to be utilized. This provision allowed the president, in the interest of national security, and after consultation with the attorney general and the chairman of the Federal Trade Commission (FTC), to exempt from the antitrust laws "those (enterprises) entering into voluntary agreements and programs to achieve the objectives of the act."[20]

Over the course of the next two months PAD, in cooperation with representatives of nineteen U.S. oil companies doing business abroad, put together a set of procedures under which the participating companies could take cooperative action to prevent, eliminate, or alleviate shortages of petroleum without being subject to antitrust action by the Justice Department. The PAD/company draft was extensively revised and narrowed in scope by the Justice Department and the FTC, and a final Voluntary Agreement was approved by the attorney general on June 26. With the stoppage of Iranian exports in late June, the Voluntary Agreement was utilized to establish a specific plan of action to deal with the situation. Under Plan of Action Number 1, dated 26 July 1951, the nineteen participating companies were permitted to take such actions as allocating markets, scheduling supplies, regulating production, and adjusting imports and exports. The Voluntary Agreement and Plan of Action were explained as being necessary to meet the oil needs of the non-Communist world during the Korean War emergency. They had other effects if not intentions, however. The arrangements also provided a mechanism whereby the gap in world oil supplies left by the sudden withdrawal of Iranian oil from world markets could be filled without disrupting established patterns of marketing, pricing, and distribution. As Roy Prewitt of the FTC pointed out, they made it possible for U.S. oil companies, in cooperation with Shell and AIOC, to protect the markets formerly supplied by AIOC against encroachments by outsiders and independents.[21]

The Voluntary Agreement and the Plan of Action achieved both of their objectives. Losses of crude from the Iranian shutdown were offset by increases in production in the rest of the Middle East, and the shortfall resulting from the closing of the Abadan refinery was made up by an expansion of refinery capacity in the Eastern Hemisphere and increased exports of refined products from the Western Hemisphere. As a result, the only major problems experienced were a shortage of aviation gasoline and a temporary "tightness" of fuel oil supplies in the Indian Ocean area. In addition, AIOC was able to maintain supplies to its established customers by increasing its production in Iraq and Kuwait, and by procuring necessary supplies from other oil companies. Its profits before taxes fell from £84 million in 1950 to £47 million in 1952, but as a result of much lower British income taxes its net profits declined only from £33 million to £25 million. Dividend payments were actually maintained at the 1951 level.[22]

Search for a Settlement

The expulsion of AIOC forced the United States to reexamine its policy. The major U.S. oil companies recommended backing the British. In a meeting with Acheson and McGhee on October 10, the top executives of Jersey, Socony, Gulf, SOCAL, and the Texas Company stressed that giving Iran better terms than those received by other producing countries would threaten the international oil industry. In their opinion, loss of Iran to the West would be preferable to the instability that successful nationalization would create. The situation transcended the oil industry, for what was at stake was the "sanctity of contractual relations" upon which all U.S. investment abroad depended. Acheson, however, explained that the U.S. government also had to keep in mind the strategic and political conse-quences of the loss of Iran to the West.[23]

In contrast, the military felt that maintaining Iran's orientation toward the West was more important than backing AIOC. In a report to the National Security Council on October 18, the Joint Chiefs of Staff (JCS) warned that Soviet control of Iran would have serious consequences for the security of the United States. It would mean immediate loss of Iranian oil and eventual loss of all Middle East oil, which would greatly increase the West's deficiency in oil resources. Loss of Iran would also outflank Turkey and thus provide the Soviets with a springboard for the domination of the entire Middle East and Eastern Mediterranean. Whereas the JCS in August had felt that U.S. objectives in Iran could be attained "without serious prejudice to British interests," they now felt that "strictly from the United States military point of view, Iran's orientation towards the United States in peacetime and maintenance of the British position in the Middle East now transcend in importance the desirability of supporting British oil interests in Iran."[24]

The State Department believed that U.S. security and economic interests in Iran were complementary rather than mutually exclusive. The coop-eration of the international oil industry was essential for the efficient operation of Iran's oil industry and hence for Iran's prosperity and stability. Moreover, a settlement in Iran that undermined U.S. oil concessions and investments elsewhere was not in the national interest. The State Department was willing to urge the British to reach a settlement with Iran, but the department would not support any settlement that had the "effect of injuring seriously the fabric of the world oil industry, which is so vital to the whole free world." Therefore, the United States should continue to work for a settlement in Iran that was equitable to all the interests involved.[25]

The State Department's conclusions required renewed efforts to reach an agreement with Iran. Mossadegh had traveled to New York in early October to argue Iran's case in the United Nations, and the department took advantage of his presence in the country to try to work out a solution to the impasse. After two weeks of discussions, the United States and Iran agreed on the outline of a tentative settlement. The key element involved the massive refinery complex at Abadan, which Mossadegh offered to turn over to a neutral foreign company to operate, with the company assuming responsibility for paying compensation to AIOC. The NIOC would take over all other facilities as well as the oil fields, and would sell oil to AIOC under a long-term contract. Mossadegh offered to forgo all counterclaims against AIOC in lieu of paying compensation, and proposed that the NIOC be run by a technical director of neutral nationality. No final agreement was reached on price, however. The State Department stressed that Iran could not expect a higher price for its oil than the other countries in the Middle East. Plenty of crude oil was available, and if it cost the companies more to buy oil from Iran than to buy or produce elsewhere, they would look elsewhere.[26]

Acheson seemed to think that Mossadegh did not understand the complexities of oil pricing, and tried to explain them to the prime minister by drawing an analogy with wholesale and retail beef prices. Acheson's analogy was incorrect. The price the United States was urging on Iran was the price that producing companies such as ARAMCO paid to host governments. Mossadegh felt that Iran deserved the higher price at which producing companies sold oil to distributing companies. The problem was that in cases like ARAMCO the same parent companies owned both the producing and distributing companies so high transfer prices did not affect overall profits because the higher profits of the producing company eventually made their way back to the parent company. What Mossadegh wanted would have interfered with this arrangement.[27]

In essence, the U.S. position meant that the companies would be able to deny Iran the fruits of nationalization by offering no more for its oil than they paid other countries, even though Iran now owned the production company and should theoretically receive the higher prices at which producing companies sold oil to their parent companies. The Iranians were understandably reluctant to accept this reality of the international oil business, and it is unlikely that the companies would have given in on the price issue, given the profits of controlling the production end of the business. In any event, the British refused to resume negotiations.

By this time, the British had come to the conclusion that it would be impossible to reach a satisfactory agreement with Mossadegh. Sir Donald Fergusson, permanent under secretary at the Ministry of Fuel and Power, pointed out in early October that even though Iranian oil was important to Britain it was not so important as all the rest of Britain's foreign oil interests. Moreover, Fergusson warned that it would be disastrous for the general British position overseas if foreign governments came to the conclusion that they could unilaterally repudiate contracts with British companies, seize British assets, and pay only as much compensation as they themselves unilaterally decided. In view of these considerations, Fergusson recommended that Britain stand firm on its rights until the Iranians "came to their senses" and replaced Mossadegh with a more reasonable government.[28]

Fergusson's line of reasoning gained in popularity as elections in late October returned the Conservatives to power. The decline in British power and prestige had been a major issue in the campaign and the Conservatives were in no mood to compromise over Iran. Churchill told Truman in early January 1952 that, had he been in office, "there might have been a sputter of musquetry, but [Britain] would not have been kicked out of Iran." Aside from wounded pride, the Conservative position was based on the importance of overseas investments to Britain's economic survival. Following Fergusson's reasoning, the new government argued that Iran's nationalization of AIOC threatened to undermine the last vestige of confidence in British power and hence in its whole overseas position. If Britain acquiesced in Iran's action by cooperating in the transfer of control to Iran, all of Britain's overseas holdings would be in jeopardy. Iran simply could not receive better terms than countries that kept their contracts.[29]

Acheson felt that British policy was "depressingly out of touch with the world of 1951." A conservative Democrat who admired the British and their empire, Acheson explained that the United States was willing to let existing circumstances operate to Mossadegh's detriment, but it would not let matters go so far as to threaten seriously Iran's orientation toward the West. Acheson agreed that it was probable that no settlement could be reached with Mossadegh. Negotiations, however, would serve to demonstrate that there were alternatives to Mossadegh's policies and would strengthen the shah. The U.S. position was based on a hierarchy of priorities. The "fundamental objective" was to prevent the loss of Iran to the "free world." Thus while the United States wanted a settlement of the oil controversy that would meet legitimate British interests, if such a settlement was not possible, it would consider other means of preventing a "tragedy which would do incalculable damage to the entire West." With these goals in mind, the United States continued to provide aid to Iran, though only enough to prevent collapse and not so much as "unduly to relieve the pressure on the Iranian government to reach a settlement."[30]

Meanwhile, another attempt to find a solution to the Iranian crisis was underway. At the suggestion of the Pakistani ambassador to the United States, Robert L. Garner, vice president of the World Bank, had met with Mossadegh on November 10. Out of this meeting came a plan for the bank to operate the Iranian oil industry as a trustee during the period it took for Iran and Britain to reach a permanent settlement. After consultation with the governments of the United States and Great Britain, the bank formulated a plan that called for it to assemble a management team to run the industry and arrange for a bulk export contract for the sale of Iranian oil, presumably with AIOC. Revenues would be divided among the operating company, Iran, and an escrow account for possible compensation to AIOC. Independent oilman Torbild Rieber, former head of the Texas Company, and Hector Prud'homme, the bank's loan officer for Iran, were named to deal for the bank. After Rieber and Prud'homme made preliminary soundings in Iran in early January 1952, they returned to Iran with Garner in early February. To help bolster the bank's efforts, Truman, in a February 11 reply to an Iranian appeal for U.S. loans, made it clear to Mossadegh that the United States would not loan Iran funds since Iran could obtain the revenues it needed by reaching an oil agreement. Negotiations between the bank and Iran began on February 12, but broke down after seven days of intense discussion over the issues of price, the use of British technicians, and the distribution of oil income. Further attempts to reach agreement proved fruitless. At the end of March, the United States publicly turned down an Iranian request for additional assistance, noting that the United States could not justify additional aid because Iran had the opportunity to receive adequate revenues from its oil industry. On April 3, the bank announced the end of its efforts.[31]

Failure to reach a settlement on the oil issue increased divisions within Iran. In July the shah tried to replace Mossadegh, who had been named "Man of the Year" by *Time*, with Qavam Saltaneh, the "hero" of the 1946 crisis. To bolster the new government, the United States, through Ambassador Loy Henderson, who had replaced Henry Grady in September 1951, promised military and financial assistance. Qavam fell before this aid could arrive, however, and the United States found itself in a worse situation than before. In response to a request by Mossadegh for aid, Acheson proposed a plan whereby the United States and Britain would provide interim aid to Iran without undermining a final settlement of the oil issue. Acheson proposed that the United States make an emergency grant of $10 million to Iran out of the $26 million that had been approved for aid to Qavam. The British would buy oil NIOC had in storage at a suitable discount. This aid would be conditioned on Mossadegh's agreement to accept arbitration of the compensation issue by an international body. The British resisted, arguing that Mossadegh was no longer a bulwark against communism and that only a coup could save Western interests in Iran. The United States held to its position that there was no alternative to Mossadegh. After some pulling and tugging, and the intervention of Prime Minister Churchill, the British finally agreed to the plan. At the end of August President Truman and Prime Minister Churchill dispatched a joint proposal to Iran.[32]

The Iranians read the joint proposal in the worst possible light, and charged that it returned AIOC to its position of dominance in Iran. Mossadegh stressed that compensation would have to be limited to the company's physical facilities. While rejecting the International Court of Justice as a forum for settling compensation, Mossadegh suggested that the ICJ could rule on the claims Iran and the company had against each other as well as determine the damages due Iran on account of the British-led boycott of Iranian oil. Before negotiations could start, however, Mossadegh insisted that Britain pay Iran the £49 million shown on the company's balance sheet as being due Iran. The United States and Britain ignored Iran's counterproposal, and claimed in similar messages on September 24 that Iran had misunderstood the joint proposal. Mossadegh accepted their explanation of the intent of the proposal, and repeated his offer to negotiate with the British on the basis of his September 24 counterproposal. He stipulated, however, that the British would have to place £20 million of the £49 million owed Iran at the country's disposal before negotiations could begin. Mossadegh's demands were too much for the British, who, in any event, had little interest in negotiating with him. In a strongly worded note, the Foreign Office rejected Iran's proposals and reiterated the British government's intention to support AIOC in its demand for compensation not merely for loss of property but also for the "unilateral termination of the 1933 concession agreement." Iran had no right to compensation for its inability to sell oil since AIOC was well within its rights in blocking the sale of its oil. The Foreign Office termed the £49 million claimed by Iran as a result of the never ratified 1949 agreement a "fictitious debt." In response the Iranians broke diplomatic relations with Britain on October 22.[33]

With relations between Iran and the West rapidly deteriorating, the United States moved to take a more active role in the search for a settlement of the dispute. According to Acheson, he and his colleagues had come to the conclusion that the British were so obstructive and determined on a "rule-or-ruin policy" in Iran that the United States would have to strike out on an independent policy or run the risk of having Iran disappear behind the "Iron Curtain" and the whole military and political situation in the Middle East change adversely. In an October 8 meeting with the secretaries of defense and treasury, the attorney general, and the chairman of the Joint Chiefs of Staff, Acheson raised the possibility of asking the major U.S. oil companies to move Iranian oil. The rationale behind the plan involved a recognition of the "realities" of the world oil economy. Acheson explained that "one of the concrete problems in securing a resumption of the flow of Iranian oil is to determine whom it is we can call on, and who is able in fact, to move Iranian oil in the volume which is required to save Iran. The independents are not in a position to give us any real help." Only the major oil companies were able to move Iranian oil in volume. Likewise, AIOC would have to get the lion's share of the oil. Attorney General James McGranery, however, pointed out that "it would be most difficult to work out a program involving the majors and at the same time maintain the present antitrust action."[34]

The antitrust action to which McGranery referred grew out of an investigation of the world oil industry by the staff of the Federal Trade Commission (FTC). In June 1952, President Truman, over the objections of the State Department, had authorized the Justice Department to begin a grand jury investigation of the major international oil companies. Utilizing the FTC report, "The International Petroleum Cartel," and other material, Justice charged that the major oil companies controlled the world oil industry through a series of agreements to divide markets, fix prices, control production, and monopolize reserves, and convened a special grand jury to look into the matter. Leonard Emmerglick, chief of the Foreign Commerce Section of the Antitrust Division, was named to conduct the case, and subpoenas were served on twenty-one oil companies directing them to produce thousands of documents on their foreign and domestic operations. The companies fought back with an extensive public relations campaign which argued that the investigation was undermining their position abroad, and hence the national security of the United States. Although FTC Commissioner Stephen J. Spingarn undertook a personal campaign in defense of the report, he found little support either within the government or in the press because the oil companies' definition of the issue as an attack on business proved more compelling than the intricacies of antitrust law. In addition, Britain and other foreign governments challenged the proceedings. AIOC and Shell were removed from the investigation, but the court denied other motions attacking its jurisdiction, and ordered that the documents sought by the government be produced.[35]

Because of the Justice Department's objections to Acheson's plan, the State Department proposed a different plan a few days later. Under the new plan, a group of U.S. oil companies not involved in the Middle East would form a joint company and contract with Iran to buy around 130 million barrels of oil and oil products a year for ten to fifteen years. The joint company would pay Iran $100 million in advance, with half of the money coming from the Export-Import Bank; $37.5 million from AIOC as advance payment for oil which it would buy from the joint company; and $12.5 million from the companies themselves. The compensation issue would be settled by Iran delivering to AIOC 32.5 million barrels of oil a year for six years. As an integral part of the arrangement, the joint company would sell at least 75 percent of the oil it bought from Iran to AIOC: AIOC had transportation and marketing facilities to handle the large quantity involved; unless AIOC were taken care of, the British government would not agree to a settlement and lift its blockade of Iranian oil; and if any U.S. company tried to sell Iranian oil in markets previously regarded as belonging to AIOC, the British would take reprisals, and relations between the United States and Britain would be damaged. Thus the plan had the effect of protecting AIOC's markets against encroachment by companies selling Iranian oil. Even though the joint company would be free to dispose of the remaining 25 percent any way it chose, the major companies' control of tankers meant that this oil would probably have to be sold either to AIOC or to one or more of the other majors.[36]

The Antitrust Division of the Justice Department objected that the State Department's new plan would be "illegal under the antitrust laws, and particularly the Sherman Act." The problem was that the plan had the "necessary and direct effect" of preventing U.S. oil companies from competing to purchase at least 75 percent of Iranian output. This denial of opportunity to buy Iranian oil constituted "unreasonable restraint" upon the foreign commerce of the United States, and it would be illegal for U.S. companies to participate in such a plan. As an alternative, the Antitrust Division suggested a plan whereby the United States would make Iran a direct loan to cover its immediate needs on the condition that Iran cancel its claims against AIOC and reach a settlement on compensation. Britain could cancel its claims against Iran and lift its blockade of Iranian oil, freeing Iran to sell its oil to whomever it chose. Such a plan would meet the "legitimate interest" of the United States in preventing Iran from coming under Soviet domination, without requiring "monopolistic rights" over Iran's oil for AIOC.[37]

To avoid these antitrust problems, the National Security Council recommended that in the event of increased crisis in Iran, the United States "be prepared to avail itself of the authority of Section 708 (a) and (b) of the Defense Production Act of 1950, as amended." At the same time, Acheson secured President Truman's approval for a plan whereby the United States would advance Iran up to $100 million against the future delivery of Iranian oil, and U.S. oil companies (including majors), alone or in conjunction with AIOC, would purchase and market Iranian oil. In a November 26 letter, the president authorized Acheson to discuss these plans with representatives of U.S. oil companies and AIOC. This letter, Acheson later wrote, was designed to protect the oil companies from the Antitrust Division and to give them "heart" to meet with the State Department.[38]

Before going to the companies, Acheson met with British Foreign Secretary Anthony Eden in New York City on November 20, and told him that a new and "in all probability . . . determinative" crisis was about to occur. The United States feared the imminent loss of Iran to the West and planned to take action to prevent it. The British, however, still insisted that the compensation issue had to be settled to their satisfaction before negotiations on purchasing Iranian oil could begin. In addition, they were reluctant to commit themselves to take the volume of oil the United States felt was necessary to pay compensation, repay U.S. advances, and keep the Iranian economy from collapsing. Acheson pointed out that while the United States hoped that a solution acceptable to Britain could be found, it would "not remain idle while the Iranian problem drifts into a steadily worse situation. . . . In the last analysis, the U.S. Government may have no alternative but to move forward in a manner best designed in its opinion to save Iran." To spur the British along, the State Department announced on December 6 that it no longer objected to U.S. companies purchasing oil from Iran.[39]

At a pair of meetings in early December the State Department explained its plans to representatives from Jersey, Socony, Gulf, SOCAL, and the Texas Company. AIOC had tentatively agreed to take 65 million barrels of Iranian oil per year, but for Iran to have enough revenues to pay compensation and "to survive as a nation" it needed to sell at least twice that amount. Thus U.S. companies would have to take around 65 million barrels. The companies were not eager to take Iranian oil since they had increased production in their concessions and were concerned about possible reaction to cutbacks that would be necessary to accommodate Iranian production. They also stressed that before they took any oil, there would have to be an agreement on compensation; that the British would have to agree to the plan; and that prices would have to be readjusted in order to conform to the fifty-fifty profit-sharing formula now prevalent in the Middle East. It was finally agreed that the possibilities of something being worked out along the lines of the State Department plan were sufficiently good to continue discussions with the British.[40]

While the State Department worked out the final details with the British, Ambassador Henderson began talks with Mossadegh. On the key issue of compensation, Mossadegh repeated his earlier offer to accept arbitration on the basis of the principles used to determine compensation for nationalized properties in Britain. Once arbitration was agreed on, the United States planned to make large and immediate advances to Iran against future delivery of oil. In addition, the United States was prepared to furnish Iran picket boats for its navy and to revive a previously approved

$25 million Export-Import Bank loan. An arrangement would then be worked out whereby AIOC with the help of U.S. companies could lift around 130 million barrels of Iranian oil a year. These terms were presented formally on January 15, 1953, as part of a joint U.S.-British package proposal.[41]

Mossadegh rejected the joint proposal. He objected to British insistence that compensation take into account the prospective earnings lost by AIOC because of the "premature termination" of its concession. Although he had earlier agreed that "any" British nationalization statute could be used in determining compensation, he backtracked upon learning that British laws such as the Coal Mines Nationalisation Act of 1946 took the loss of future profits into account in setting compensation. Mossadegh considered the 1933 agreement, which had extended AIOC's concession to 1993, as invalid, and felt that compensation should be limited to physical properties. The British, on the other hand, were adamant on the issue of compensation. From the British standpoint, a generous compensation settlement was necessary not only to help make up for the losses caused by nationalization, but more importantly to serve as a deterrent to nationalist groups in other countries who might be considering nationalization of British properties. Iran had to be punished; otherwise, Britain's entire overseas investment position would be jeopardized.[42]

At U.S. insistence, the joint proposals were modified to limit compensation to twenty years, and resubmitted to Iran on February 20. Mossadegh was warned that this was the final offer Britain and the United States were going to make. Nevertheless, he rejected the new proposals on March 20 because of his continued objection to compensating AIOC for lost profits. Mossadegh's action brought the United States closer to the British position. Henderson summed up the situation in April: "There's no use, so far as Mossadegh is concerned. He will not agree to anything that the oil companies could possibly accept or for that matter that the American government would accept. I am sure that the British government would not approve of anything that would meet Mossadegh's approval." The British had already drawn a similar conclusion. The United States and Britain, Anthony Eden argued after the Iranian rejection, "should be better occupied looking for alternatives to Mossadegh rather than trying to buy him off."[43]

Finding an Alternative

The search for alternatives to Mossadegh had already begun. In early November 1952, British intelligence contacted Kermit Roosevelt, a CIA operative in the Middle East and a grandson of Theodore Roosevelt, and raised the possibility of jointly sponsored covert action to remove Mossadegh. Roosevelt passed these plans to CIA head Walter Bedell Smith, who was eager to cooperate. Deputy Director Allen W. Dulles, however, was concerned that President Truman and Secretary of State Acheson might oppose such action, and recommended keeping the plans secret until after the newly elected Republican administration took office. Dulles was slated to become director of Central Intelligence, and his older brother, John Foster, had been named secretary of state by President-elect Dwight D. Eisenhower. Planning could, and did, begin immediately. In February 1953, a British delegation came to Washington and reviewed their plans with the Dulles brothers and Smith, who had moved to the State Department. Kermit Roosevelt was selected to be in charge of the operation.[44]

The United States decided to intervene in Iran because it was persuaded to accept British arguments that Iran under Mossadegh was slipping under Soviet control. The pro-Soviet Tudeh party was, in this view, growing more powerful and Mossadegh was becoming more and more dependent on its support. The arguments had their effect on U.S. policymakers. Allen Dulles has even claimed that Mossadegh was a Communist "stooge," and "Communism . . . achieved control of the governmental . . . apparatus . . . for a time." While careful not to claim that Mossadegh was personally a Communist, Ambassador Henderson has argued that "Iran would have fallen under Soviet domination if the CIA had not intervened." The political and economic situation was rapidly deteriorating, and the government was without funds to pay salaries. A U.S. "bail-out" of Mossadegh would have damaged relations with the British, and encouraged other developing countries to cancel "mutually beneficial economic arrangements with the West." According to Henderson, Mossadegh would have dethroned the shah only to have been overthrown by the Tudeh shortly thereafter. Iran would have become a "Soviet satellite," and the "strategic situation in the whole Middle East . . . fundamentally altered— in a way that would have had world-wide implications." Thus it was accepted in Washington that Mossadegh, if not the "stalking horse" of a Communist takeover, was, through his refusal to reach a settlement on the oil issue, creating the conditions for eventual Communist control. Whatever Mossadegh's intentions, the United States became persuaded that it could not stand by and risk letting Iran drift into a situation in which the Soviets could gain control and thereby establish themselves on the Persian Gulf.[45]

The British desire to find an alternative to Mossadegh was due in part to a weakening of their boycott of Iranian oil. With the assistance of PAD and the major U.S. oil companies, AIOC had been able to meet the needs of its former customers. The threat of legal action had initially prevented most independent oil companies from buying Iranian oil. In addition, the British Treasury prohibited use of sterling to buy Iranian oil. British military power was also involved. In the summer of 1952, Royal Air Force fighters forced the oil tanker *Rose Mary*, carrying Iranian oil to

Italy, into port in British-controlled Aden. The cargo was impounded, and in January 1953, a British court ruled that the oil still belonged to AIOC. Moreover, behind the scenes maneuvering apparently convinced Cities Service head W. Alton Jones, a close friend and supporter of President Eisenhower, not to become involved in Iran. Soon after the *Rose Mary* case, however, Italian and Japanese companies were able to import Iranian oil successfully. Iran would not have been able to reach the level of its former exports, but it is possible that other small deals might, in time, have brought in sufficient revenue to make nationalization a qualified success.[46]

Oil policy considerations continued to play a large, if indirect, role, in U.S. policy toward Iran. On May 28, Mossadegh again appealed for aid. Ironically using the same approach that the British were employing to enlist U.S. support for his removal, Mossadegh warned that without assistance Iran could fall to the Communists. President Eisenhower delayed his response until June 29 and then replied that it would not be fair to U.S. taxpayers for the U.S. government to extend any considerable amount of economic aid to Iran since Iran could have access to funds derived from the sale of oil and oil products and the large-scale marketing of Iranian oil could be resumed if Iran would only reach a reasonable agreement with the British on compensation. Eisenhower explained that there was "considerable sentiment" in the United States that payment of compensation only for the losses of physical assets was not a reasonable settlement, and that agreement to such terms "might tend to weaken mutual trust between free nations engaged in friendly economic intercourse." For these reasons, the United States would not extend more aid to Iran or purchase Iranian oil.[47]

Eisenhower's response was made public and broadcast to Iran over the Voice of America. The U.S. refusal to extend aid to Iran intensified pressure on Mossadegh, who was facing increasing opposition because of his desire to eliminate the monarchy and the military from politics. Matters came to a head in early August when Mossadegh announced his intention to dissolve the Majlis. Under the revised 1949 constitution, dissolving the Majlis had become the prerogative of the shah, and Mossadegh's action provided the monarch and his supporters with legal grounds to force the popular prime minister from office. On August 12, the shah issued a decree replacing Mossadegh with Gen. Fazollah Zahedi, former chief of police of Tehran and, for a short period in 1951, Mossadegh's minister of interior before becoming a leader of the opposition to the prime minister.[48]

The initial attempt to remove Mossadegh failed, however, and the shah fled the country. The antishah forces failed to consolidate their initial advantage, however. While Mossadegh and his supporters discussed what to do, Tudeh mobs began destroying all symbols of imperial power, and in some cases attacked mosques. The Tudeh's actions alienated many of Mossadegh's allies and convinced many Iranians, including the military, that a Communist takeover was a real possibility. In addition, Ambassador Henderson who had returned to Iran after several weeks' absence when the coup failed, threatened to pull out U.S. aid missions unless law and order were restored. Henderson's threat convinced Mossadegh to order the security forces to clear the streets. At this point, a massive demonstration, organized and financed by the CIA, galvanized the security forces and other Iranians concerned over the Tudeh's activities into action. In a brief but bloody battle, the forces still loyal to Mossadegh were defeated and Zahedi was installed in power.[49]

The U.S. role in the overthrow of Mossadegh is difficult to assess accurately in the absence of documentation, though it is clear that the United States was instrumental in bringing the anti-Mossadegh forces to power. According to Kermit Roosevelt, the shah greeted him after the coup with the words "I owe my throne to God, my people, my army—and to you." This view is shared by British intelligence operative C. M. Woodhouse who has written that "without Roosevelt's presence, I doubt if the revolution would have succeeded." *New York Times* correspondent Kennett Love, who witnessed the coup, also believes that the U.S. role was decisive; that without U.S. cooperation there would have been no coup, and without U.S. assistance the coup would not have been successful.[50]

Whatever the precise nature of its role in Mossadegh's downfall, the United States moved quickly to support the new government. After Zahedi carefully linked a request for aid to a pledge to settle the oil conflict with Britain "in accordance with accepted principles of international intercourse," the United States provided Iran an emergency grant-in-aid of $45 million on September 5. In addition, the State Department began looking for a solution to the oil controversy that would enable Iran to resume oil exports on a scale sufficient to provide the shah with the resources to build a stable political base.[51]

Establishing the Consortium

The main problem standing in the way of large-scale exports of Iranian oil, apart from the continuing British boycott, was how to fit Iranian oil back into the world oil economy. Iranian oil had been completely replaced in world markets by large increases in production in the other countries of the Middle East, and a worldwide surplus productive capacity of more than 1.5 million bpd was pressing for market outlets. The resumption of Iranian oil operations on a substantial scale would require other producing countries and companies either to reduce their output, or at least to abstain from increasing it so that Iranian oil could fill the expected increase in demand. In addition, the potential market for Iranian output was largely limited to the Eastern Hemisphere because of U.S. import policies and strategic considerations. Many independent oil companies were capable of running the Iranian oil industry, but they lacked market outlets, especially in the Eastern Hemisphere, and would probably have to cut prices sharply to sell the oil. Slashing prices, in turn, could lead to disruptive price wars and destabilizing declines in revenues for the oil-producing countries. British control of the bulk of Iranian exports would avoid the currency problems that would result if U.S. companies took a large share. Therefore, if Iranian oil was to be integrated into world markets without causing disruption, the cooperation of the major oil companies, particularly AIOC, would be needed.[52]

On the other hand, it was also clear that nationalization was still overwhelmingly popular in Iran, and that it would be impossible for AIOC to resume its old position. Faced with the need to restore Iranian government revenues, the State Department engaged Herbert Hoover, Jr., son of the former president and partner in the prominent oil-consulting firm of Hoover and Curtice, to find a way to get Iranian oil back into world markets in large quantities. After meeting with officials from various government agencies and representatives of the major oil companies, Hoover proposed that a consortium of British and U.S. oil companies be established to run the Iranian oil industry. The British share would be divided between AIOC and Shell, and the U.S. share would be held by the five majors and any other U.S. company with sufficient resources.[53]

Such a solution ran up against the Justice Department's cartel case, since it would give government sanction to the type of joint ownership and control that the Justice Department was trying to dissolve. In early January 1953, faced with warnings from the departments of State, Defense, and Interior, and most importantly from Gen. Omar Bradley, that criminal antitrust prosecution of the major oil companies was detrimental to the national interest, President Truman had ordered the grand jury investigation terminated. As a condition for dropping the criminal charges, however, Truman stipulated that the companies agree to turn over to the Justice Department the documents already subpoenaed by the grand jury. In addition, the oil companies would have to pledge to hold off any motion to amend or dismiss the new suit until the documents were received and the details of the government's case had been worked out.[54]

On the basis of Truman's decision, the Justice Department filed a civil complaint in the U.S. District Court in the District of Columbia on April 21, 1953. The complaint charged that the five U.S. majors along with AIOC and Shell "have been and now are engaged in an unlawful combination and conspiracy to restrain interstate and foreign commerce of the United States in petroleum and products, to increase domestic prices of petroleum and products imported into the United States, and to monopolize trade and commerce in petroleum and products between the

United States and foreign nations." This "combination and conspiracy" consisted of agreements and other arrangements to control foreign production, restrict domestic production, divide and allocate markets, and control imports and exports of oil into and from the United States. Significantly, joint ownership of companies organized to produce, refine, and/or market crude oil or products was considered a "vehicle for price fixing and market allocation, and the monopolization of foreign crude supplies."[55]

In view of this history, the Justice Department felt that Hoover's proposal would require antitrust immunity for the U.S. companies involved and would thus affect the cartel case. Justice officials argued that the Iranian problem could be dealt with in a manner consistent with the antitrust laws, by setting up a new company composed of U.S. companies not involved in the Middle East. Defense and State objected to the Justice Department proposal. Defense representatives pointed out that such an approach would present serious problems from the standpoint of national defense since the new company would only be able to market its oil in the United States, which would "deter incentives" for domestic producers to develop U.S. reserves. The State Department warned that the British would object to being excluded from Iran. Moreover, if the new company tried to force Iranian oil into European markets, the result would be price cutting as well as production cuts in the other Middle East producing countries. The resulting decline in revenues, in turn, could have a destabilizing effect on the whole region. These national security considerations prevailed, and the NSC took direction of the matter away from the Justice Department and gave it to the State Department.[56]

Armed with this new mandate, Hoover traveled to Tehran and London to sound out the other potential participants in his plan. Talks in Iran confirmed his opinion that AIOC would not be able to return to Iran alone. This "ideal solution" being impossible, Hoover convinced Sir William Fraser, the crusty chairman of AIOC, that his company would only be able to resume operations in Iran as part of a consortium. Accordingly, Fraser sent out invitations to the heads of Shell, Jersey, Socony, Gulf, SOCAL, and the Texas Company to meet in London to discuss the establishment of a multinational consortium to produce and market Iran's oil. After clearing the matter with the State Department, the U.S. companies met with the British in London in late December. With the assistance of Hoover, they worked out a plan whereby the seven major companies would run the Iranian oil industry on behalf of the NIOC. The consortium would provide capital, pay compensation to AIOC "for such interest as it might not retain," and agree to take Iranian crude oil and products in quantities to be agreed on by the members. The Iranian oil industry would be run "in accordance with sound commercial practices," which meant that in practice the group would "expand or contract production, transportation and refining in Iran as economic conditions indicated." Article IV of the plan noted that any U.S. oil company that participated might find it necessary to reduce its taking of crude oil and refined products from other sources in order to absorb its share of Iranian oil. For this and other reasons, the companies wanted an opinion from the attorney general on the legality of the proposed arrangements before proceeding further.[57]

Aware that the State Department's interest in finding a solution to the Iranian problem put them in a strong position, the companies renewed their attack on the cartel case. In a meeting with the secretaries of state, defense, and treasury on January 6, 1954, company representatives warned that further progress in the London discussions would be prejudiced if the antitrust suit were pressed. Hoover, who had stayed in London, noted that he had been putting pressure on the British to solve the Iranian problem and had undertaken obligations to the new government of Iran for the same purposes. If the U.S. government were to decide that it could not approve of the plan because of antitrust considerations, "the consequences might well be of the gravest kind." Dulles, Secretary of Defense Charles E. Wilson, and Secretary of the Treasury George M. Humphrey pressed these concerns on Attorney General Herbert Brownell on January 8, arguing that it was crucial that the U.S. companies be given sufficient legal clearance to participate in the proposed Iranian consortium. These national security arguments put Brownell, a New York corporation lawyer who had managed Eisenhower's election campaign, on the spot. If the consortium fell through because of antitrust problems and Iran subsequently "went Communist," he would be blamed. Therefore he agreed to give some kind of legal clearance to the consortium if so requested by the NSC. Nonetheless, he insisted that the lawsuit would have to be continued. An arrangement along these lines was worked out, and on January 14, the NSC formally advised Brownell that the security interests of the United States required that U.S. oil companies participate in the Iranian consortium.[58]

Accordingly, Brownell wrote the president on January 21 that in his opinion the proposed consortium "would not in itself constitute a violation of the antitrust laws, nor create a violation of antitrust laws not already existing." This opinion was limited to the actions described in the revised consortium plan and did not include any other agreements among the participating companies. Brownell also reserved his right to prosecute the existing civil complaint. The NSC approved the arrangements the same day, and the following day they were explained to a select group of congressional leaders. Perhaps because of the heavy military emphasis in the administration's presentation, the congressmen raised no objections to the proposed course of action. Concerns over possible imports of Iranian oil into the United States were quieted by Secretary of the Navy

Robert B. Anderson, a former Texas oilman, who reassured the congressmen that sale of Iranian oil would be limited to the Eastern Hemisphere.[59]

Negotiations on the consortium resumed, and after two months of hard bargaining the companies reached agreement on the terms of the proposed agreement. AIOC was to receive a 40 percent share, Shell 14 percent, Compagnie Française des Pétroles 6 percent, and the five U.S. majors 8 percent apiece. In an agreement to be kept secret until negotiations with Iran were completed, the five U.S. companies agreed to relinquish 1 percent of their holdings apiece to make a 5 percent share available for U.S. independent oil companies. (See table 3.) In return for giving up 60 percent of its former holdings, AIOC was to receive £32.4 million from the others. In addition, the other participants agreed to pay AIOC an overriding royalty of ten cents per barrel on all oil exported from Iran until an additional £188 million had been paid.[60]

While discussions among the oil companies were going on in London, the British government, on behalf of AIOC, began negotiations in Iran on the question of compensation. Previously the main barrier to a settlement, the issue had receded in importance since the question of future profits had been partly solved by AIOC's retaining an interest in the consortium and by payments from the other consortium members. Nevertheless, formal ownership was to be retained by Iran, and AIOC wanted heavy compensation for "damages." The United States was not pleased with the British desire to punish Iran, and Secretary of State Dulles warned the British that the United States was "not going to let Iran go down the drain because of Fraser's position." If the British terms were unfair the United States would not be concerned about British interests.

Table 3. Participation in the Iranian Consortium

Company	Nationality	Share(%)
Anglo-Iranian Oil Company	British	40
Royal Dutch/Shell	British/Dutch	14
Standard Oil Company (New Jersey)	U.S.	7
Socony-Vacuum Oil Company	U.S.	7
The Texas Company	U.S.	7
Standard Oil Company of California	U.S.	7
Gulf Oil Corporation	U.S.	7
Compagnie Française de Pétroles	French	6
Iricon Agency[a]	U.S.	5

Source: U.S. Congress, Senate, Committee on Foreign Relations, Subcommittee on Multinational Corporations, *Multinational Corporations and United States Foreign Policy*, part 7: 244 – 246. The consortium agreement was initialed on 15 August 1954, and approved by Iran on 29 October 1954.

[a] The Iricon Agency, a group of nine U.S. independent oil companies added to the consortium in 1955, was composed of Richfield Oil Corporation, 25 percent; American Independent Oil Company (a consortium of ten U.S. independent oil companies, the most important of which were Phillips, Signal, and Ashland), 16.66 percent; Standard of Ohio, Getty Oil Company, Signal Oil and Gas Company, Atlantic Refining, Hancock Oil Company, Tidewater Oil Company, and San Jacinto Petroleum Company, each with 8.33 percent.

Secretary of the Treasury Humphrey agreed, noting that "what [the Iranians] tried was a complete failure and having to pay out some money for it is rubbing salt into the wound." Hoover and Ambassador Henderson also felt that there had to be a limit on British demands. After prolonged negotiations, the Iranian government, in late July, agreed to pay AIOC £25 million over a ten-year period in settlement of all claims.[61]

Negotiations between the consortium and Iran over the terms of their new relationship were also long and arduous, lasting from early April until mid-August. According to the terms of the agreement, NIOC retained title to Iran's oil industry. Actual operation of the industry was placed in the hands of two operating companies, one to handle exploration and production and the other refining. Both operating companies were wholly owned by the consortium, though the NIOC had the right to name two of each company's directors. The operating companies, though legally exercising their powers on behalf of the NIOC, were given full and effective management and control of all operations, and their rights and powers could not be modified during the twenty-five-year term of the agreement. The vitally important matter of production level was reserved to the consortium owners. Iran was to receive 50 percent of the net profits from the industry's operations, and royalties and taxes were adjusted so that U.S. companies could count their payments to Iran as tax credits against their U.S. income taxes, as ARAMCO did in Saudi Arabia.[62] Although not pleased with the stiff terms, the shah pushed the agreement through the Majlis and the Senate, and Iranian exports resumed at the end of October 1954.[63]

The establishment of the Iranian consortium completed the emasculation of the cartel case. Aware that approval of the consortium was inconsistent with the cartel case, Attorney General Brownell decided to change the focus of the case from production control to marketing. This decision seriously weakened the government's case. As Burton Kaufman has pointed out, "joint marketing and price-fixing agreements were only the most burdensome and difficult to maintain of the many parts of the scaffolding by which the majors controlled the world's flow of oil." Such agreements were becoming superfluous, since once the vertically integrated majors had secured control over almost all foreign oil reserves, production, refining, and transportation, their control of markets was also assured. Not enough "free oil" was available to disturb their control of markets or to disrupt prices. As for competition with each other, their common interests as oligopolists, as well as their many joint ventures, served as sufficient checks on inclinations to enlarge market shares through price competition.[64]

The solution to the Iranian crisis not only maintained Iran's orientation to the West but also helped the major oil companies maintain their control over the world oil economy. The costs, however, were high: The necessity of providing U.S. oil companies with a compatible international environment ultimately led the United States, in the case of Iran, to violate the principle of nonintervention in the affairs of other nations.[65] Moreover, reliance on private companies as vehicles of the national interest in foreign oil resulted in the United States sanctioning and supporting, in the name of national security, private control over one of the most important natural resources in the world.

Conclusion

The Political Economy of Foreign Oil Policy

By the eve of the World War II, U.S. oil companies had acquired rights to extensive oil reserves all over the world. The U.S. government had aided in this expansion by insisting on the Open Door principle of equal opportunity for U.S. companies. Moreover, the State Department interpreted the Open Door in oil to mean noninterference in the business affairs of the companies, and thus took no action as the leading international oil companies, with Jersey, Shell, and AIOC in the forefront, sought to control the world oil economy by allocating markets, setting prices, regulating production, and monopolizing reserves.

The stability of this informal system of control had been threatened at the end of the 1930s. In 1937, Bolivia nationalized the holdings of Standard Oil of Bolivia. In March 1938, Mexico nationalized the leading U.S. and British oil companies operating within its borders. In both cases, a government-owned oil company assumed responsibility for operations after nationalization. The seizure of control of oil operations by producing countries directly threatened the very existence of the international oil companies. A more subtle threat to the position of the international majors was the 1938 discovery in Saudi Arabia of oil in commercial quantities. The two U.S. companies holding the concession in Saudi Arabia were not closely aligned with the other majors, which were bound together in varying degrees through a series of agreements. How Saudi Arabian oil would be fitted into world markets was a key issue facing the world petroleum economy when World War II broke out.

Changes in the U.S. domestic oil economy forced U.S. policymakers to consider the importance of foreign oil to the nation's security and economic well-being. The tremendous growth in oil consumption in the interwar period had been more than matched by increases in production and reserves, but by the eve of World War II the United States had become concerned that it would not be able to continue to find oil from domestic supplies at the rate needed to meet steadily increasing demand. U.S. entry into World War II heightened this fear as the voracious appetite of modern warfare for petroleum products threatened to overtake the nation's ability to supply its needs from domestic sources. The end of the war did little to alleviate the concern as consumption continued at high levels and the emerging cold war further underlined the strategic importance of foreign oil to the United States. Moreover, postwar increases in production and reserves were achieved only through greatly expanded drilling and development efforts, an indication that new reserves were becoming more difficult to find. These developments posed potential threats to the nation's strategic position and to the long-term health of the U.S. economy and led to plans to secure access to foreign oil.

Early efforts to formulate and implement foreign oil policy were marked by sharp conflict between contrasting conceptions of the proper relationship of the government to an industry whose operations were increasingly vital to the nation's security and economic well-being. One approach, that espoused by State Department oil adviser Max W. Thornburg, sought to revive and institutionalize the informal public-private partnership in foreign oil that had evolved during the 1920s. This approach involved reliance on the private initiatives of the major U.S. oil companies coupled with strong diplomatic support for the companies in their foreign operations. In exchange for public support, Thornburg wanted an end to companies' competitive practices that had created problems in the past and, in general, closer coordination between public and private policies.

Close cooperation between the government and the oil companies, though of a more informal nature than Thornburg envisaged, proved successful in Venezuela because the government and the companies shared the same goal—maintaining U.S. control of Venezuelan oil. Another factor working for success was that no government action was required beyond informal mediation efforts by Thornburg and limited diplomatic support from the State Department. Thus, neither Congress nor other executive agencies were involved. Corporatist policies did not work in Mexico, however. No U.S. oil companies that could serve as the instruments of U.S. policy existed. Moreover, Thornburg's strategy for reversing nationalization and reopening Mexico to U.S. oil companies was undercut by Secretary of the Interior Harold L. Ickes who supported efforts by independent oil companies to enter Mexico and held out the prospect of U.S. government assistance for Mexico's nationalized oil industry. Ickes's intervention strengthened the hand of those elements within Mexico that opposed the re-privatization of the national patrimony and offset efforts to alter the balance of forces within Mexico in favor of those more sympathetic to foreign involvement in the Mexican economy. As a result, the political consequences of denationalization continued to be too high for the Mexican government to risk. Mexico's oil industry remained nationalized.

Thornburg's plans involved a fairly open use of public power for private benefit, and therefore were opposed by New Dealers like Ickes, who believed that the private interests of the oil companies would not always coincide with the national interest. When the focus of foreign oil policy shifted to Saudi Arabia in early 1943 because the United States was concerned about running short of oil, Ickes got the opportunity to put his ideas into action. Drawing on the statist strand in New Deal thought, Ickes argued that U.S. government ownership of Saudi Arabian reserves was necessary to protect the national interest and to assure public support for whatever measures might be necessary to secure U.S. access to the desert kingdom's oil. The emergence of oil as a national security issue provided Ickes with vital support from elements of the armed forces who shared his concerns. Equally important, SOCAL and Texas were willing to accept a degree of government participation to secure their somewhat tenuous position in Saudi Arabia and provide capital for further development of their potentially rich concession. Thus, Ickes was able to gain President Roosevelt's support for a plan to buy the Saudi Arabian concession held by SOCAL and the Texas Company.

The State Department opposed Ickes's plans for direct government participation in the oil business. Many in the department objected to New Deal–style government intervention in the economy, and feared similar opposition from the oil industry and Congress. More important, U.S. government ownership of oil properties could provoke nationalization measures by other oil-producing nations, particularly in Latin America, where the Mexican example was still fresh in everyone's mind. The main reason for the failure of the "New Deal" solution, however, was the structure and power of the U.S. oil industry. SOCAL and the Texas Company were willing to accept a limited government role because they needed assistance and because they feared that the government might plead "wartime necessity" and seize their property, but they resisted giving up their concession entirely. As word of Ickes's intentions leaked out, the rest of the oil industry began to mobilize to oppose another attempt by the Interior secretary to inject the federal government into oil matters. Faced with a major conflict with the oil industry in the midst of the war, the pragmatic Ickes gave up the purchase plan.

Ickes quickly rebounded from this initial setback with a plan for the government to build and own a pipeline from the Persian Gulf oil fields to the Mediterranean. The pipeline, Ickes argued, would give the government sufficient control over Arabian oil to protect U.S. interests, and would, in addition, provide the government with the tangible stake in Middle East oil necessary for public support. SOCAL, Texas, and Gulf supported the idea since government construction of the pipeline would save them the expense of building one themselves, provide low-cost access to European markets, and enhance the security of their investments by giving the government a concrete interest in the future of their concessions. The other two U.S. majors, Jersey and Socony, vigorously opposed the

202 / *Oil and the American Century*

plan because it would give their competitors significant advantages. They were joined by domestic producers, who feared that the U.S. market would be flooded with cheap foreign oil as Arabian oil took away European markets from Caribbean oil. Congress, increasingly anti–New Deal and ever receptive to appeals cast in terms of defense of free enterprise, joined the majority of the oil industry in defeating the pipeline plan.

The State Department desired a solution to the problem of assuring U.S. access to foreign oil that meshed with its overall plans for the postwar world. State Department policy was marked by a commitment to the traditional liberal objective of an Open Door for U.S. interests. Regarding political freedom and economic freedom as inseparable, department policymakers believed that an open world was essential for peace and prosperity. An Open Door in oil could provide a model for other raw materials and would best serve U.S. interests because U.S. companies were favorably positioned to gain maximum advantage from equality of access. As an alternative to government ownership, the State Department proposed that the United States protect and promote its interests in Middle East oil through an oil agreement with Great Britain that would provide guarantees for existing concessions, grant equal opportunity for U.S. companies to compete for new concessions, and remove the existing restrictions on the development of Middle East oil. Because of the special characteristics of oil production, a joint petroleum commission would be needed to perform the allocation function previously carried out by private agreements and to help integrate Middle East oil into world markets with a minimum of disruption. "Orderly" expansion of Middle East production would, in turn, enhance U.S. security by reducing the drain on U.S. and other Western Hemisphere reserves.

The structure of the U.S. oil industry prevented a solution along the liberal lines proposed by the State Department. The State Department's strategic goal of promoting increased Middle East production in order to conserve Western Hemisphere oil reserves was compatible with the interests of the five U.S. international majors, all of which held reserves in the Middle East. The majors initially favored the agreement because they wanted an international allocation mechanism to assimilate Middle East oil without disrupting markets and believed that government action was needed to overcome company rivalries and avoid running afoul of the U.S. antitrust laws. The domestic oil industry, on the other hand, had worked out a system of production control in the 1930s, centered on the Texas Railroad Commission, which protected the interests of marginal producers. Domestic producers feared that the proposed petroleum commission would allow cheap foreign oil to flood the U.S. market. Unlike the Texas Railroad Commission, with its elected membership and commitment to the welfare of small, independent producers, a joint Anglo-

American petroleum commission would not assure protection of their interests. The State Department had no intention of destroying the domestic oil industry, but it favored using the commission to increase oil imports into the United States in order to conserve U.S. domestic supplies for future defense needs. This strategy conflicted with the immediate financial interests of the independent oil companies, and with those of the more than twenty states involved in oil production. The concerns of domestic producers were loudly voiced in Congress.

The only foreign oil policy on which all segments of the industry could agree was that the government should limit its involvement in foreign oil matters to providing and maintaining an international environment in which private enterprise could operate with security and profit. U.S. foreign oil policy gradually evolved in this direction with the departure of Max Thornburg, the collapse of the plan to buy CASOC, the overwhelming opposition to the pipeline plan, and the dim prospects for the Anglo-American oil agreement. As in the 1920s, entrenched opposition from private interests in conjunction with strong ideological objections prevented a policy of public control. And, as in the 1920s, a laissez-faire policy of government noninvolvement was also unacceptable because of the economic and strategic importance of assured access to foreign oil reserves.[1]

By the end of the war the main elements of a renewed public-private partnership in foreign oil policy had emerged. First, the policy opposed economic nationalism. This aspect was clearest in U.S. policy toward Mexico which aimed at reversing nationalization, or at least preventing nationalization from being successful. At the same time, with national security considerations transforming U.S. relations with the Middle East, efforts to maintain the security and stability of the region could be justified without being explicitly tied to oil. Finally, instead of directly intervening in oil matters, U.S. policymakers increasingly looked to private arrangements by the major oil companies to secure the nation's interests in foreign oil.

Having failed in their efforts to obtain government assistance for their operations, SOCAL and the Texas Company decided that a secure if subordinate status in world oil was preferable to the risks involved in forcing their way into world markets. They therefore brought Jersey and Socony into ARAMCO in order to provide the capital and markets needed to ensure the rapid development of the concession. Similarly, Gulf secured its position in Kuwait by arranging to share its output with Shell. In addition, Jersey and Socony contracted to buy large quantities of oil from AIOC, thus easing the potential pressure of British competition. The result was a private, but nonetheless government-sanctioned, system of worldwide production control, which allowed Middle East oil to be integrated into world markets without disrupting markets and prices. Moreover, the deals contributed to stability by adjusting company shares in the region's oil and by extending joint ownership of oil reserves by the major companies. They also realized the strategic goals the United States had sought with the PRC and the Anglo-American Oil Agreement: securing U.S. concession rights in the Middle East, ending restrictions on the development of Middle East oil, facilitating its entry into world markets, and thus reducing the drain on Western Hemisphere reserves.

Utilizing private oil companies as vehicles of the national interest in foreign oil did not mean that the government had no role to play. On the contrary, the logic of a policy of relying on foreign oil to meet the nation's increasing energy needs required that the United States take an active interest in the security and stability of the Middle East. This was especially the case in Iran, where fear of Soviet expansion and determination to secure access to the region's resources transformed U.S. policy from one of relative indifference to one of deep concern for Iranian independence and territorial integrity. The principal interest of the United States in Iran was strategic: Iran constituted a buffer between the Soviet Union and the oil fields of the Middle East. To secure Iran's orientation to the West, the United States provided economic and military assistance and gradually assumed Britain's role as a barrier to the expansion of Russian influence in the Middle East.

The imperatives of the emerging foreign oil policy also affected U.S. policies on the Palestine issue. In this case, the apparent conflict between the nation's economic and strategic interests in Middle East oil on the one hand and its emotional support for a Jewish homeland in Palestine on the other led the United States to follow a policy of minimal involvement. Ironically, the Palestine problem enhanced the status of the oil companies as vehicles of the national interest in Middle East oil. While U.S. prestige suffered somewhat because of the government's ambivalent policies, the oil companies managed to maintain a degree of distance from the official U.S. position, and thus escaped the burden of Arab displeasure. The result was greater reliance on the major oil companies to achieve U.S. objectives in the Middle East.

The doubtful security of the Middle East coupled with growing U.S. requirements for oil renewed interest in Latin American oil. The United States was firmly convinced that oil development, on the scale it believed necessary, could only be accomplished through the international oil companies. Thus, the United States opposed efforts by the Venezuelan government to increase national control of the oil companies by taking part of its royalties in oil and establishing a government oil agency. On the other hand, the United States did not oppose Venezuelan efforts to gain increased revenues from the oil companies because the State Department, for strategic reasons, wanted to avoid any dispute that might disrupt Venezuelan oil production. Both the department and the major U.S. oil companies were ready to accept such reforms as higher taxes and wages that did not challenge corporate control of Venezuela's oil.

In Mexico, however, reliance on private interests presented the United States with a dilemma. On the one hand, U.S. security interests called for the rapid and extensive development of Mexico's nearby reserves. On the other hand, U.S. assistance to Mexico to achieve this goal could be seen as a reward for nationalization and thus encourage other nations to nationalize their oil industries. Therefore, the main objective of State Department policy continued to be the reversal of nationalization. Congressional opposition, divisions within the oil industry, and the reluctance of President Truman to pressure Mexico on behalf of the major companies hindered the implementation of the department's strategy. Nevertheless, the department was able to maintain a policy of providing no assistance to Mexico's nationalized oil industry.

Reliance on private oil corporations to protect and promote U.S. national interests in foreign oil also imposed important constraints on the ability of the United States to secure its stake in the Middle East. While Marshall Plan aid helped maintain vital markets for U.S. oil companies in the Middle East by providing Europe with the dollars to buy oil from U.S. companies, the major companies were able to maintain the administrative pricing system they had designed to avoid price competition. The position of the major companies as vehicles of the national interest in foreign oil also forced the Economic Cooperation Administration to protect them by working to limit Marshall Plan aid to the European oil refinery program. Moreover, the sheltered position of the domestic oil economy, which was an integral part of the oligopolistic structure of the world oil industry, drove the problem of the dollar shortage into the international arena. Finally, by utilizing private interests as agents of public policy the U.S. government supported a solution to Saudi Arabia's demands for higher revenues which transferred the cost of higher payments from the oil companies to the U.S. Treasury. Although at first opposed to such a solution, the State Deparment changed its position after ARAMCO claimed that increased payments were necessary to preserve U.S. control of Saudi Arabian oil.

The policy constraints inherent in a corporatist foreign oil policy significantly shaped the U.S. response to the nationalization of the Iranian oil industry. The Iranian action presented the United States with a dilemma. On the one hand, the United States feared that if Iran succeeded in gaining control over its oil other oil-producing states would follow its example. Thus the Iranians, like the Mexicans before them, could not be allowed to succeed. On the other hand, the Truman administration feared that the use of force to reverse nationalization would drive Iranian nationalism into the waiting arms of the Soviet Union. The Eisenhower administration, however, feared that inaction was leading to chaos and a possible

Communist takeover. It directed covert action to overthrow nationalist leader Mohammed Mossadegh and reinstate the shah, who was willing to settle the oil dispute on Western terms. Moreover, in order to fit Iranian oil back into world markets without disruptive price wars and destabilizing reductions in revenues for other oil-producing countries, the Eisenhower administration enlisted the major U.S. oil companies in an international consortium to run Iran's oil industry. The antitrust exemption required for this strategy undercut efforts by the Justice Department to loosen the major companies' control over the world oil economy.

By 1955, the U.S. stake in foreign oil was secure. Through their own efforts, and with significant support from the U.S. government, U.S. oil companies were firmly entrenched in the great oil-producing areas of the world. The threat of economic nationalism in these areas had been successfully contained, at least for the time being, and the U.S. government was actively involved in maintaining the security and stability of the Middle East. Finally, the government had made its peace with the arrangements devised by the major oil companies to manage the world oil economy.

At the peak of their influence in the early 1950s, the seven major oil companies controlled over 90 percent of oil reserves outside the United States, Mexico, and the centrally planned economies; accounted for almost 90 percent of world (defined as above) oil production; owned almost 75 percent of world oil-refining capacity; and provided about 90 percent of the oil traded in international markets. The major companies, in conjunction with the Texas Railroad Commission and other state prorationing bodies in the United States, were thus able to exercise an informal control over the world oil economy, matching supply with demand and maintaining prices at profitable though not unreasonable levels. Behind this system of private control stood the power of the U.S. government.[2]

Because of the divided structure of the U.S. oil industry, U.S. foreign oil policy guaranteed access to foreign oil reserves to all U.S. oil companies, not just the majors. In the decade after 1955 the major companies' control of the world oil economy was undermined by the entry of a large number of smaller, mostly U.S.-owned, oil companies into the international field. Drawn by the lure of high profits and the security provided by the *Pax Americana*, the independents were determined to carve out a niche in world markets. Unconcerned about the effect of their actions on oil prices in other countries, the independents cut prices in order to sell their oil. Growing production from these companies exerted a steady downward pressure on world oil prices from the late 1950s to the early 1970s. Moreover, the U.S. domestic oil industry was never completely in accord with the corporatist system. Concern over the impact of rising oil imports on the domestic oil economy led the United States, in March 1959, to establish a Mandatory Oil Import Program, which set quotas on oil imports. This action limited U.S. demand for foreign oil, accelerated the exhaustion of domestic reserves, and exerted downward pressure on world oil prices.[3]

Reliance on private initiative to secure the nation's energy future also contributed to the decline of U.S. dominance in world oil. During the 1950s and 1960s U.S. domestic oil investment declined as oil companies sought more profitable outlets for their capital in the Middle East and other overseas areas. More important, the unchecked rise in oil consumption in the United States decimated the U.S. reserve position, as the United States intensified its embrace of patterns of social and economic organization premised upon high levels of oil use.[4]

The final element in the transformation of the political economy of international oil was the resurgence of economic nationalism in the producing countries. Declining oil prices reduced the income of the oil-producing countries since their share of profits was calculated on the basis of posted prices. In September 1960, the oil ministers of Saudi Arabia, Iraq, Iran, Kuwait, and Venezuela met in Baghdad and formed the Organization of the Petroleum Exporting Countries (OPEC). Although OPEC achieved limited success in its first decade, the continuing transformation of the world oil economy and declining U.S. hegemony enabled the oil-producing countries to exercise considerable influence over world oil prices in the 1970s.[5]

The history of U.S. foreign oil policy during World War II and the early cold war offers important insights into the relationship between private power and public policy that are crucial to understanding the nature and development of the political economy of the United States in the twentieth century. The structure of the U.S. petroleum industry played a key role in determining the content and conduct of foreign oil policy. Like much of U.S. industry, the oil industry was divided between a mass of small- and medium-sized domestic companies and a handful of large multinational firms. Within these divisions the competing strategies of different firms often led to intense conflict and efforts to enlist the government as an ally in the competitive struggle. Any public policy that seemed to favor one group of companies was certain to be opposed by the rest of the industry. Divisions within the industry were at the base of much of the ideological opposition to government involvement in oil matters. In addition, the oil industry shared the general distrust of the democratic state that prevailed throughout U.S. business.[6]

Even though these divisions effectively blocked certain types of corporatist cooperation, the split did not reduce the industry's influence.[7] All segments of the industry agreed on policies aimed at creating and maintaining an international environment in which all U.S. oil companies could operate with security and profit. The ultimate impact of business divisions was not a free hand for public initiatives but rather, as in the 1920s, a policy of public support for private control of one of the most important natural resources in the world. Moreover, reliance on the major oil companies as vehicles of the national interest in foreign oil facilitated control of the world oil economy by the most powerful private interests.[8]

The U.S. government itself was divided into competing bureaucracies and institutions, each with its own set of organizational interests.[9] While these divisions played a major role in shaping the specific contours of foreign oil policy, behind them lay the divisions in the U.S. oil industry. The Department of State, for both organizational and ideological reasons, represented the interests of the major U.S. oil companies. Congress, in turn, played its traditional role of protector of small business interests and backed the numerically dominant independent companies. Ickes's influence was largely based on his personal relationship with President Roosevelt. His institutional base, the Department of the Interior, was divided between New Dealers and industry advocates. With the oil industry strongly entrenched in the wartime mobilization structure, the immense powers of the Petroleum Administration for War were unavailable for reformist purposes.[10]

Divisions within the government masked overall agreement that some form of U.S. control over world oil reserves was necessary. As Clayton Koppes has noted, the idea that the United States had a preemptive right to the world's oil resources was well entrenched by World War II.[11] The struggle over foreign oil policy during the war years revolved around the nature and degree of government action needed to assure this hold. Viewed in this light, the various approaches—Thornburg's institutionalized business-government cooperation, Ickes's statism, the liberalism advocated by elements in the State Department, and the informal public-private partnership that eventually emerged—were all strategies for creating coordination between the public and private spheres to ensure U.S. control of world oil. Although the antitrust approach of the Justice Department, which limited the government's role to enforcing competition in the oil industry, was largely outside this consensus, even Justice defended its approach as the best means for gaining U.S. access to foreign oil.[12]

Once the issue was defined in terms of access to additional supplies, the interests of the major oil companies, which possessed extensive foreign petroleum reserves and the means to discover, develop, and deliver this oil, coincided with the national interest. In these circumstances, the major international oil companies were vehicles of the national interest in foreign oil, not just another interest group.

National security concerns were also an important source of U.S. policies, but their impact on oil policy varied.[13] During the war, when public support for government policies was paramount, national security was used to justify government ownership of foreign oil properties. Later, when the crisis was over, reliance on private interests to protect and promote the U.S. national interest in foreign oil produced a symbiosis between U.S. national security interests and the private interests of the major U.S. oil companies. Thus to maintain an international environment in which private companies could operate with security and profit, the U.S. government became actively involved in maintaining the stability of the Middle East, in containing economic nationalism, and in sanctioning and supporting private arrangements for controlling the world's oil.

Moreover, definitions of national security and national interest were not shaped in isolation from the nature of the society they were meant to defend.[14] Foreign oil policy was shaped not only by the structure of the oil industry but also by the "privileged position of business" in the United States.[15] Corporate power influenced the outcome of specific decisions, and, more importantly, significantly shaped the definition of policy objectives.[16] In the case of oil, the realization that U.S. oil consumption threatened to outpace domestic reserves led to plans to assure access to foreign oil reserves. The alternative of reducing, or at least slowing, the growth of rapidly rising consumption was not seriously considered. Part of the reason for a supply-side focus was that the key decisions were made during wartime, when no one wanted to risk running out of oil. Even more important, however, the consideration and adoption of alternative policies for the production and consumption of oil clashed with well-organized political and economic interests, deep-seated ideological beliefs, and the "structural weight of an economic system in which most investment decisions are in private hands."[17] In the oil industry the U.S. government faced one of the most modern and well-organized sectors of the corporate economy. Domestic producers opposed conservation, arguing that greater incentives for domestic production were the answer. Companies with interests overseas were anxious to secure government protection and markets for their oil.

Demand-side planning, in contrast, involved end-use and other restrictions which clashed with the interests of the oil industry and other industries based on high petroleum use. Planning for publicly defined purposes, such as limiting demand for oil products, required a role for public authority—supplanting the market in many areas—that was unacceptable to the dominant political culture of the United States.[18] Conservation also went against the ideology of growth and the desire, born out of the experience of depression and war, to escape redistributionist conflicts by expanding production and the absolute size of the "economic pie."[19] Moreover, the patterns of social and economic organization—in particular the availability of inexpensive private automobiles, the consequent deterioration of public transportation, and the trend toward increased suburbanization—were premised upon high oil consumption and were regarded as natural and not subject to conscious control rather than as the results of identifiable, and reversible, social, economic, and political decisions.[20]

Closing the Door

A battle between ARAMCO's old and new partners over the price at which the company sold oil to its parent companies highlighted the effect of the most important of the "great oil deals" on the "competitive balance of power in the world oil industry."

In March 1947, just before the merger, the off-take price charged by ARAMCO for crude oil transferred to the California-Texas Company (Caltex), the marketing subsidiary of SOCAL and the Texas Company, was $1.02 per barrel, on the basis of an estimated cost of $0.40 per barrel plus an amount designed to build up a cash and investment reserve within ARAMCO. This price was essentially an internal bookkeeping matter since profits made by Caltex or other downstream subsidiaries on the sale of this oil ultimately found their way back to the same parent corporations. This arrangement did not affect the Saudi Arabian government since its royalties were tied to production (at 4 shillings per ton, roughly $0.23 per barrel) rather than to profits.[1]

Since this practice of establishing a low transfer price was basically the same as that followed by the other companies in the Middle East, SOCAL and Texas apparently assumed that it would remain unchanged by the entry of Jersey and Socony into ARAMCO. At the first meeting of the new ARAMCO board of directors on March 12, 1947, however, the Jersey and Socony directors announced that ARAMCO should sell its oil on the basis of "competitive market prices." To do otherwise would violate Article IV of the Off-Take Agreement which specified that the price to be paid to ARAMCO should be established by the board "in accordance with the principle that Seller (ARAMCO) should be run for its own benefit as a separate entity." Thus the terms of the agreement required that the transfer price be raised to around $1.48 per barrel, the current "competitive market price" for Middle East oil. SOCAL and Texas, whose lawyers had overlooked the significance of the "separate entity" clause, argued that the existing price was in the best interests of the company and refused to change it. Since together they held a majority of votes on the ARAMCO board, their view prevailed. Nevertheless, a special subcommittee was appointed to study the question and report back to the board with recommendations.[2]

In a series of memorandums over the following year, SOCAL and Texas laid out their reasons for opposing a high off-take price for ARAMCO crude. First, there was the problem of "coat-tail riding." The regular off-take agreement, which would go into effect when the Trans-Arabian Pipeline was completed, divided

ARAMCO production on the same percentage basis as stock ownership—30 percent each for SOCAL, Texas, and Jersey, and 10 percent for Socony. In the meantime, an interim off-take agreement divided ARAMCO production roughly on a basis of 74 percent for Caltex and 26 percent for Jersey and Socony in order to give the senior partners time to adjust to the new situation. With ARAMCO's profits divided on the basis of stock ownership, Jersey and Socony were entitled to 40 percent of ARAMCO's profits, and the higher the transfer price the larger ARAMCO's profits. According to a Jersey study, the net effect of coat-tail riding would be $20.1 million transferred from Caltex to Jersey and Socony at the $1.02 price and $33.8 million at the $1.48 price. Calculations by Caltex, on the basis of a different division of production, produced an even greater transfer to Jersey and Socony—$37.4 million at $1.02 and $63.1 million at $1.48. Thus SOCAL and Texas would lose and Jersey and Socony gain a considerable sum of money if the transfer price were increased.[3]

In addition to the effect on their financial position, SOCAL and Texas were concerned about the impact of a higher transfer price on their basic position in the industry. Caltex pointed out that Middle East oil was divided between two unequal groups of companies. One group was composed of Jersey, Socony, Gulf, Shell, and AIOC, and the other was made up of SOCAL and Texas. The first group, with the exception of Gulf, had worked closely together for many years and had, "to all intents and purposes, pooled their Middle East oil production." Together they controlled some 20 billion barrels of reserves, which was sufficient to provide all the members of the group with all the crude oil they needed for the foreseeable future "without calling on Saudi Arabian supplies at all." Moreover, the oil shared by this group was sterling oil, which would remain worldwide in its salability. Saudi Arabian oil, in contrast, was dollar oil, and its salability once sterling resources were adequately developed would probably be "greatly circum-scribed."[4]

The "most powerful weapon" that this group had forged for possible use against Caltex was the low cost of their Middle East supplies. In general, the members of this group were able to obtain crude oil at $0.50 to $0.75 per barrel. As long as SOCAL and Texas could obtain oil from ARAMCO at a low price, they could compete with members of the first group. According to calculations made by Caltex in June 1947, a transfer price of $1.15 per barrel would allow Caltex to compete in Western Europe, and the existing $1.02 price was "approximately right" for Caltex to ship oil to the United States without loss against competition from Venezuela. Sales to the United States were particularly important in view of the uncertain future for sales of dollar oil elsewhere in the world. A high transfer price, on the other hand, would restrict the area in which ARAMCO crude could be resold by Caltex without loss. The problem had to do with the location of profits. Profits under a low off-take price went to the company making the sale. With a high off-take price, profits were moved inside ARAMCO and were shared by the partners on the basis of stock ownership. If Jersey and Socony decided, either because of the dollar shortage or because of other obligations, not to take their percentage share of ARAMCO production, they would suffer no penalty, but would still share in 40 percent of the profits made by ARAMCO. Thus a high transfer price, by moving profits inside ARAMCO, discouraged Caltex from cutting prices to expand its markets since 40 percent of the profits resulting from greater sales of ARAMCO crude would go to Jersey and Socony. For this reason, SOCAL and Texas suspected that a high transfer price was part of an overall strategy by Jersey and Socony to limit competition from Caltex.[5]

SOCAL and Texas also warned that increasing the ARAMCO off-take price could lead to political problems. The people of Europe regarded low-cost Middle East oil as essential to recovery, and a high price could "produce political consequences of very far reaching effects." In addition, a high transfer price, by restricting the free movement of Arabian oil in world trade, was contrary to assurances given U.S. government agencies and congressional committees that the ARAMCO deal would not lead to restrictions on the production and sale of Arabian oil. ARAMCO's directors were also directors of four large oil companies doing business in the United States, and any agreement by them on ARAMCO's prices could be construed as an agreement in restraint of U.S. trade. Determination of the transfer price on a cost-plus basis, on the other hand, would leave the stockholders in the same competitive position as before the merger and thus would be less likely to raise antitrust issues.[6]

Finally, the restriction of production resulting from a high transfer price would also affect Saudi Arabia's revenues. Raising the transfer price above a minimum level would result in higher profits showing up within ARAMCO, and this would invite further pressure from Saudi Arabia for higher royalties. At projected production levels, Saudi Arabia would receive $65 million in royalties over the period 1947–1949, and the U.S. government would receive $44 million in income taxes if the transfer price remained at $1.02 per barrel. Each ten-cent increase in the transfer price would increase U.S. income taxes by $18 million so that a $1.48 per barrel transfer price would result in U.S. taxes of $80 million. Since Saudi income was tied to production and not to profits, it would stay at $65 million.[7]

Jersey and Socony held to their insistence that Article IV of the ARAMCO Off-Take Agreement required that the transfer price be set at the level of "competitive world market prices." Article IV had been inserted by Jersey President Orville Harden, who was fully aware of the competitive implications of a high transfer price. A low transfer price was not in Jersey's interests. One Jersey study revealed that if ARAMCO crude were priced so as just to cover costs, it would be $1.00 per barrel by 1949 and $0.60 per barrel two years later. Another Jersey study warned that "if an arbitrarily low price is set for Arabian crude to the partners, say 90 cents per barrel, it would in all probability result in price cutting to obtain business." For a company with interests all over the world, including relatively high-cost production in Venezuela, this was not a welcome prospect.[8]

Negotiations by the price subcommittee in the spring and summer of 1947 went nowhere as each side stuck stubbornly to its position. In October, the ARAMCO board voted along straight company lines to retain the $1.02 transfer price for the remainder of the year. Jersey and Socony thereupon informed ARAMCO that, although they would continue to accept deliveries of oil at the lower price, they did so under protest and did not waive their demand that the transfer price be set in accordance with the off-take agreement. In addition, Jersey obtained an opinion from the prestigious law firm of Sullivan and Cromwell that supported its argument that the Off-Take Agreement legally required that the transfer price be set in accordance with world market prices. Any other course could subject the majority stockholders and directors "not only to serious criticism, but to possible liability for mismanagement."[9]

After several months of battles, a compromise was reached whereby Jersey and Socony were allowed to lift their pro rata 40 percent share of ARAMCO products and crude. This took care of the problem of coat-tail riding. In July 1948, SOCAL and Texas agreed to an increase in the transfer price to $1.40 per barrel. The new price was set on the basis that it would allow Arabian crude to be competitive with crude from other sources delivered to U.S. East Coast markets. (The price of oil in the United States had risen considerably since Caltex's earlier calculations that an off-take price of $1.02 was necessary to compete in the U.S. market.) Tying the ARAMCO price to any other world market, a SOCAL analyst argued, would unduly restrict ARAMCO production. Because of the pattern of world refinery location and ownership, the United States would be an important market for Arabian crude oil even as refined products from the Caribbean and the United States continued to move to Western Europe. The new price allowed Caltex to ship oil to the United States without loss, but it was high enough to discourage price cutting to gain market shares and to prevent Caltex from selling significant amounts of crude oil to the medium-sized U.S. oil companies that desired foreign oil. In this regard, the higher price protected Jersey and Socony from low-cost Arabian oil.[10]

A variety of factors led both sides to compromise. ARAMCO's management had asked for a higher price to finance development of the concession though SOCAL and Texas believed that ARAMCO could get by with less money. In addition, SOCAL had some second thoughts about price cutting to gain markets for Arabian oil, especially in the United States where domestic producers were certain to raise a potent protest. Moreover, both SOCAL and Texas had significant holdings within the United States, and thus had no reason to precipitate a drop in the U.S. price level by importing large amounts of oil. Similarly, Jersey and Socony probably decided that it would be easier to maintain equilibrium in world markets by keeping Iraqi rather than Arabian production down. Fear of disorder arising out of the unrest in Palestine probably led all the parties to be more cautious. Finally, the advent of Marshall Plan assistance in the spring of 1948 broadened the market for Arabian oil since the Marshall Plan provided potential customers in Western Europe with the necessary dollars. This factor not only lessened the urgency for Caltex to find dollar outlets for its share of ARAMCO production, but also further reduced the incentive for Jersey and Socony to draw on their sterling supplies.[11] The July decision led to the end of the price controversy as well as the threat of Arabian oil disrupting the international petroleum order.

The United States and World Oil

Table A. U.S. Production, Imports, Exports, and Demand for Petroleum, 1920–1975 (Millions of Barrels)

Year	Domestic Production			Imports		Exports		Net imports[b]	Domestic demand[c]	Net imports ÷ demand
	Crude oil	Natural gas liquids[a]	Total	Crude oil	Total	Crude oil	Total			
1920	442.9	10.9	453.9	106.2	108.8	9.3	79.6	29.2	483.1	6.04
1921	472.2	12.1	484.2	125.4	128.8	9.6	71.8	57.0	541.2	10.53
1922	557.5	13.7	571.3	127.3	136.0	10.8	74.6	61.4	632.7	9.70
1923	732.4	21.8	754.2	82.0	99.6	17.5	102.2	-2.6	751.6	-0.35
1924	713.9	24.4	738.4	77.8	94.6	18.2	117.5	-22.9	715.5	-3.20
1925	763.7	28.7	792.4	61.8	78.2	13.3	114.2	-36.0	756.4	-4.76
1926	770.9	34.6	805.4	60.4	81.3	15.4	132.3	-51.0	754.4	-6.76
1927	901.1	41.6	942.8	58.4	71.7	15.8	142.0	-70.3	872.5	-8.05
1928	901.5	46.0	947.5	79.8	91.6	19.0	155.3	-63.7	883.8	-7.21
1929	1007.3	56.2	1063.6	78.9	108.7	26.4	163.5	-54.8	1008.8	-5.43
1930	898.0	55.3	953.3	62.1	105.6	23.7	156.9	-51.3	902.0	-5.69
1931	851.1	45.4	896.5	47.2	86.1	25.5	124.7	-38.6	857.9	-4.50
1932	785.2	37.3	822.5	44.7	74.5	27.4	103.5	-29.0	793.5	-3.47
1933	905.6	35.2	940.8	31.9	45.4	36.6	107.0	-61.6	879.2	-7.01
1934	908.1	38.3	946.3	35.6	50.5	41.1	114.8	-64.3	882.0	-7.29
1935	996.6	41.2	1037.8	32.2	52.6	51.4	129.4	-76.8	961.0	-7.99
1936	1099.7	45.3	1145.0	32.3	57.1	50.3	132.4	-75.3	1069.7	-7.04
1937	1279.2	52.0	1331.1	27.5	57.2	67.2	173.3	-116.1	1215.0	-9.56
1938	1214.4	53.1	1267.5	26.4	54.3	77.2	194.1	-139.8	1127.7	-12.40
1939	1265.0	54.1	1319.1	33.1	59.1	72.1	189.5	-130.4	1188.7	-10.98
1940	1353.2	58.9	1412.1	42.7	83.8	51.5	130.5	-46.7	1365.4	-3.42
1941	1402.2	84.3	1486.5	50.6	97.1	33.2	108.8	-11.7	1474.8	-0.79
1942	1386.6	85.7	1472.4	12.3	36.0	33.8	116.9	-80.9	1391.5	-5.81
1943	1505.6	90.1	1595.7	13.8	63.4	41.3	150.0	-86.6	1509.1	-5.74
1944	1677.9	102.4	1780.4	44.8	92.3	34.3	207.6	-115.3	1665.1	-6.93
1945	1713.6	114.9	1828.5	74.3	113.6	33.0	183.0	-69.4	1759.1	-3.95
1946	1733.9	117.8	1851.7	86.1	137.7	42.4	153.1	-15.4	1836.3	-0.84
1947	1857.0	132.9	1989.8	97.5	159.4	46.4	164.5	-5.1	1984.7	-0.26

Year										
1948	2020.2	147.1	2167.3	129.1	188.1	39.7	134.7	53.4	2220.7	2.40
1949	1841.9	157.3	1999.2	153.7	235.6	33.1	119.4	116.2	2115.4	5.49
1950	1973.6	182.1	2155.7	177.7	310.3	34.8	111.3	199.0	2354.7	8.49
1951	2247.7	205.0	2452.7	179.1	308.2	28.6	154.0	154.2	2606.9	5.92
1952	2289.8	223.9	2513.7	209.6	348.5	26.7	158.2	190.3	2704.0	7.04
1953	2357.1	239.1	2596.2	236.4	377.5	19.9	146.6	230.9	2827.1	8.17
1954	2315.0	252.6	2567.6	239.5	384.0	13.6	129.7	254.3	2821.9	9.01
1955	2484.4	281.9	2766.3	285.4	455.6	11.6	134.2	321.4	3087.7	10.41
1956	2617.3	293.2	2910.5	341.8	525.6	28.6	157.4	368.2	3278.7	11.23
1957	2616.9	295.2	2912.1	373.2	574.6	50.2	207.2	367.4	3279.5	11.20
1958	2449.0	295.2	2744.2	348.0	620.6	4.3	100.6	520.0	3264.2	15.93
1959	2574.6	321.1	2895.7	352.3	649.6	2.5	77.1	572.5	3468.2	16.51
1960	2574.9	340.4	2915.4	371.6	664.1	3.1	73.9	590.2	3505.6	16.84
1961	2621.8	361.8	2983.6	381.5	669.7	3.2	63.6	636.1	3619.7	17.57
1962	2676.2	372.8	3049.0	411.0	759.8	1.8	61.4	698.4	3747.4	18.63
1963	2752.7	401.0	3153.7	412.7	774.7	1.7	75.9	698.8	3852.5	18.14
1964	2786.8	422.5	3209.3	438.6	826.7	1.4	73.9	752.8	3962.1	19.00
1965	2848.5	441.6	3290.1	452.0	900.8	1.1	68.3	832.5	4122.6	20.19
1966	3027.8	468.7	3496.4	447.1	939.2	1.5	72.4	866.8	4363.2	19.87
1967	3215.7	514.5	3730.3	411.6	926.0	26.5	112.1	813.9	4544.2	17.91
1968	3160.9	707.7	3868.6	472.3	1039.4	1.8	84.5	954.9	4823.5	19.80
1969	3204.0	748.0	3953.0	514.1	1155.6	1.4	84.9	1070.7	5023.7	21.31
1970	3350.7	772.7	4123.4	483.3	1248.1	5.0	94.5	1153.6	5277.0	21.86
1971	3296.6	775.1	4071.7	613.4	1432.9	0.5	81.7	1351.2	5422.9	24.92
1972	3284.3	798.3	4082.5	811.1	1735.3	0.2	81.4	1653.8	5736.3	28.83
1973	3206.0	789.3	3995.3	1184.0	2283.5	0.7	84.4	2199.1	6194.4	35.50
1974	3056.9	761.7	3818.6	1269.2	2230.9	1.1	80.5	2150.4	5969.0	36.03
1975	2922.6	730.0	3652.6	1498.2	2210.3	2.1	76.4	2133.9	5786.5	36.88

Source: Douglas R. Bohi and Milton Russell, *Limiting Oil Imports: An Economic History and Analysis* (Baltimore: Johns Hopkins University Press, 1978), pp. 22–23. Table A is based on 1920–1967: American Petroleum Institute, *Petroleum Facts and Figures*, 1971 ed., pp. 283–288. 1968–1975: American Petroleum Institute, *Basic Petroleum Data Book*, sec. 7, table 3; sec. 9, table 2; sec. 10, table 1; except crude exports: U.S. Bureau of Mines, Crude Petroleum, Petroleum Products, and Natural Gas Liquids, annual issues 1969–1975, table 1.

[a] Prior to 1968, benzol and other hydrocarbons were included; thus 1968 and later data are not strictly comparable with those of prior years.

[b] Total imports minus total exports.

[c] Total domestic production plus net imports.

Table B. Liquid Fuels Production: Percentage Distribution by Major World
Regions and by Selected Countries, Selected Years, 1925–1965

Region and Country	1925	1938	1950	1955	1960	1965
	Regions as Percent of world production					
North America	70.9%	61.1%	53.5%	47.5%	37.3%	30.2%
Western Europe	0.2	0.3	0.7	1.2	1.4	1.3
Oceania	–	–	–	–	–	–
U.S.S.R. & Communist						
E. Europe	6.8	13.5	8.3	10.7	15.1	16.6
U.S.S.R.	4.7	10.8	7.2	9.1	13.8	15.6
Communist E. Europe	2.1	2.7	1.1	1.6	1.3	1.0
Communist Asia	–	0.1	–	0.1	0.5	0.6
Latin America	15.4	15.5	18.9	17.9	18.0	15.6
Caribbean	14.0	13.8	17.8	16.9	16.4	14.0
Other Latin America	1.4	1.8	1.1	1.0	1.6	1.6
Asia	6.5	9.3	18.0	22.3	26.4	28.8
Middle East	3.0	5.7	15.8	20.0	23.9	26.7
Other Asia	3.5	3.6	2.2	2.3	2.5	2.1
Africa	0.1	0.1	0.5	0.3	1.3	6.9
North Africa	0.1	0.1	0.5	0.2	1.1	5.9
Other Africa	–	–	–	–	0.2	1.0
World	100.0	100.0	100.0	100.0	100.0	100.0
World (mill. metric tons coal equiv.)	(228.0)	(421.3)	(813.2)	(1,205.8)	(1,641.2)	(2,342.8)
World (mill. metric tons original units)	(152.0)	(280.9)	(542.1)	(803.9)	(1,094.1)	(1,561.9)
	Selected countries[a] as percent of world production					
United States	70.9%	60.8%	52.7%	45.3%	34.9%	27.4%
U.S.S.R.	4.7	10.8	7.2	9.1	13.8	15.6
Venezuela	1.9	9.8	14.5	14.1	13.7	11.7
Kuwait	–	–	3.2	6.8	7.5	7.0
Saudi Arabia	–	–	4.9	5.9	5.7	6.5
Iran	3.1	3.7	6.0	2.0	4.8	6.0
Iraq	–	1.6	1.2	4.2	4.3	4.1
Libya	–	–	–	–	–	3.8
Canada	–	0.3	0.7	2.2	2.4	2.7
Algeria	–	–	–	–	0.8	1.7
Indonesia	2.1	2.7	1.2	1.5	1.9	1.5
Kuwait Neutral Zone	–	–	–	0.2	0.7	1.2
Mexico	11.6	2.0	1.9	1.6	1.4	1.2
Total above 13	94.3	91.7	93.5	92.9	91.9	90.4
All other countries	5.7	8.3	6.5	7.1	8.1	9.6
World	100.0	100.0	100.0	100.0	100.0	100.0

Source: Joel Darmstadter, Perry D. Teitelbaum, and Jaroslav G. Polach, *Energy in the World Economy: A Statistical Review of Trends in Output, Trade, and Consumption Since 1925* (Baltimore: Johns Hopkins Press, 1971), p. 125.

[a] The thirteen countries shown comprise all those accounting for 1 percent or more of world liquid fuels production in 1965; the countries are listed in descending order according to their 1965 percentage shares.

Abbreviations Used in the Notes

Age of Energy	Harold F. Williamson, Ralph L. Andreano, Arnold R. Daum, and Gilbert C. Klose, *The American Petroleum Industry: The Age of Energy 1899–1959* (Evanston, Ill.: Northwestern University Press, 1959).
ANPB Records	Record Group 334. Records of Interservice Agencies, Army-Navy Petroleum Board. National Archives. Washington, D.C.
Arabian Oil Hearings	U.S. Congress, Senate, Special Committee Investigating the National Defense Program, *Investigation of the National Defense Program: Part 41: Petroleum Arrangements with Saudi Arabia*, 80th Cong., 1st sess., 1948.
CCS	Combined Chiefs of Staff.
Consortium Documents	U.S. Congress, Senate, Committee on Foreign Relations, Subcommittee on Multinational Corporations, *The International Petroleum Cartel, the Iranian Consortium and U.S. National Security* (Washington, D.C.: U.S. Government Printing Office, 1975).
Davies Diary	Ralph K. Davies Papers. Harry S. Truman Library. Independence, Mo.
Davies Papers	Ralph K. Davies Papers. Harry S. Truman Library. Independence, Mo.
DIA	Denise Folliot, ed., *Documents on International Affairs, 1951–1954* (London: Royal Institute of International Affairs, 1954–1957).
DSB	Department of State *Bulletin*.
ECA Records	Record Group 286. Records of the Agency for International Development. Federal Records Center. Suitland, Md.
ECEFP Records	Record Group 353. Records of the Interdepartmental and Intradepartmental Committees of the Department of State, Executive Committee on Economic Foreign Policy. National Archives. Washington, D.C.
FDRL	Franklin D. Roosevelt Library. Hyde Park, New York.
FO	Foreign Office (Great Britain).
FO 371	British Foreign Office Political Correspondence.

Forrestal Diary   James V. Forrestal Diary (Xerox from Princeton University). Operational Archives, Naval Historical Division, Navy Yard. Washington, D.C.

Forrestal Diaries   Walter Millis, ed., *The Forrestal Diaries* (New York: Viking Press, 1951).

FR   U.S. Department of State, *Foreign Relations of the United States* (Washington, D.C.: U.S. Government Printing Office, 1959–1984).

HSTL   Harry S. Truman Library. Independence, Mo.

Ickes Diary   The Diary of Harold L. Ickes. Manuscript Division, Library of Congress. Washington, D.C.

Ickes Papers   Harold L. Ickes Papers. Manuscript Division, Library of Congress. Washington, D.C.

Interior Records   Record Group 48. Records of the Office of the Secretary of the Interior. National Archives. Washington, D.C.

International Petroleum Cartel   U.S. Congress, Senate, Select Committee on Small Business, *The International Petroleum Cartel: Staff Report to the Federal Trade Commission* (Washington, D.C.: U.S. Government Printing Office, 1952).

Iranian Documents   Iranian Embassy, *Some Documents on the Nationalization of the Oil Industry in Iran* (Washington, D.C.: Iranian Embassy, 1951).

JCS Records   Record Group 218. Records of the Joint Chiefs of Staff. National Archives. Washington, D.C.

JPD   Joint Petroleum Discussions.

LC   Manuscript Division, Library of Congress. Washington, D.C.

MemCon   Memorandum of Conversation.

MNC Documents   U.S. Congress, Senate, Committee on Foreign Relations, Subcommittee on Multinational Corporations, *Multinational Corporations and United States Foreign Policy*, Part 8, 93rd Cong., 2d sess., 1975.

Multinationals Report   ———, *Multinational Corporations and United States Foreign Policy*, 93d Cong., 2d sess., 1975.

New Horizons   Henrietta M. Larson, Evelyn H. Knowlton, and Charles S. Popple, *New Horizons, 1927–1950*. Vol. 3 of *History of Standard Oil Company (New Jersey)* (New York: Harper & Row, 1971).

NSC Documents   University Publications of America, Inc., *Documents of the National Security Council, 1947–1977* (Washington, D.C.: University Publications of America, 1980).

NSRB Records   Record Group 304. Records of the National Security Resources Board. National Archives. Washington, D.C.

NYT   *New York Times.*

OF   Office File.

Oil Lift Hearings	U.S. Congress, Senate, Committee on the Judiciary and Committee on Interior and Insular Affairs, *Emergency Oil Lift Program and Related Oil Problems*, Joint Hearings, 85th Cong., 2d sess., 1957.
OSD Records	Record Group 330. Secretary of Defense Office Files. National Archives. Washington, D.C.
OSS Records	Record Group 226. Records of the Office of Strategic Services, Research and Analysis Branch. National Archives. Washington, D.C.
OSS/INR Records	Record Group 59. Records of the Department of State, Research and Analysis Branch (OSS) and the Bureau of Intelligence and Research (State Department). National Archives. Washington, D.C.
OWMR Records	Record Group 250. Records of the Office of War Mobilization and Reconversion.
PAW History	John W. Frey and H. Chandler Ide, *A History of the Petroleum Administration for War, 1941–1945* (Washington, D.C.: U.S. Government Printing Office, 1946).
PAW Records	Record Group 253. Records of the Petroleum Administration for War. National Archives. Washington, D.C.
PED Records	Record Group 59. Records of the Department of State, Petroleum Division. National Archives. Washington, D.C.
Petroleum Committee Records	RG 46. Records of the U.S. Senate. Special Committee Investigating Petroleum Resources.
PRC Documents	U.S. Congress, Senate, Committee on Foreign Relations, Subcommittee on Multinational Corporations, *A Documentary History of the Petroleum Reserves Corporation, 1943–1944* (Washington, D.C.: U.S. Government Printing Office, 1974).
PSF	President's Secretary's File.
RG	Record Group.
Roosevelt Papers	Franklin D. Roosevelt Papers. Franklin D. Roosevelt Library. Hyde Park, New York.
SANACC	State-Army-Navy-Airforce Coordinating Committee.
SecDef	Secretary of Defense.
SECNAV/CNO Records	Records of the Secretary of the Navy/Chief of Naval Operations Operational Archives, Naval Historical Division, Navy Yard. Washington, D.C.
SecState	Secretary of State.
Senate Records	Record Group 46. Records of the U.S. Senate. National Archives. Washington, D.C.
SWNCC Records	Record Group 59. Records of the Department of State, State-War-Navy Coordinating Committee. National Archives. Washington, D.C.
Truman Papers	Harry S. Truman Papers. Harry S. Truman Library. Independence, Mo.

Note on the Department of State Decimal Files. Record Group 59. Records of the Department of State. National Archives. Washington, D.C.

The central files of the Department of States are indexed according to a decimal number system, for example, 812.6363/3-145. The first digit indicates the general subject matter (in the example, political affairs); the next two digits indicate country or region (in the example, Mexico); the four-digit number indicates specific subject matter (in the example, petroleum); and the numbers after the slash, the date (i.e., March 1, 1945.) Prior to mid-1944, the numbers after the slash were sequence numbers and were not hyphenated.

Author-date method of citation. To save space secondary works are cited according to the author-date method of citation. Full bibliographic information on all works cited is given in the bibliography.

Notes

Introduction: Corporatism and U.S. Foreign Oil Policy

1. Fusfeld 1972; see also Chandler and Galambos 1970.
2. This conclusion was arrived at independently. Kaufman 1977a, 937; and *Multinationals Report*, 14–15, make essentially the same argument.
3. I wish to acknowledge the assistance of Melvyn P. Leffler in helping me clarify this point.
4. Hawley 1978 provides a comprehensive guide to the literature. Schatz 1983, 11 n., argues that the term *corporatism* is more accurate and more comprehensive than *corporate liberalism* and has the added advantage of placing developments in the United States in comparative perspective. Shonfield 1965, Wolfe 1977, and Maier 1981 are good examples of the benefits of a comparative perspective. For a discussion of these developments from the standpoint of political theory, see Wolin 1960, chap. 10.
5. Panitch 1980; McCormick 1982, 323–326. On the differences between the *state corporatism* of prepluralist, authoritarian societies and the *societal corporatism* of pluralistic, democratic societies, see Schmitter 1974, especially 105–108; Schmitter links the differences between the two varieties of corporatism to "different stages in the institutional development and international context of capitalism."
6. Exceptions include Collins 1981; Griffith 1982 and 1984; and Hogan 1985.
7. Notably, Wilson 1974; Hogan 1977; and Leffler 1979.
8. Griffith 1982, 97; McQuaid 1978, 364–366, makes much the same point regarding the impact of the war; on the Truman period, see Griffith 1984 (I wish to thank Dr. Griffith for providing me with a copy of this unpublished paper).
9. Schmitter 1974, 107. On the ways the economic structure of a society sets limits on policy formation, see Offe 1975.
10. On these points, see Mills 1959, 159–182; McCormick 1982; and Hogan 1984. The quotation is from Falk 1978, 144.
11. Readers familiar with the literature will note the similarity with the argument in Hawley 1975, 51–52. Wolfe 1977, 108–145, distinguishes between the *societal corporatism* of the 1920s and what he terms the *franchise state* arrangements of the 1930s and 1940s.
12. *Age of Energy*, part 1; and Pratt 1980. Cook 1941 provides details on the structure of the industry after the breakup of the Standard Oil trust. In the United

States, the term *majors* usually applied to the twenty largest integrated companies; in international oil, the *majors* were the seven largest companies which by the eve of World War II controlled most of the non-Communist world's oil. See the definitions in the glossary.

13. Nash 1968, 1–48; DeNovo 1955. Galambos 1970 discusses these trends in the context of the larger political economy and provides citations to the literature. Koistinen 1967 and Cuff 1973 discuss the wider significance of the war experience for business-government relations.

14. Chandler and Galambos 1970; Hawley 1974. On the oil industry in the 1920s, see Pratt 1981; Nordhauser 1973; and *Age of Energy*, part 2.

15. The conflict also extended to Latin America and the Netherlands East Indies (NEI); on Latin America, see Tulchin 1971; on the NEI, see Reed 1958. On relations between the British government and British oil companies, see Jones 1977.

16. The literature on international oil in the 1920s is extensive. See Gibb and Knowlton 1956, 291–308; DeNovo 1956 and 1963, chap. 6; Hogan 1974; Stivers 1981; and Mejcher 1984. Cooperation extended beyond oil; see Hogan 1977.

17. On the Red Line, see Stivers 1983; and Cowhey 1985, 84–93.

18. Sampson 1975, 69–74 provides a readable account. See also *New Horizons*, 303–314; and *International Petroleum Cartel*, 197–208, 228–274, 353–355; and on pricing, Walter Levy, "The Past, Present and Future Price Structure for the International Oil Trade." Wilkins 1976, 160–161 argues that these arrangements failed to achieve their goals and soon broke down under the impact of continuing competition and rising international tensions.

19. Nordhauser 1973; *Age of Energy*, 299–338; on East Texas, see Prindle 1981, 21–27.

20. On Hoover, see Nordhauser 1979, 37–42. See also Hawley 1974a and 1974b; and Williams 1973, 201–212.

21. *New Horizons*, 84–91; Nash 1968, 98–127.

22. This account of the NRA and oil draws on Brand 1983; see also Nash 1968, 128–145; Nordhauser 1979, chaps. 6–8; and Hawley 1966, 213–216.

23. Nash 1968, 146–151; Hawley 1966, 216–219. On the workings on the Texas Railroad Commission, see Prindle 1981.

24. This interpretation is largely based on Hawley 1966 and 1975; and Griffith 1984.

25. *International Petroleum Cartel*, 129–134; Sampson 1975, 91–94; DeNovo 1963, 204–206.

26. *International Petroleum Cartel*, 74–84, 114–117; Anderson 1981, 21–29; Miller 1980, 16–21; Leatherdale 1983, chap. 8, discusses Saudi Arabian oil from a British perspective. The currently preferred form of the king's name is Abd al-Aziz. This work uses Ibn Saud, the form in U.S. documents of the period.

27. On the Mexican revolution, see Smith 1972; on relations between Mexico and the oil companies, see Meyer 1977; Koppes 1982, 61–62, 66–68.

28. On Bolivia, see Klein 1964; on Iran, see Beck 1974; and Ferrier 1977, 93–101.

29. DeGolyer & McNaughton, *Twentieth Century Petroleum Statistics*; *Age of Energy*, 760, 775; U.S. percentage of world reserves calculated from Fanning 1954, 359. These reserve figures were calculated by the traditional method of

accounting for the discovery of new reserves. Other methods yielded strikingly different results; see the discussion in chap. 1.

30. *Age of Energy,* 442–462, 648–672; *PAW History,* 9–12.

31. I am indebted to Clayton Koppes for the term *Anglo-American petroleum order;* see Koppes 1983. (I wish to thank Dr. Koppes for providing me with a copy of this unpublished paper.)

Chapter 1 Origins of Foreign Oil Policy, 1941–1943

1. *Age of Energy,* 747–748.

2. DeNovo 1977 (I wish to thank Dr. DeNovo for granting permission to cite this unpublished paper); *Age of Energy,* 747–748, 753, 757.

3. DeNovo 1977; Roosevelt to Ickes, 28 May 1941, printed in *PAW History,* 374–375; Stoff 1980, 18.

4. See the testimony of William R. Boyd, Jr., president of the API, 17 Feb. 1942, in U.S. Congress, House, Committee on Interstate and Foreign Commerce, *Petroleum Investigation (Gasoline and Rubber),* 76. On Ickes's background, see Lear 1981; see Stoff 1980, 12–16 for the characterization of Ickes as a Bull Moose Progressive; see also Randall 1983, 367–368; on Ickes and the oil industry in the 1930s, see Nash 1968, 128–156; and Brand 1983; *Energy Resources and National Policy: Report of the Energy Resources Committee to the Natural Resources Committee,* 1–3; on U.S. antitrust policy in the later New Deal, see Hawley 1966, 420–455.

5. DeNovo 1977; Ickes testimony, *Arabian Oil Hearings,* 25206–25207; on Davies's background, see U.S. Congress, Senate, Committee on Naval Affairs, *Nomination of Edwin W. Pauley for Appointment as Under Secretary of the Navy,* 61–62 (transcript in the Ickes Papers); see also Randall 1983, 369–370 and Stoff 1980, 18–19.

6. *PAW History,* 33–34; Ickes testimony in *Arabian Oil Hearings,* 25207–25209; Harold L. Ickes, *Fightin' Oil,* 73–80; Ickes Diary, 12 Apr. and 14 June 1942; see Ickes's defense of these arrangements in Ickes to I. F. Stone, 28 Sept. 1942, Ickes Papers, Box 220, "Oil (23)"; and Ickes to Kirchwey, 7 Jan. 1944, Ickes Papers, Box 221, "Oil (28)."

7. DeNovo 1977; Ickes, *Fightin' Oil,* 70–73; *PAW History,* 17–18, 56, and the Justice Department letters printed in *PAW History,* 392–394.

8. *PAW History,* 25–32, 56–69; Engler 1961, 278–289; Ickes, *Fightin' Oil,* 74–85; *New Horizons,* 459–464; Randall 1983, 370–372.

9. *Arabian Oil Hearings,* 25207; Ickes Diary, 4 and 18 Jan. 1942. On the initial clash between OPC and State, see Davies's testimony in *Pauley Hearings,* 1977–1978; and the discussion in Stoff 1980, 22–32.

10. DeNovo 1956; Hogan 1974; Stivers 1981; Wilkins 1974, 238–241; Walter J. Levy, "The Progressive Threat to the U.S. Position in Foreign Oil," 25 July 1952, Oscar Chapman Papers, Box 102, "Petroleum Cartel File," HSTL.

11. G. M. Richardson Dougall, "The Petroleum Division," Oct. 1944, RG 59, War History Branch Study, Box 22 (hereafter PED History); Thornburg, "Preliminary Memorandum for Discussion on Relations between This Government and American Oil Companies Operating Abroad," 15 Jan. 1943, PED Records, Box 1, "Foreign Oil Policy 1942." See also Thornburg, "Implications of the Atlantic Charter with Regard to Oil," 19 Apr. 1943, Frank P. Corrigan Papers, General

Correspondence, "Max Thornburg," FDRL; two papers by Thornburg entitled "Notes on the Proposed International Oil Agreement," one dated 22 Nov. 1944, Acheson Papers, Assistant Secretarv Correspondence, 1941–1945; and the other dated 8 Feb. 1945, 800.6363/5-1845; and Thornburg's testimony in *Pauley Hearings*, 80–86.

12. Thornburg to Welles, 12 Jan. 1943, 811.6363/525; Thornburg to Bernays, 17 Feb. 1941, Davies Papers, Box 15, "Confidential Correspondence, 1940–1948"; see also Thornburg to Henry F. Grady, 5 Mar. 1941; and Thornburg to Winthrop Rockefeller, 4 Mar. 1941; both in Davies Papers, Box 15, "Confidential Correspondence, 1940–1948." On British oil policy in the interwar period, see Kent 1984.

13. See the papers by Thornburg cited in note 11. The quotation is from the "Preliminary Memorandum", 15 Jan. 1943.

14. Thornburg, "Preliminary Memorandum", 15 Jan. 1943.

15. Feis testimony, *Arabian Oil Hearings*, 25272; Thornburg to Acheson, 18 May 1942, 800.6363/690 1/2, summarizes the work of the first nine months; the petroleum attaché program is discussed in the PED History.

16. PED History; Thornburg Memo, 11 Oct. 1941, printed in PED History; Thornburg, "Foreign Oil Policy of the United States," 20 Oct. 1941, Interior Records, Central Classified File, 1-188, part 14; Ferris, "Long-Term American Foreign Oil Policy and Its Relation to the Conservation of Domestic Reserves of Crude Oil," 24 Nov. 1941, 800.6363/1-2142; see also Ferris to Thornburg, 24 Mar. 1942; and Ferris, "Post-War Arrangements Concerning Oil and Commitments of the United Nations," 18 Mar. 1942, both in 800.6363/579 1/2; Ferris, "The Foreign Oil Policy of the United States," 31 Aug. 1942, PED Records, Box 8.

17. Stinebower to Feis, 26 Dec. 1941, 811.6363/12-2641; Hawkins to Acheson, 9 Jan. 1942, 800.6363/514 1/2; see also Hawkins to Thornburg, 15 Jan. 1942, PED Records, Box 8, "Foreign Oil Policy, 1941–1942."

18. Thornburg to Acheson, 18 Nov. 1942, PED Records, Box 1, "Foreign Oil Policy 1942"; Thornburg to Acheson, 23 Nov. 1942, Interior Records, Central Classified File, 1-188, part 14.

19. On the prewar development of Venezuelan oil, see Rabe 1982, 22–57; Lieuwen 1954, 33–89; Wilkins 1974, 209–210; *The International Petroleum Cartel*, 163–193; and McBeth 1983.

20. Wood 1961, 260–271; *New Horizons*, 479–480; Rabe 1982, 57–65, 74–80; Tugwell 1975, 40–43.

21. Baloyra 1974, 43–45; *New Horizons*, 479–480; Salazar-Carrillo 1976, 68, 72, 76, 77, 81; Rabe 1976, 128–129; Wood 1961, 260–270, discusses earlier problems between the companies and the Venezuelan government.

22. Wood 1961, 270–271; *New Horizons*, 481.

23. FR 1942 6: 743–749; Wood 1961, 271–273. On Welles, see May 1974, 148–149.

24. FR 1942 6: 751–752; see also 745–749; Welles to Roosevelt, 30 Dec. 1942, 811.6363/497A.

25. Welles to Roosevelt, 30 Dec. 1942, 811.6363/497A; Wood 1961, 274; FR 1942 6: 750; see also 751–753; Thornburg to Manrique Pacanins, 30 Oct. 1942, Interior Records, Central Classified File, 1-188, part 11.

26. *New Horizons*, 481–483.

27. Ibid., 483–484; on Shell, see Philip 1982, 62–64; *FR* 1942 6: 753–755; *FR* 1943 6: 807–809, 812–813; Wood 1961, 275–277. The Hoover in Hoover and Curtice was Herbert Hoover, Jr., the former president's son.

28. Rabe 1982, 89; Wood 1961, 276; Tugwell 1975, 43–44; Hellinger 1984, 40–41; *New Horizons*, 479, 482, 484; Suman (Jersey) to Clayton, 27 Aug. 1946, 831.6363/8-2746, points out that the law did not require the companies to convert their old concessions to its terms, but rather made it in their interest to do so.

29. *New Horizons*, 474, 485–488; Salazar-Carrillo 1976, 75–77; Suman to Clayton, 27 Aug. 1946, 831.6363/8-2746.

30. Rabe 1982, 90–91; see also Tugwell's assessment in his introduction to Betancourt 1979, xiv.

31. Wood 1971, 277–282; Rabe 1982, 87–93.

32. Feis to Welles, 2 Jan. 1943, 811.6363/498 1/2; and A. A. Berle to Hull, Welles, Feis, and Thornburg, 6 Jan. 1943, 800.6363/1125; Feis MemCon, 11 Jan. 19.;3, 800.6363/1091. On Thornburg's plans, see Thornburg to Welles and Duggan, 24 May 1943, 812.6363/7959; see later in this chapter for an account of Thornburg's departure from the State Department.

33. Koppes 1982, 62, 66–68; Cronon 1960, 122–129.

34. Cronon 1960, 208–211, 238–269; Koppes 1982, 69–71; Meyer 1974, 193–200, 217–223. On U.S. policy toward Mexico from the expropriation to the settlement, see the excellent study by Ring 1974.

35. Cronon 1960, 267–270; Ring 1974, 312–321, 329–330; Meyer 1977, 223–224; Rabe 1982, 83–84. Shell did not settle until 1947, when it received an award of $130,338,868; see Bermúdez 1963, 26–27.

36. The historian quoted is Ring 1974, 316. The clearest statement of this argument is in Koppes 1982, 73–74; see also Welles to Roosevelt, 17 Feb. 1943, Roosevelt Papers, PSF: Confidential, "Office of the Petroleum Coordinator"; and the unpublished memoirs of Messersmith in the University of Delaware Library, Vol. 2, part 17. Messersmith had served as ambassador to Austria, 1934–1937, and to Cuba, 1940–1942.

37. Bursley to Duggan, 3 Sept. 1941, 812.6363/7354 10/11; Thornburg, "Mexican Oil Settlement Memorandum," 10 Sept. 1941, 812.6363/7365 11/21; Cronon 1960, 264–267.

38. Allen to Davies, 29 Sept. 1941, PAW Records, Box 666, "Correspondence with President"; Davies to Ickes, 15 Oct. 1941, Ickes Papers, Box 221, "Oil-Octane (100);" Ickes to Hull, 11 Oct. 1941, Interior Records, Central Classified File, 1–188, part 3.

39. Ickes to Roosevelt, 18 Oct. 1941, Ickes Papers, Box 221, "Oil-Octane (100)." See the explanation of the OPC methodology in *Wartime Petroleum Policy under the Petroleum Administration for War*, 84; *Age of Energy*, 796. See also the pamphlet issued by Standard Oil of New Jersey in 1944 entitled *Some Facts on the Search for Oil in the United States*, which concluded that the "success of oil-finding effort has not changed substantially in recent years as compared with previous experience."

40. Ickes to Roosevelt, 1 Dec. 1941, 8 Dec. 1941, Interior Records, Central Classified File, 1–188, part 3.

41. Ickes to Roosevelt, 20 Feb. 1942; and Roosevelt to Ickes, 28 Feb. 1942, both in Interior Records, Central Classified File, 1–188, part 4; Welles to Duggan, 24 Feb. 1942, 812.6363/7616.

42. Ickes Diary, 7 and 15 Feb. 1942; Department of State Press Release no. 148, 7 Apr. 1942. On U.S.-Mexican cooperation during World War II, see Cline 1953, 265–278; and Vazquez and Meyer 1982, 182–188.

43. *Pauley Hearings,* 64–73, 90–96, 115–116, 187–192, 600–610, 654–665. The testimony of Henry D. Collier, chairman of the board of SOCAL, *Pauley Hearings,* 1308–1340, 1348–1399, contains some interesting information about the close business ties between SOCAL and Pauley. (I wish to thank Linda Qaim-Maqami for drawing my attention to this testimony.) See also Thornburg MemCon, 21 Jan. 1943, 812.6363/7852; and Duce to Davies, 22 Jan. 1943, PAW Records, Box 667, "Mexico–100 Octane Gasoline Plant."

44. Thornburg to Welles and Duggan, 4 Apr. 1942, 812.6363/7672; *FR* 1942 6: 525–526; Thornburg to Duggan, 16 June 1942, 812.6363/7709; Bonsal MemCon, 23 Apr. 1942, 812.6363/7678; Thornburg to Bonsal, 28 Apr. 1942, 812.6363/7675. Thornburg may have "fudged" his figures somewhat in order to portray Pauley's project in the worst possible light; see *Pauley Hearings,* 218–219, 683–684, 1976. Green 1971, 192–194, accepts the State Department version of these events.

45. *FR* 1942 6: 528–533; Thornburg to Duggan, 17 Nov. 1942, 812.6363/7800.

46. Ickes to Roosevelt, 17 Apr. 1942, Interior Records, Central Classified File, 1–188, part 7; Ickes Diary, 28 Mar., 19 Apr., 24 May, and 28 June 1942; Ickes to Perkins, 1 June 1942, PAW Records, Box 668, "Secretary Misc."; Ickes to Hull, 4 Aug. 1942, Interior Records, Central Classified File, 1–188, part 10.

47. *FR* 1942 6: 526–527, 533–535; "Report of Mexican Technical Mission," 27 Oct. 1942, 812.6363/7766. The mission was headed by noted oil geologist Everett Lee DeGolyer.

48. Roosevelt to Hull, Jones, and Ickes, 23 Nov. 1942; Hull to Roosevelt, 27 Nov. 1942; Roosevelt to Hull, 7 Dec. 1942; all in Roosevelt Papers, PSF: Confidential, "Office of the Petroleum Coordinator for War." In addition, Ickes began trying "to smoke out" the State Department on its delays; see Ickes Diary, 13 Dec. 1942.

49. Thornburg to Duggan and Bonsal, 18 Jan. 1943, 812.6363/7968; see also Thornburg MemCon, 21 Jan. 1943, 812.6363/7852; and Duce to Davies, 22 Jan. 1943, PAW Records, Box 667, "Mexico–100 Octane Gasoline Plant;" Duggan MemCon, 12 Dec. 1942, 812.6363/7799 1/2; Thornburg to Welles, 12 Jan. 1943, 811.6363/525;

50. J. F. McGurk, "Present State of the Mexican Petroleum Industry," 12 Jan. 1943, PED Records, Box 19, "Petroleum Policy Study Group;" Welles to Roosevelt, 17 Feb. 1943, Roosevelt Papers, PSF: Confidential, "Office of the Petroleum Coordinator for War"; Messersmith to Welles, 3 Nov. 1942, 812.6363/7811; Messersmith to Bonsal, 4 Dec. 1942, 812.6363/7814; *FR* 1943 6: 449–452; Messersmith to Welles, 7 Jan. 1943, 812.6363/7821. On the political situation in Mexico, see Meyer 1974, 110–112, 127–128; and Cline 1953, 261–265. On Messersmith, see Koppes 1982, 75.

51. Ickes Diary, 20 Feb. 1943; Ickes to Roosevelt, 16 Feb. 1943, Roosevelt

Papers, PSF: Departmental Correspondence, "Ickes, Harold L., 1942–45"; Roosevelt to Welles, 16 Feb. 1943; and Welles to Roosevelt, 17 Feb. 1943; both in Roosevelt Papers, PSF: Confidential, "Office of the Petroleum Coordinator for War"; Duggan Memo, 18 Feb. 1943, 812.6363/7933; *FR* 1943 6: 455.

52. Thornburg to Welles, 8 Feb. 1943, Frank P. Corrigan Papers, General Correspondence, "Max Thornburg," FDRL; *FR* 1943 6: 457–459; Thornburg to Messersmith, 10 Mar. 1943, Interior Records, Central Classified File, 1–188, part 15.

53. Ickes Diary, 14 Mar. 1943; Ickes to Roosevelt, 15 Mar. 1943, Interior Records, Central Classified File, 1–188, part 15.

54. *FR* 1943 6: 459–460; State Department to Mexican government, 30 Mar. 1943, 812.6363/7897A; see also the memorandums by Thornburg on the negotiations, 812.6363/7893, /7895, /7896, /7934, AND /7935.

55. Ickes Diary, 3 Apr. 1943; Thornburg to Feis, Bonsal, and McGurk, 9 Apr. 1943, 812.6363/7992; Ickes to Roosevelt, 20 Apr. 1943, Roosevelt Papers, PSF: Departmental Correspondence, "Ickes, Harold L., 1942–45."

56. Ickes Diary, 22 and 23 May 1943; Acheson MemCon, 17 May 1943, 812.6363/7930; Davies Notes, 12 May 1943; and Brown Memo for File, 28 May 1943; both in PAW Records, Box 667, "Mexico–100 Octane Gasoline Plant"; *Pauley Hearings*, 128–130, 202–203, 695–697.

57. Messersmith to SecState, 2 Apr. 1943, 812.6363/7900; Messersmith to Welles, 6 May 1943; Duggan to Bonsal and Thornburg, 12 May 1943; Thornburg to Duggan and Bonsal, 18 May 1943; all three in 812.6363/7969; Messersmith to SecState, 8 June 1943, 812.6363/7956; Koppes 1982, 78.

58. For the text of the president's address, see *DSB*, 24 Apr. 1943, 348–349; see also Cline 1953, 271–272.

59. DeNovo 1977; The executive order establishing PAW and defining its functions is printed in *PAW History*, 375–377.

60. Roosevelt to Welles, 30 June 1943, Roosevelt Papers, PSF: Departmental, "State Department, June-December 1943;" see also Berle MemCon, 10 June 1943, Berle Diary, A. A. Berle Papers, FDRL. Pauley played a key role in Thornburg's dismissal, see *Pauley Hearings*, 133–135; and Ickes Diary, 10 Apr. 1943. Hull and Feis later claimed that Thornburg was dismissed because they discovered that he was still receiving a salary from his former oil company; see Cordell Hull, *Memoirs*, 1517; and Feis testimony in *Arabian Oil Hearings*, 25273. Thornburg disputed this version; see his testimony in *Pauley Hearings*, 138–139; and Thornburg to Senator Owen Brewster, 10 Feb. 1948, Acheson Papers, Box 27. Thornburg's personnel file is empty. According to Linda Qaim-Maqami, who had access to Thornburg's personal papers, Thornburg voluntarily resigned on June 24, "out of recognition of the tenuous nature of his continued employment." Letter to the author, 2 Oct. 1984.

Chapter 2 Mr. Ickes's Arabian Nights

1. Miller 1980, 33–38; *Arabian Oil Hearings*, 25389–25393, 25409–25411.

2. *Arabian Oil Hearings*, 24804–24809, 25393–25397; Miller 1980, 38–39; *FR* 1941 3: 624–627. See also an interview with Moffett conducted in June 1948 in the James Landis Papers, Box 31, "1947–49 ARAMCO Case," LC. What was to be sold was the royalty oil of the king and not oil belonging to the company.

Essentially the proposal allowed CASOC to make its advances to the king in oil rather than in cash.

3. *FR* 1941 3: 627–629, 634–636, 638–640; *Arabian Oil Hearings*, 25415; Miller 1980, 40–44. From August 1939 until April 1942 when a one-man legation was established in Jidda, the U.S. minister to Egypt was concurrently accredited to Saudi Arabia.

4. *Arabian Oil Hearings*, 25371, 25434, 25445; *FR* 1941 3: 642–643; Jones to Harry Hopkins, 22 July 1941, RG 234, Reconstruction Finance Corporation/War Assets Corporation Records, Box 1 "PRC Organization" (hereafter RFC/WAC Records); Alling MemCon, 3 June 1941, 890F.51/32; Miller 1980, 45, 234. In 1941, the British provided Saudi Arabia $5,285,500 in aid and CASOC advanced the king $2,433,222 above rentals and royalties. The figures for 1942 were $12,090,000 from the British and $2,307,023 from CASOC; see *Arabian Oil Hearings*, 25381.

5. See the reports by Roy Lebkircher, CASOC representative in Saudi Arabia, dated Feb. 1942 and 10 May 1942, in 890F.6363 Standard Oil Company/138 and 890F.6363 Standard Oil Company/143. Davies is quoted in Miller 1980, 50.

6. Miller 1980, 45–54; Anderson 1981, 33.

7. *Age of Energy*, 757–760, 762–766. Wilson later became the chairman of the board of Standard Oil of Indiana; see Engler 1961, 282.

8. *Truman Committee Report*, 4–11; Davies to Ickes, 15 October, 1941, Ickes Papers, Box 221, "Oil-Octane (100)"; "The Importance of Foreign Oil Reserves to the United States," 29 Dec. 1942, enclosed in Reginald Stoner to Rear Admiral H. A. Stuart, 8 Jan. 1943, ANPB Records, ANPB 39, "EF-4"; Stoner was head of the Production Division of SOCAL. See also the PIWC report printed in *Petroleum Investigation (Gasoline and Rubber)*, 85–95.

9. Herbert Feis, *Seen from E.A.*, 102; "The Importance of Foreign Oil Reserves to the United States," 29 Dec. 1942, cited in note 8. The figures on CASOC production are from Anderson 1981, 28.

10. Paul Alling to Dean Acheson and A. A. Berle, 14 Dec. 1942, 890F.24/21. Acheson and Berle approved the recommendation. "The Importance of Foreign Oil Reserves to the United States," 29 Dec. 1942, cited in note 8.

11. For the company's view, see the testimony by W.S.S. Rodgers, president of the Texas Company in *Arabian Oil Hearings*, 24805; *FR* 1943 4: 856–857. See also Miller 1980, 66–67 and Anderson 1981, 46–47. On Kirk, see Baram 1978, 73.

12. Ickes Diary, 7 and 13 Feb. 1943; *Arabian Oil Hearings*, 24854–24856, 25232–25236, 25385–25387.

13. "Memorandum on Petroleum Reserves—Saudi Arabia," 12 Feb. 1943, ANPB Records, ANPB 40, Box 74, January-March 1943; Ickes Diary, 24 Feb. 1943.

14. Feis MemCon, 11 Feb. 1943, PED Records; see also the minutes of the second meeting of the Committee on International Petroleum Policy, 10 Feb. 1943, PED Records, Box 19, "IPP Minutes." Rodgers later observed that Feis took some time to convince whereas the Near East Division was sympathetic from the outset; *Arabian Oil Hearings*, 24856.

15. Ickes Diary, 20 Feb. 1943.

16. *FR* 1943 4: 854–855, 859; Hopkins to Stettinius, 11 Feb. 1943; and

Stettinius to Hopkins, 17 Feb. 1943; both in Harry L. Hopkins Papers–Sherwood Collection, Book 7, "Lend-Lease in Operation: Middle East (Folder 20)", FDRL. Anderson 1979, 413–423 argues that the lend-lease decision had little to do with corporate lobbying efforts. It is true that the State Department's lend-lease recommendation predated the February 1943 corporate initiatives, but Anderson's argument ignores the close relationship between CASOC and the State Department so ably described in Miller 1980, 45–61.

17. Feis, *Seen from E.A.*, 129.

18. On State Department plans, see Thornburg, "Memorandum for discussion concerning the advisability and means of United States Government acquiring rights to oil in Saudi Arabia for current delivery or reserve," 16 Feb. 1943, PED Records, Box 28, "IPPC-F. Reserves"; "Minutes of Fifth Meeting", 22 Feb. 1943, PED Records, Box 19, "IPP Minutes"; Sappington, "Desirability of This Government Obtaining Foreign Petroleum Reserves," 23 Feb. 1943, PED Records, Box 19, "CIPP Memos 1943"; Committee on International Petroleum Policy to Hull and Welles, 24 Feb. 1943; and Thornburg Memo, 25 Feb. 1943; both in PED Records, Box 28, "IPPC-F. Reserves."

19. "Report of the Committee on International Petroleum Policy," 22 Mar. 1943, 800.6363/3-2243; see the explanation of the plan in Thornburg Memo, 27 Mar. 1943, PED Records, Box 1, "IPPC-F. Reserves."

20. On concern within the State Department, see Bonsal to Feis, 1 Mar. 1943, PED Records, Box 19, "CIPP Memos 1943"; Hawkins to Feis, 4 Mar. 1943, PED Records, Box 1, "PRC Organization 2/22–7/2/43;" Minutes of Meeting, 5, 9, 15, and 19 Mar. 1943, PED Records, Box 19, "IPP Minutes"; Hawkins to Hull, 23 Mar. 1943; and Berle to Hull, 24 Mar. 1943, both in 800.6363/1145. The report is attached to Hull to Stimson, Knox, and Ickes, 31 Mar. 1943, 811.6363/524A. For Ickes's reaction, see Ickes Diary 25 Apr. 1943.

21. Carter to SecNav, 9 Apr. 1943; Horne to SecNav, 10 Apr. 1943; Carter to Gingrich, 23 Apr. 1943; all in ANPB Records, ANPB 40, Box 74, "April-June 1943."

22. Bullitt's undated paper is printed in *PRC Documents*, 3–6; on Bullitt's background, see Gardner 1970, 3–25.

23. Knox to Hull, 24 May 1943, SecNav/CNO Records, "JJ7 Secret 1943"; JCS 342, 31 May 1943; and JCS, 91st Meeting, 8 June 1943; both in JCS Records, CCS 463.7 (5-31-43); *FR* 1943 4: 921.

24. JCS, 92d Meeting, 15 June 1943, JCS Records, CCS 463.7 (5-31-43); *FR* 1943 4: 921–924; Berle MemCon, 10 June 1943, Berle Diary, FDRL.

25. Ickes Diary, 25 Apr., 9 and 29 May, and 12 and 13 June 1943; Ickes's letter is printed in *Arabian Oil Hearings*, 25237–25238; see also "Suggested Practical Method for Acquisition and Administration of U.S. Governmental Participation in Crude Oil Reserves," enclosed in Ickes to Bullitt, 25 May 1943, Ickes Papers, Box 221, "Oil Rationing."

26. Feis to Hull, 10 June 1943, 890F.6363/80; Hull to Roosevelt, 14 June 1943, PED Records, Box 1, "PRC Organization 2/22–7/2/43." See also Hull, *Memoirs*, 81, 1518–1519.

27. Feis to Hull, 11 June 1943, 890F.6363/79; *PRC Documents*, 6–8; Ickes Diary, 20 June 1943.

28. *FR* 1943 4: 924–930; compare *PRC Documents*, 9–12. Careful reading of

these documents answers the questions about the origin of the PRC raised by Keohane 1982b, 169.

29. *FR* 1943 4: 924–930.

30. Feis to Hull, 3 July 1943, 890F.6363/52; "The Position of the Department on the Petroleum Reserves Corporation," 6 Feb. 1944, 800.6363/2-544; Ickes Diary, 26 June, 3, 11, 20, and 25 July, and 1 Aug. 1943; Ickes to Roosevelt, 26 July 1943; and Roosevelt to Ickes, 30 July 1943; both in Davies Papers, Box 21, "PRC"; *FR* 1943 4: 933–934; Jesse H. Jones, *Fifty Billion Dollars: My Thirteen Years with the RFC (1932–1945*, 482–484.

31. Ickes Diary, 8 Aug. 1943.

32. Ibid.; Feis to Hull, 4 Aug. 1943, 890F.6363/9-943.

33. Feis to Hull, 4 Aug. 1943, PED Records, Box 1, "PRC Activities 7/3/43–1/1/45"; Ickes Diary, 22 and 29 Aug. and 5 Sept. 1943; *PRC Documents*, 23–26.

34. Feis to Hull, 9 Sept. 1943, 890F.6363/66; Feis to Hull, 10 Sept. 1943, 890F.6363/67. The JCS approved the refinery project on August 3, and on August 10 the ANPB assigned responsibility for the project to the PRC. After negotiations with CASOC, Ickes issued a letter of intent authorizing CASOC to begin work on the project. The letter of intent left the question of the future ownership of the project open; see John P. Frank, "PRC—Brief Chronology," 19 Apr. 1944, Davies Papers,; *PRC Documents*, 23–26.

35. Ickes Diary, 12 and 25 Sept. 1943; Feis to Hull, 16 Sept. 1943, 890F.6363/69; *PRC Documents*, 29–33.

36. Feis to Hull, 9 Sept. 1943, 890F.6363/66; *PRC Documents*, 26–29; Ickes Diary, 17 Sept. 1943; Feis to Hull, 15 Sept. 1943, 890F.6363/65; Rodgers testimony, 27 Apr. 1944, Petroleum Committee Records, *Transcript of Executive Sessions*, vol. 6.

37. *PRC Documents*, 29–39; Feis to Hull, 25 Sept. 1943, 890F.6363/70; Wirtz to Ickes, 25 Sept. 1943, 3 Oct. 1943, RFC/WAC Records, Box 1 of Item 225, "Refinery File"; Ickes Diary, 10 and 18 Oct. 1943.

38. Ickes Diary, 18 Oct. 1943; *PRC Documents*, 40–41; *Arabian Oil Hearings*, 25240–25243, 24867–24869; Anderson 1981, 61.

39. *Arabian Oil Hearings*, 25242; Ickes Diary, 18 Oct. 1943. Anderson 1981, 65 n.–66 n., points out that Ickes had the minutes of the 3 November meeting of the PRC Board changed to omit reference to the meeting with Brown. Ickes Diary puts the meeting with Brown after his decision to break off negotiations, whereas Rodgers's testimony indicates that it was before Ickes ended the talks; compare Ickes Diary, 18 Oct. 1943 with *Arabian Oil Hearings*, 24868. For yet another version of these events, see Randall 1983, 376–377.

40. *PRC Documents*, 60–68.

41. Ickes Diary, 7 Nov., 4, 5, and 11 Dec. 1943; *PRC Documents*, 80–81; Harden to Ickes, 5 Nov. 1943, Ickes Papers, Secretary of the Interior File, Box 159, "Foreign Oil Policy," explains the position of SOCAL, Texas, and Gulf on the PIWC report.

42. "Report of the Committee on International Petroleum Policy," 22 Mar. 1943, 800.6363/3-2243; see also the minutes of the Committee on International Petroleum Policy for 28 Apr., 5 and 14 May 1943, in PED Records, Box 19, "IPP-Minutes of Meetings;" Thornburg to Feis, 26 May 1943, PED Records, Box

6, "Saudi Arabia-Proposed Loan." The purpose of the Red Line Agreement was to ensure that Middle East oil development took place in a cooperative rather than a competitive manner. On the Red Line Agreement, the As-Is Agreement, and other market-sharing arrangements among the major oil companies, see the discussion in chapter 1.

43. Duce, Aide Memoire, 9 Aug. 1943, Roosevelt Papers, PSF: Diplomatic Correspondence, "Saudi Arabian Pipeline." Ickes forwarded a copy of Duce's memorandum to President Roosevelt on August 18, commenting that "next to winning the war, the most important matter before us as a nation was the world oil situation." Jackson later met with State Department Economic Adviser Herbert Feis; see Feis MemCon, 1 Oct. 1943, 800.6363/10-143.

44. See Stoff 1980, 95–97 on British concerns.

45. Sheets to Feis, 1 Sept. 1943, PED Records, Box 5, "Anglo-Iranian Agreement."

46. *PRC Documents*, 44–46; Feis to SecState, 17 Sept. 1943, PED Records, Box 3, black binder; see also the Minutes of the Special Committee on Petroleum for 21 and 28 Sept. 1943, PED Records, Box 28, "Special Committee on Petroleum (Minutes)."

47. Stoff 1980, 97–100; Ickes Diary, 29 Aug. and 17 Sept. 1943. For the views of the five senators, see *Truman Committee Report*.

48. Stoff 1980, 104–109; DeNovo 1984, 86–87. (I wish to thank Dr. DeNovo for providing me with a copy of his paper.)

49. *FR 1943* 4: 6–10, 943–947; *PRC Documents*, 41–44; James C. Sappington, "Memorandum on the Department's Position on the Proposed PRC Purchase of CASOC," 1 Dec. 1943, PED Records, Box 1, "PRC Activities 7/3/43–1/1/45."

50. See the documents cited in note 49.

51. *FR 1943* 4: 943–950; *FR 1944* 5: 10–12, 15–16; James C. Sappington, "Memorandum on the Department's Position on the Proposed PRC Purchase of CASOC," 1 Dec. 1943, PED Records, Box 1, "PRC Activities 7/3/43–1/1/45."

52. Ickes Diary, 5 Dec. 1943. The "cookie pusher" story is found in Alling to Murray, Stettinius and Hull, 6 Dec. 1943; Ickes to Roosevelt, 27 Dec. 1943; both in PED Records, Box 3, black binder. The source of the story was Max Thornburg. Ickes had also been conducting talks with the British; see Anderson 1981, 73–74.

53. *FR 1944* 5: 13–17; Ickes to Roosevelt, 4 January 1944, Roosevelt Papers, PSF: Diplomatic Correspondence, "Saudi Arabian Pipeline"; Ickes Diary, 9, 23, and 29 Jan. and 12, 19, and 20 Feb. 1944; Stoff 1980, 122–127.

Chapter 3 Obstacles to a Liberal Petroleum Order

1. E. L. DeGolyer, "Preliminary Report of the Technical Oil Mission to the Middle East," 1 Feb. 1944, Roosevelt Papers, OF 4226-D. There was no final report. A State Department report on "The Petroleum Situation in the Middle East" estimated that proved reserves in the Middle East as of 1 January 1943 were in excess of 14 billion barrels. Moreover, the potential reserves in the Persian Gulf area were "substantially in excess" of the figures for proved reserves; "The Petroleum Situation in the Middle East," 13 Jan. 1944, RG 59, Harley Notter Files, Box 83, E-251 (hereafter Notter Files).

2. *FR 1944* 5: 17–20; Carter testimony, 14 Apr. 1944, Petroleum Committee Records, *Transcript of Executive Sessions*, vol. 3.

3. Ickes Diary, 22 Jan. 1944; *Arabian Oil Hearings*, 25243–25246; PRC Minutes, 27 Jan. 1944, PED Records, Box 20, "PRC Minutes."

4. Ickes Diary, 6 Feb. 1944; Ralph K. Davies Diary, 19, 28, 30, and 31 Jan. and 1 Feb. 1944, Davies Papers, Box 15.

5. Ickes Diary, 22, 23, 29, 30, and 31 Jan. 1944; *FR* 1944 5: 20–21. See the testimony of Duce and Rodgers, 19 Apr. 1944, in Petroleum Committee Records, *Transcript of Executive Sessions*, vol. 4. The text of the agreement is printed in Mikesell and Chenery 1949, 192–197.

6. Ickes Diary, 29 Jan. and 6 Feb. 1944; *Arabian Oil Hearings*, 25387–25388; PRC Minutes, 27 Jan. 1944, PED Records, Box 20, "PRC Minutes"; *FR* 1944 5: 21–25; Miller 1980, 96.

7. Ickes testimony, *Arabian Oil Hearings*, 25244; *FR* 1944 3: 94; *NYT*, 10 Feb. 1944; "America in the Middle East," *Economist* 146 (11 Mar. 1944): 328–329; Feis, *Seen from E.A.*, 146–147; Beaverbrook is quoted in Anderson 1981, 85; see also DeNovo 1984, 90–91.

8. Stoff 1980, 142–144.

9. *FR* 1944 3: 97–100; Stoff 1980, 129–132; *NYT*, 17 Feb. 1944; *Truman Committee Report*, 516.

10. *FR* 1944 3: 100–105; see also Woodward 1975, 395–396, 397–398.

11. *PRC Documents*, 83–85, 88–119; Ickes Diary, 6 Feb. 1944; Pogue's criticisms, dated 17 Feb. 1944, were given to Ickes by Davies, PAW Records, Item 11, Deputy Administrator's Personal File, Box 666, "Arabia and National Oil Policy"; *PAW History*, 278–279.

12. "Mr. Ickes' Arabian Nights," *Fortune* 24 (June 1944): 124–125, 274; Pogue statement, cited in note 11; "The Relation of the Proposed Saudi Arabian Pipeline to United States Interests in Middle East Oil," 5 May 1944, OSS Records, Report. No. 109155; Randall 1983, 380–383.

13. *NYT*, 10 Feb. and 17 Mar. 1944; Ickes Diary, 21 Mar. 1944; Stoff 1980, 139–140. On the conservative mood in Congress, see Brody 1975, 272–274; and Moore 1967.

14. Compare *Newsweek*, 20 Mar. 1944 with William R. Boyd (chairman of the PIWC) to Senator Maloney, Petroleum Committee Records, SF–Box 2, "Boyd, W. R., Jr.;" *PM*, 6, 7, and 9 Feb. 1944; *Chicago Sun*, 10 Feb. 1944; *Christian Science Monitor*, 20 Mar. 1944; *Time* was enthusiastic about the project; see Randall 1983, 379.

15. Herbert Feis, *Petroleum and American Foreign Policy*, 45–47; *NYT*, 16 Apr. 1944.

16. Carter to Boyd, 18 Feb. 1944, ANPB Records, ANPB 40, Box 75, "Jan.-Mar. 1944"; Ickes to Zook, 3 Mar. 1944, Interior Records, Central Classified File, 1–188, PRC, part 1; Knox is quoted in the *NYT*, 10 and 22 Mar. 1944.

17. On the State Department's position, see Wright Memos, 11 and 17 Feb. 1944, PED Records, Box 1, "PRC Activities 1/1–7/1/44"; Kirk to SecState, 23 Feb. 1944, 800.6363/1500. For the split within the ranks of the project's defenders, see Ickes Diary, 17 Mar., 2, 17, 23, and 29 Apr. 1944; PRC Minutes, 25 Apr. 1944, PED Records, Box 20, "Record of Meeting of the PRC held 25 April 1944"; Ickes to Forrestal, 19 May 1944, Davies Papers, Box 21, "PRC"; Davies Diary, 15 Dec. 1943 and 2 Feb. 1944; Fortas to Ickes, 29 Mar., 8, 12 (two) April 1944; Davies to Ickes, 10 and 24 Apr. 1944; Ickes to Fortas, 24 Apr. 1944; Ickes

1944; Ickes to Davies, 26 Apr. 1944; all in Ickes Papers, Secretary of the Interior File, Box 161, "Friends-Fortas." On Fortas, see "Mr. Ickes' Arabian Nights," 125.

18. "Mr. Ickes' Arabian Nights," 274; *NYT*, 15 June 1944; for the hearings, see Petroleum Committee Records, *Transcripts of Executive Sessions*; and the correspondence between Ickes and the committee in Petroleum Committee Records, Subject File, Box 6, "Ickes, Harold L."; Ickes Diary, 28 May and 3, 4, and 18 June 1944; Stettinius MemCon, 1 June 1944, PED Records, Box 1, "PRC Activities 1/1–7/1/44"; Forrestal and Stimpson to Ickes, 30 June 1944, ANPB Records, ANPB 40, Box 75, "April-June 1944"; *FR* 1944 5: 33–34. For the fate of the pipeline project see, Shwadran 1973, 335–340.

19. *FR* 1944 5: 27–33; see also the earlier studies, "United States Petroleum Policy in the Middle East," 13 Jan. 1944, Notter Files, Box 83, E-252; Military Intelligence Division, "The Military Significance of Persian Gulf Petroleum," 12 Mar. 1944, PAW Records, Item 11, Deputy Administrator's Personal File, Box 666, "Arabia and National Oil Policy."

20. The British and U.S. delegations are listed in *PAW History*, 280. On the change in the U.S. position, see the analysis by John Loftus, "Suggested Memorandum of Information for the Secretary," 9 Nov. 1944, PED Records, Box 1, "Memoranda Re Petroleum, 1943–44." Davies's paper is "Trends in World Supply and Demand for Oil," PED Records, Box 3, "Petroleum Discussions with the British/Joint Minutes."

21. Joint Minutes, 20, 22, and 24 Apr. 1944, PED Records, Box 3, "Petroleum Discussions with the British/Joint Minutes."

22. *FR* 1944 3: 112–115.

23. Joint Minutes, 25 and 26 Apr. 1944, PED Records, Box 3; Rayner to SecState (draft), 26 Apr. 1944, PED Records, Box 3, black binder.

24. In addition to the documents cited in note 23, see Stinebower to Acheson, 1 May 1944, 800.6363/1629.

25. On the British side, see Woodward 1975, 399–401; Stoff 1980, 169–171; and DeNovo 1984, 95–96. *FR* 1944 3: 117; *NYT*, 15 June 1944; Churchill to Roosevelt, 20 June 1944; Roosevelt to Churchill, 23 June 1944; Winant to Roosevelt, 21 June 1944; all in Roosevelt Papers, Map Room 604 (2), Section 1 Oil Conference; Achilles MemCon, 24 July 1944, PED Records, Box 3, black binder. The British and U.S. delegations are listed in *FR* 1944 3: 119–120.

26. The industry advisers are listed in *PAW History*, 279–280. The British delegation included the heads of AIOC and Shell. *NYT*, 14, 20, 26, and 27 Apr. 1944; "Report on Comments of the Industry Advisers," 20 May 1944; Brown Memo, 26 Apr. 1944; Minutes of Meeting of U.S. Technical Group IX, 3 May 1944; all in PED Records, Box 3, "Petroleum Discussions with the British"; Stinebower to SecState, 25 Apr. 1944, 800.6363/1628; Brown to Rayner, 29 Apr. 1944, 800.6363/1668; Hill to Rayner, 29 Apr. 1944; and Rayner to Hill, 12 June 1944; both in 800.6363/1724; English to Rayner, 3 May 1944, PED Records, Box 20, "Petroleum Agreement U.S.-U.K., Memos, Letters"; Zook to Rayner, 5 May 1944, 800.6363/1666; Jacobsen to Rayner, 9 May 1944, PED Records, Box 3, black binder.

27. Petroleum Committee Records, *Transcript of Executive Sessions*, vols. 7 and 8; Maloney to Hull, 24 May 1944; Acheson to Maloney, 21 June 1944; both

in Petroleum Committee Records, Subject File, Box 5, "Hull, Cordell"; Report of the Expert Technical Group, 10 May 1944, PED Records, Box 3, "Petroleum Discussion with the British"; Rayner to SecState, 21 and 24 July 1944, PED Records, Box 1, "Memoranda Re Petroleum, 1943–44."

28. See the summary of the Minutes of the Joint Subcommittee, 26 July–1 Aug. 1944; Digest of Minutes of Plenary Sessions; Minutes of the Plenary Sessions; all in Davies Papers, Box 13; Minutes of the Joint Subcommittee Meetings, PED Records, Box 3, "Petroleum Discussions with the British/Joint Minutes." See also the Minutes of the Meeting of the Cabinet Committee on Petroleum, 3 Aug. 1944, 800.6363/8-344; Stettinius to SecState, 15 Aug. 1944, 800.6363/8-1544; and Stoff 1980, 172–177.

29. See the minutes cited in note 28.

30. In addition to the minutes cited in note 28, see the analyses by Rayner and Levy in PED Records, Box 17, "Memo of Understanding with the British"; Haley to Rayner, 27 June 1944, 800.6363/6-2744; Haley and Phillips, "Commodity Policy Issues in Petroleum Discussions," 11 July 1944, ECEFP Records, Subject File 5.19E; Haley to SecState, 5 July 1944, 800.6363/7-544; Loftus to Rayner, 11 July 1944, PED Records, Box 8, "CBR Working Papers"; and the later analysis by Loftus, "Suggested Memorandum of Information for the Secretary," 9 Nov. 1944, PED Records, Box 1, "Memoranda Re Petroleum, 1943–44."

31. See the minutes cited in note 28; Ickes Diary, 6 Aug. 1944. The minutes of the final session and the text of the final agreement are printed in Anderson 1981, 218–224.

32. *NYT*, 10 and 21 Aug. and 26 Oct. 1944; Giebelhaus 1980, 267–270; Zook Address before the Interstate Oil Compact Commission, PED Records, Box 20, "Anglo-American Oil Agreement/IPPA Report"; *PAW History*, 281–282, summarizes oil industry opposition.

33. Hill, "Discussion of the Anglo-American Oil Agreement," PED Records, Box 20, "Anglo-American Oil Agreement/IPPA Report"; Long MemCon, 15 Aug. 1944, 800.6363/8-1644; Rayner MemCon, 17 Aug. 1944, 800.6363/8-1744; Long MemCon, 24 Aug. 1944, 800.6363/8-2444; Minutes of Executive Session, 11 Aug. 1944, Petroleum Committee Records, Box 25; Fraser to Connally, 14 Aug. 1944, 800.6363/8-1644. The Justice Department was inclined to agree with the Senate's interpretation of the language of the agreement; see Biddle to Hull, 26 Sept. 1944, 800.6363/9-2644; MemCon with Justice Department Officials, 27 Oct. 1944, 800.6363/10-2744; and Biddle to Hull, 25 Nov. 1944, 800.6363/11-2544.

34. See the statements by Zook and Hill cited in notes 32 and 33 respectively; Sappington (drafted by Loftus) to Rayner, 13 Nov. 1944, PED Records, Box 1, "Memoranda Re Petroleum, 1943–44."

35. Joseph Pogue, "The Purpose of an International Oil Agreement," 7 Oct. 1944, PAW Records, Box 666, "Foreign Oil Policy"; see also Pogue, "Must an Oil War Follow This War?" *Atlantic Monthly* 173 (Mar. 1944): 41–47; "Preliminary Report of the Technical Oil Mission to the Middle East," 1 Feb. 1944, cited in note 1 of this chapter; Loftus MemCon, 14 Sept. 1944; Loftus to Rayner, 30 Oct. 1944; both in PED Records, Box 1, "Memoranda Re Petroleum, 1943–44."

36. Ickes Diary, 17 Sept., 15 Nov. and 2 Dec. 1944; Rayner Memo, 24 Oct. 1944, PED Records, Box 20, "Petroleum Agreement, U.S.-U.K., Memos, Letters";

Sappington to Rayner, 13 Nov. 1944, PED Records, Box 1, "Memoranda Re Petroleum, 1943–44;" Ickes to Roosevelt, 29 Nov. 1944, Interior Records, Central Classified File, 1–188, "PRC, part 2"; Ickes to Roosevelt, 1 Dec. 1944, Roosevelt Papers, OF 56.

37. Loftus to Rayner, 30 Oct. 1944; Loftus, "Proposed Report to the Policy Committee," 3 November 1944; Loftus, "Suggested Memorandum of Information for the Secretary," 9 November 1944; all in PED Records, Box 1, "Memoranda Re Petroleum, 1943–44"; Rayner, "Recommended Action Regarding the Anglo-American Agreement," 13 Nov. 1944, Notter Files, "Policy Committee Documents, ECA-11"; Minutes of 91st Meeting of the Policy Committee, 29 Nov. 1944, Notter Files, "Policy Committee Minutes, 81–91;" Rayner MemCon, 30 Nov. 1944, PED Records, Box 2, "United States National Oil Policy"; Acheson MemCon, 1 Dec. 1944, 800.6363/12-144.

38. *NYT*, 3 and 7 Dec. 1944 and 11 Jan. 1945; Acheson MemCon, 26 Dec. 1944, 800.6363/12-2644; a copy of the PIWC revision is in Interior Records, Fortas Papers, Box 11, "Oil (Foreign) PRC"; *FR* 1944 3: 126–127.

39. Ickes Diary, 13 and 21 Jan. 1945; Rayner MemCon, 11 Jan. 1945, PED Records, Box 20, "Petroleum Agreement, U.S.-U.K. Memos, Letters."

40. Ickes Diary, 21 Jan., 24 and 25 Feb 1945; Ickes to Roosevelt, 23 Jan. 1945, Roosevelt Papers, PSF: Departmental Correspondence: State Department 1945; Mason to Clayton, 21 Feb. 1945 and 22 Feb. 1945, 800.6363/1-2145. Senator Connally is quoted in the *NYT*, 23 Feb. 1945.

41. Charles Darlington, "The Anglo-American Agreement," 2 Feb. 1945; and Darlington Memo, 27 Feb. 1945; both in PED Records, Box 4, "Anglo-American Oil Agreement 1945"; Ickes Diary, 3 Mar. 1945; Minutes of Mtg. of President's Oil Committee, 27 Feb. 1945, SecNav/CNO Records, "JJ7 Secret 1945"; the quotations are from Grew to Roosevelt, 21 Mar. 1945, Roosevelt Papers, PSF: Departmental: Anglo-American Oil Agreement.

42. Coordinating Committee-28 (with annexes), Mar. 1945, PED Records, Box 4, "Anglo-American Oil Agreement 1945"; the quotation is from Annex 5. Rayner disagreed with his colleagues, see Annexes 9 and 10.

43. Ickes Diary, 4 Mar. 1945; Ickes to President Roosevelt, 12 Mar. 1945, Roosevelt Papers, PSF: Departmental: Anglo-American Oil Agreement.

44. Ickes to Roosevelt, 26 Mar. 1945, PED Records, Box 4, "Anglo-American Oil Agreement 1945"; Ickes to Roosevelt, 28 Mar. 1945, Roosevelt Papers, PSF Subject File: Anglo-American Oil Agreement; Ickes Diary, 7, 24, and 29 Apr. 1945; for the text of the Justice Department revision see, Haley to Clayton, 12 Apr. 1945, PED Records, Box 4, "Anglo-American Oil Agreement 1945"; Ickes to Stettinius, 12 Apr. 1945, 800.6363/4-1245.

45. Ickes Diary, 9 and 26 June, 8 and 21 July, 19 Sept., and 7 Oct. 1945; Meeting of the President's Committee on Oil, 6 June 1945, 800.6363/6-1345; Minutes of 1st Meeting of U.S. Delegation and Industry Advisers, 17 Sept. 1945, Davies Papers, Box 22, "U.S. Government/Industry Advisers."

46. Stoff 1980, 185–187; Ickes Diary, 24 Sept. 1945; Minutes of Joint Official Subcommittee, 19 Sept. 1945; and Minutes of Plenary Session 3, 22 Sept. 1945; both in Davies Papers, Box 13, "Anglo-American Oil Treaty, 1945–47, F. 1."

47. Stoff 1980, 187–189; Minutes of Plenary Sessions I, II, and III, 19, 20, and 22 Sept. 1945; Minutes of Joint Official Subcommittee Meetings 1, 2 and 3,

19, 20 and 22 Sept. 1945; both in Davies Papers, Box 13, "Anglo-American Oil Treaty, 1945–47, F. 1."

48. See the documents cited in note 47; the text of the revised agreement is printed in Anderson 1981, 224–228.

49. Ickes Diary, 24 Sept. 1945; Zook to Kerr, 6 Aug. 1946, Davies Papers, Box 13, "Anglo-American Oil Treaty, 1945–47, F. 3;" Goodloe to Henderson, 12 Dec. 1945, RG 234, Reconstruction Finance Corporation/War Assets Corporation Records, Box 1, "PRC Organization;" Jones 1952, 482–484.

50. Wilcox to Clayton, 29 Jan. 1946, Clayton-Thorp Papers, Box 5, Subject File, "Petroleum", HSTL; on U.S. attitudes toward control of international cartels, see Taylor 1981.

51. P. C. Spencer (Sinclair), "Memorandum in Opposition to the Anglo-American Oil Agreement," 25 Jan. 1946, Senate Petroleum Committee Records, Box 27, "Treaty, Anglo-American Oil;" Heroy to Davies, 8 Feb. 1946, Davies Papers, Box 13, "Anglo-American Oil Treaty, 1945–1947, F. 5"; East Texas Oil Association, Watch Your Purse Uncle Sam, n.d., 800.6363/2-1646; Knight to Byrnes, 15 Mar. 1946, 800.6363/3-1546.

52. Ickes to Truman, 12 Feb. 1946, Truman Papers, OF 6; Ickes Diary, 3 Feb. 1946; secondary accounts are Engler 1961, 341–348; and Donovan 1977, 177–184. On the attitude of Ickes's successors, see Gardner to Chapman, 1 Mar. 1946; Gardner to Krug, 12 Apr. 1946; both in Warner Gardner Papers, Box 4, "Legal-Interior File," HSTL; Chapman to Krug, 12 Mar. 1946, Ickes Papers, Box 40, "Krug, Julius A., 1946–47."

53. Loftus, "Policy Decisions Required on the Anglo-American Petroleum Agreement," 18 Feb. 1946, PED Records, Box 28, "Anglo-American Oil Agreement, 1946"; Wilcox to Clayton, 19 Feb. 1946, 800.6363/2-1946.

54. FR 1946 1: 1379–1381; Acheson Note, 20 May 1946, 800.6363/5-946.

55. Stoff 1980, 193–195; Ickes Diary, 21 Dec. 1946 and 4 Jan. 1947; University of the Air Transcript, 17 Aug. 1946, 800.6363/10-246; Platt's Oilgram (6 Nov. 1946); U.S. Congress, Senate, Committee on Foreign Relations, Executive Session Hearings, vol. 1, 51–52.

56. Ickes Diary, 7 June 1947; U.S. Congress, Senate, Committee on Foreign Relations, Petroleum Agreement with Great Britain and Northern Ireland, 80th Cong., 1st sess., 1947; and the Executive Session hearing cited in note 55; Stoff 1980, 194.

Chapter 4 Return to Open Door Diplomacy

1. The text of the treaty is in DSB, 21 Mar. 1942, 249–252. For information concerning the Anglo-Soviet invasion, see Woodward 1971, 23–27, 54–57; Motter 1952, 7–12; Wright 1985, 213; and Kuniholm 1980, 138–143.

2. See the documents printed in Yonah Alexander and Allan Nanes, eds., The United States and Iran: A Documentary Record, 76-81 (hereafter U.S. and Iran); Motter 1952, 13–23; "Conflicts and Agreements of Interest of the United Nations in the Near East," 10 Jan. 1944, OSS/SD Records, Report No. 1206; McFarland 1980, 338; Kuniholm 1980, 147–156; FR 1941 3: 377–378; Parker MemCon, 12 Jan. 1942, 740.0011 European War 1939/19113 (I wish to thank Aaron Miller for drawing my attention to this document); Ferris to Thornburg, 24 Mar. 1942,

800.6363/579 1/2; Acheson to Dreyfus, 16 May 1942, 891.6363/277A; Ernst to Mattingly, 27 Dec. 1943, 891.6363/812 1/2; *FR* 1942 4: 184–185, 227–228.

3. *U.S. and Iran*, 90–93; Motter 1952, 28–29, 213–214; Kuniholm 1980, 145–146. Nearly 8 million tons of supplies crossed Iran into the Soviet Union between 1941 and 1945; McFarland 1980, 336–337; Ramazani 1975, 70–72; for British views, see Lytle 1973, 36–41, 82–85; on U.S. motives see, *FR* 1941 3: 475; *FR* 1942 4: 247–248. On the U.S. advisers, see George V. Allen, "American Advisers in Persia," *DSB*, 23 July 1944, 88–93; Motter 1952, 161–173; *U.S. and Iran*, 108–127; Lytle 1973, 191–229. The head of the main U.S. advisory mission Arthur Millspaugh wrote of his experiences in *Americans in Persia*.

4. "The Three Power Problem in Iran," 13 July 1944, OSS/SD Records, Report No. 2201; *FR* 1943 4: 331–336; Woodward 1962, 161.

5. *FR* 1943 4: 330–331 and 377–379.

6. The text of the Tripartite Declaration is in *U.S. and Iran*, 142–143. The Tehran Declaration was largely the work of Gen. Patrick J. Hurley, who had been on two fact-finding missions to the Middle East for President Roosevelt during 1943; see Buhite 1973, 125–133. On the background to the Tehran Declaration, see *FR* 1943 4: 363–370, 420–426; Hull, *Memoirs*, 1505–1506; and Kuniholm 1980, 149–150, 159–160, 164–171.

7. Feis, *Seen from E.A.*, 178; Morris to SecState, "Chronological Summary," 11 Dec. 1944, 891.6363/12-1144; *FR* 1943 4: 625–628; and the following documents from the 891.6363 file: /808, /809, /813, /829, and /830.

8. Lytle 1973, 106–149; Pfau 1974, 104–114; "Chronological Summary," 891.6363/12-1144; on Hurley's connections with Sinclair, see Buhite 1973.

9. *FR* 1944 5: 447–452; the following documents from the 891.6363 file: /842, /850, /843, /846, /5-1544; 7-1044, and /8-544; and Rayner to Hawkins, 2 June 1944, PED Records, Box 1, "Memoranda Re Petroleum, 1943–44."

10. McFarland 1980, 342; "Chronological Summary," 891.6363/12-1144; "Survey of Iranian Oil Concessions," 8 Apr. 1944, Report. No. 1981.1; "Iranian Oil as a Source of Potential Conflict," 9 May 1944, Report No. 1971.2; both in OSS/SD Records. According to Feis, *Seen from E.A.*, 136, the Soviets were left out of Anglo-American oil negotiations because they did not own any concessions in the Middle East, and because it was feared that "they might bring the whole pattern of ownership in the region into question."

11. Leavell to SecState, 25 Aug. 1944, 891.6363/8-2644; *FR* 1944 5: 452; "Chronological Summary," 891.6363/12-1144; "Russian Proposals to the Iranian Government," 19 Oct. 1944, OSS Records, Report L47580; Mark 1975, 56, points out that the area requested by the Soviets was smaller than that actually granted to a U.S. company in 1937, and that the concession was spread across a wide area because of the uncertain nature of petroleum exploration.

12. *FR* 1944 5: 445–446, 452–455; on Bullard's views, see Kuniholm 1980, 155; and Louis 1984, 59–64; see also Lytle 1973, 147–48, 155–59.

13. Southgate to Alling, 11 Oct. 1944, 891.6363/10-1144; "Notes on the Postponement of Oil Concession," 11 Nov. 1944, OSS Records, Report No. L49311; Minor, "Oil Negotiations in Iran," 17 Nov. 1944, 891.6363/11-1744; "Chronological Summary," 891.6363/12-1144; *FR* 1944 5: 455–478; Pfau 1974, 106–107; McFarland 1980, 342.

14. Woodward 1975, 446–451; FR 1944 5: 351–352, 462–463, 467; Murray to Stettinius, 20 Nov. 1944, 891.6363/11-2044; Murray to SecState, 8 Dec. 1944, 891.6363/12-844; Lytle 1973, 161–163; Kuniholm 1980, 198–200.

15. Woodward 1975, 456–458; McFarland 1980, 342; Ramazani 1975, 103–104.

16. Morris to SecState, 11 Dec. 1944, 891.6363/12-1144; FR 1944 5: 470–471, 480–482; see also Mark 1975, 62–63.

17. Thornburg to Acheson, 24 June 1943, 800.6363/7979; FR 1943 6: 469–470. I would like to express my appreciation to Clayton R. Koppes for sharing with me his unpublished paper "Reversing Nationalization: The United States, Mexico, and Oil, 1938–1950," which was presented at the Annual Meeting of the American Historical Association, Washington, D.C., December 30, 1980.

18. Ickes Diary, 29 Aug., 27 Nov., 11 and 26 Dec. 1943, and 23 Jan. 1944; Ickes to Roosevelt, 20 Dec. 1943, Ickes Papers, Box 20, "Oil-Octane (100)"; FR 1944 7: 1339–1342.

19. FR 1943 6: 470–475; FR 1944 7: 1339–1342; Messersmith to Duggan, 16 Feb. 1944, 812.6363/2-1644; Meyer 1974, 133–135. On Mexico's oil laws, see "Constitutional and Legal Provisions Regarding the Operations of Foreign Petroleum Companies in Mexico," 5 Apr. 1949, Report No. 4917-PV, O.S.S./State Department Intelligence and Research Reports: Latin America, 1941–1961 (Washington, D.C.: University Publications of America, 1979).

20. Messersmith to SecState, 18 Apr. 1944, 812.6363/8165; Acheson to Messersmith, 5 June 1944; and Lieb to Sollenberger, 12 Apr. 1944; both in 812.6363/8146A. The various plans are described in "United States Government Policy Regarding Mexican Petroleum," 23 Aug. 1946, 812.6363/8-2746.

21. Messersmith to SecState, 18 Apr. 1944, 812.6363/8165; Loftus, "Memorandum on Mexican Petroleum Situation," 15 May 1944; and Acheson to Messersmith, 5 June 1944; both in 812.6363/8146A; see also Messersmith's earlier comments on smaller companies, Messersmith to Thornburg, 9 June 1943, 812.6363/6-943.

22. Ickes Diary, 6 Feb., 5 Mar., and 20 May 1944; Messersmith to SecState, 22 May 1944, 812.6363/8180; Carrigan Summary, 30 May 1944, 812.6363/8181; Messersmith to SecState, 8 June 1944, 812.6363/8159.

23. Sappington to Carr, 13 June 1944, 812.6363/8178; FR 1944 7: 1339–1342; Acheson to Messersmith, 27 June 1944, 812.6363/8181A; Ickes Diary, 24 June 1944.

24. Messersmith to Roosevelt, 29 June 1944, Roosevelt Papers, PSF: Diplomatic: Mexico, 1944–45; FR 1944 7: 1336–1338; Petroleum Division, "Memorandum for the President," 4 July 1944, 812.6363/5-444; Ickes Diary, 9 July 1944. On internal Mexican politics, see Meyer 1974, 136–138.

25. FR 1944 7: 1338–1339, 1342, 1346–1347; Ickes Diary, 16 July 1944; Messersmith to SecState, 21 July 1944, 812.6363/7-2144.

26. FR 1944 7: 1343–1351; Messersmith to SecState, 20 July 1944, 812.6363/7-2044; Bursley to Duggan, 8 July 1944, 812.6363/7-844; Messersmith to SecState, 21 July 1944, 812.6363/7-2144.

27. FR 1944 7: 1348–1358. Messersmith's memorandum, 1356–1358, is the only record of this important policy statement by President Roosevelt.

28. Ibid., 1358–1359; Messersmith MemCon, 29 Dec. 1944, 812.6363/1-345;

Messersmith to Roosevelt, 8 Jan. 1945, Roosevelt Papers, PSF: Diplomatic: Mexico, 1944–45; *FR* 1945 9: 1160.

29. *FR* 1945 9: 1160–1162; Messersmith to SecState, 26 Jan. 1945; and Messersmith to Carrigan, 26 Jan. 1945; both in 812.6363/1-2645; Carrigan, "Mexican Oil Negotiations," 7 July 1945, 812.6363/7-745; Byrnes, "Memorandum for the President," 11 Oct. 1945 (marked "approved Harry Truman, 10/13/45"), 812.6363/10-1345; Meyer 1974, 144–145.

30. Secretary Hull quoted in *FR* 1943 4: 941–943.

31. Miller 1980, 91, 107–109; *FR* 1944 5: 679–680; Woodward 1975, 406.

32. *FR* 1944 5: 690–691, 696–697; Miller 1980, 108–109, 111–116; Woodward 1975, 405–408.

33. For information on the aid program see, *FR* 1944 5: 719–726, 734–736, 743–744; Murray to Stettinius, 11 Nov. 1944, PED Records, Box 6, "Saudi Arabia–proposed loan to;" Miller 1980, 118–119; on Eddy, see Baram 1978, 75–77.

34. *FR* 1944 5: 748–749, 755–756, 757 n.; Memo for General Roberts, 18 Oct. 1944, RG 165, Records of the War Department General and Special Staffs, ABC 679 (5-2-43), Section 1-B; Rayner MemCon, 13 Nov. 1944, 890F.6363/11-1344.

35. On State Department concerns, see *FR* 1943 4: 547–577, 581–583, 747–751, 773–787, and 825–827. On Long's activities see, Long MemCon, 5 Feb. 1944, Breckinridge Long Papers, Box 20, "Palestine 1944," LC; and Fred Israel, ed., *The War Diaries of Breckinridge Long*, 245–275. See also the earlier warning by the president's special representative to the Middle East, Patrick J. Hurley, to Roosevelt, 9 June 1943, Roosevelt Papers, PSF Subject File: Patrick J. Hurley.

36. *FR* 1944 5: 614–616, 619, 622–626, 631–633, 643–646. On Murray, Henderson, and the other "Middle East hands," see Baram 1978, 66–96.

37. *FR* 1944 5: 757–758; *FR* 1945 8: 847; Miller 1980, 120–121, 126–127.

38. *FR* 1945 8: 2–3, 690–698; Miller 1980, 128–131; Wilson 1979, 51–53. See also William A. Eddy, *FDR Meets Ibn Saud*. Eddy served as translator at the meeting.

39. Brownell to McCoy, 30 Dec. 1944, RG 107, Records of the Office of the Secretary of War, "Plans, Policies and Agreements" (hereafter War Department Records); Rayner to Murray, 20 Jan. 1945, PED Records, Box 6, "SAUDI ARABIA–PROPOSED LOAN TO"; MILLER 1980, 127–128.

40. *FR* 1945 8: 852–860; Miller 1980, 131–132.

41. *FR* 1945 8: 861–862.

42. Collado to Clayton, 27 Mar. 1945, 890F.6363/3-2745; Collado to Clayton, 29 Mar. 1945, 890F.6363/5-647.

43. McGuire is quoted in Anderson 1981, 139–140; see also McGuire to Collado, 10 Feb. 1945, 890F.51/2-1045; McGuire to Collado, 24 Mar. 1945, 890F.51/3-2445; and Loftus to Sappington, 18 Oct. 1944, PED Records, Box 1, "Memoranda Re Petroleum, 1943–44."

44. Collado to Haley and Acheson, 20 Dec. 1944, 800.6363/12-1944; Collado to Clayton, 1 Jan. 1945, 890F.51/1-145; Collado to Clayton, 27 Mar. 1945, 890F.6363/3-2745; Collado to Clayton, 29 Mar. 1945, 890F.6363/5-647; Anderson 1981, 139–140.

45. *FR* 1945 8: 869–871. On Clayton, see Weil 1978, 203–204; and Gardner 1970, 116–118.

46. *FR* 1945 8: 866, 875–879, 882–884, 891–892, 895–896; McGowan to Bard, 4 June 1945, RG 80, General Records of the Department of the Navy, Records of the Office of Secretary of the Navy, James Forrestal General Correspondence, 36-1-30; Anderson 1981, 141–144.

47. *FR* 1945 8: 915–917; Miller 1980, 138–142; see also Gormly 1980.

48. *FR* 1945 8: 917 n., 956–958; Miller 1980, 141–142.

49. "Report on Petroleum Policy, Saudi Arabia," OWMR Records, Box 176D. The two plans are summarized in *FR* 1945 8: 940–942.

50. *FR* 1945 8: 940–942; McCloy to Vinson, 5 July 1945, War Department Records, "Plans, Policies and Agreements;" *FR* 1945 8: 917–918, 960–963.

51. *FR* 1945 8: 960.

52. Ibid., 981–983.

53. Ibid., 995; *FR* 1946 7: 740–741.

54. Anderson 1981, 120–121; *FR* 1946 7: 243–244, 746, 746 n. The U.S. Navy was an important customer. According to a 1948 Senate report, the Navy paid Caltex almost $32 million between January 1942 and June 1947. Similarly, the Navy paid ARAMCO $37.5 million between August 1945 and April 1947. See U.S. Congress, Senate, Special Committee Investigating the National Defense Program, *Navy Purchases of Middle East Oil*, Report No. 440, part 5, 80th Cong., 2d sess., 338.

Chapter 5 Public-Private Partnership

1. *Age of Energy*, 796.

2. Walter H. Voskuil and Hope M. Meyers, "Can United States Oil Reserves Meet the Postwar Demand?" *American Interests in the War and the Peace*, 4–6, provides a good summary of expert opinion in 1945; on the DeGolyer and McNaughton report see, Kenney to Hill, n.d., W. John Kenney Papers, Box 2, HSTL; see also DeGolyer's testimony in *Petroleum for National Defense*, 358–384. The reserve ratio calculations are from American Petroleum Institute, *Petroleum Facts and Figures*, Centennial Edition, 62.

3. The reserve figures are from *Petroleum Facts and Figures*, 62; see also *Age of Energy*, 812–814; Adelman 1972, 18, 30–31.

4. "Current Petroleum Import Problem," PPC D12/1, 9 Sept. 1949, ECEFP Records, Box 69, "Petroleum Policy Committee"; "Boiling Oil," *Fortune* 37 (May 1948): 107–109. Consumption statistics are from DeGolyer & McNaughton, *Twentieth-Century Petroleum Statistics*.

5. U.S. Congress, House, Committee on Armed Services, *Petroleum for National Defense*, H. Report 263, 6052–6073; Wiess to Krug, 23 Oct. 1947, NSRB Records, "J2-8, Petroleum"; J. Frederick Dewhurst and Associates, *America's Needs and Resources: A New Survey*, 260–293; Smerk 1965, 47.

6. Fanning 1950, 364–391; *Oil Lift Hearing*, 104; Truman to Short, 2 July 1948, Truman Papers, OF 56-A, "Misc. Oil Matters."

7. "The Problem of Procurement of Oil for a Major War," JCS 1741, 29 Jan. 1947, ANPB Records, ANPB 38, Box 9, Folder 5; see also Rosenberg 1976, 55–56.

8. Eakens to Brown, 9 Apr. 1948, 810.6363/4-948; Vietor 1980; Goodwin 1981, 146–167; "Synthetics: The Great Oil Reserve," *Fortune* 37 (May 1948): 110–115, 153–156; Halford L. Hoskins, *Middle East Oil in United States Foreign Policy*, 17–28, discusses synthetic fuels and the steel question.

9. "Current Petroleum Import Problem," 9 Sept. 1949, cited in note 4; Johnson to Steelman, 14 Feb. 1949, Truman Papers, OF 56; "United States Trade in Crude Oil and Petroleum Products," 12 Aug., 1949, 811.6363/8-1249; Ernest O. Thompson (head of the Texas Railroad Commission), "Middle East Oil Presents Problems for U.S. Producers," *Oil Forum* 3 (Mar. 1949): 123–126; and Thompson, "Oil Conservation Progress in Texas During 1949," *Oil Forum* 3 (Nov. 1949): 487–488.

10. In addition to the material cited in note 9, see also the numerous exchanges between the State Department and congressmen from oil states in the 811.6363 file for 1949; and T. Orchard Lisle, "Import-Export Oil Divides American Oil Industry," *Oil Forum* 3 (Mar. 1949): 115–118; and *Oil Forum* 3 (Oct. 1949): 412–414, 431.

11. Vietor 1980; Goodwin 1981, 146–167.

12. Brodie, "American Security and Foreign Oil," 298–311; see also Deale to Forrestal, 15 Oct. 1947, NSRB Records, Entry 1, Box 3, "Special Reports Petroleum, Speeches"; Biggs statement, 12 Sept. 1949, ECEFP Records, "PPC Documents 11-19/4c"; Hoskins, *Middle East Oil*, 37–41.

13. Brodie, "American Security and Foreign Oil," 307–311; DeGolyer to Forrestal, 30 Mar. 1947, enclosing letter to Congressman Dewey Short of 19 Mar. 1948, OSD Records, CD 7-1-18; Pratt, "A Study of Liquid Fuels for National Security," Aug. 1948, ECEFP Records, "PPC Documents 1-5."

14. Pratt to Forrestal, 15 Apr. 1948, OSD Records, CD 7-1-18; Hopkins Comments on Pratt Report, 18 Nov. 1948, ECEFP Records, Box 69, "PPC Documents 1–5"; "Outline of 'A National Liquid Fuels Policy' by Wallace Pratt," 24 Nov. 1948, ECEFP Records, Box 70, "PPC M-10;" *Petty's Oil Letter* 206 (25 Sept. 1948). See the astute comments on the difficulties of stockpiling in Hoskins, *Middle East Oil*, 111–112.

15. See Bachrach and Baratz 1963 for a discussion of "nondecisions."

16. *New Horizons*, 735–736; Anderson 1981, 80 n., 144–146; "The Great Oil Deals," *Fortune* 35 (May 1947): 139–143, 175–182; *Oil Lift Hearings*, 1402–1409, 2010–2011.

17. *New Horizons*, 734–736; *International Petroleum Cartel*, 47–112.

18. *New Horizons*, 736; *Multinationals Report*, 45–46.

19. *MNC Documents*, 96–111; Anderson 1981, 144 n.–145 n. Most accounts place the beginning of negotiations in the spring of 1946. According to a May 1947 article in *Fortune* magazine, however, the idea was first discussed early in the fall of 1944, when Harry Collier of SOCAL visited the downtown Manhattan offices of Socony–Vacuum Chairman Harold Sheets and "poured out his troubles and his prospects"; see "The Great Oil Deals," 176.

20. *MNC Documents*, 84–89; Anderson 1981, 149–150.

21. *MNC Documents*, 89–94, 103–111.

22. Holman quoted in *International Petroleum Cartel*, 101–102; *New Horizons*, 736.

23. MNC Documents, 94–95, 122; FR 1946 8: 31–34.

24. Multinationals Report, 50–53; MNC Documents, 112–116, 124–126; "Great Oil Deals," 176–179.

25. MNC Documents, 114–116; Loftus MemCon, 10 Dec. 1946, 891.6363 AIOC/12-1046.

26. FR 1946 8: 38–40, 44–46.

27. MNC Documents, 116–117.

28. FR 1946 8: 40–43; Loftus MemCon, 10 Dec. 1946, 891.6363 AIOC/12-1046.

29. Multinationals Report, 49; International Petroleum Cartel, 121 n.; Kaufman 1978, 27–28.

30. MNC Documents, 119–121, 124–126; New Horizons, 737–738.

31. MNC Documents, 123–128; International Petroleum Cartel, 102–104; Funkhouser, "Monthly Petroleum Report, Dec. 1946," 20 Jan. 1947, 800.6363/1-2047.

32. MNC Documents, 130–132; Funkhouser, "Monthly Petroleum Report," 20 Jan. 1947, 800.6363/1-2047; FR 1947 5: 627–629, 632–633.

33. FR 1947 5: 629–631, 634–635, 639–642; MNC Documents, 130–132; Loftus, "Memorandum for the Secretary's Staff Committee," 31 Dec. 1946, PED Records, Box 2, "Middle East Oil Deals."

34. In addition to the memorandum by Loftus cited in note 33, see International Petroleum Cartel, 12–22, 138–162; Blair 1976, 34–43.

35. FR 1947 5: 646–651; see also Eakens to Wilcox, 14 Feb. 1947, 800.6363/2-1447.

36. FR 1947 5: 651–654; MNC Documents, 133–134, 149–151, 160; Multinationals Report, 53–54.

37. FR 1947 5: 651–654; MNC Documents, 151–155, 160–163.

38. New Horizons, 738; Anderson 1981, 158, gives a final figure of $469.2 million; International Petroleum Cartel, 122–128; MNC Documents, 166–167.

39. FR 1947 5: 657–660; FR 1948 5: 64–66; International Petroleum Cartel, 104–107; Anderson 1981, 158–159.

40. Hess 1974, 123–125; Kuniholm 1980, 214–216, 272–273, 276–277.

41. For Murray's report, see FR 1945 8: 417–419; see also FR 1945 8: 393–400; and Lytle 1973, 226. On Murray's move to Iran, see Weil 1978, 194.

42. On Soviet behavior before the 1944 crisis, see "Conflicts and Agreements of Interest of the United Nations in the Near East," 10 Jan. 1944, OSS/SD Records, Report No. 1206; Mark 1975, 57–59; McFarland 1980, 340, 343–345. Kuniholm sees Stalingrad as the turning point in Soviet policy in Iran. He also takes a more traditional view of Soviet actions in Azerbaijan; see Kuniholm 1980, 147–148, 274–282. See also Hess 1974, 125–128. The United States note to the Soviets is in FR 1945 8: 448–450; see also the State Department report "Withdrawal of Allied Forces from Iran," 11 Dec. 1945, PED Records, Box 5, "Iran."

43. Kuniholm 1980, 279–282, 319; Hess 1974, 128–132. See FR 1945 8: 520–521 for an example of the new consul's reporting.

44. The shift in U.S. policy is best followed in Messer 1982, 152–180; see also Yergin 1977, 138–178; and Kuniholm 1980, 282–298, 308–313, 315–317; Truman's interpretation of the situation in Iran is in his Memoirs 1: 522–523, 551–552; see also FR 1946 7: 1–6.

Notes to Pages 113–120 / 245

45. See the accounts in Messer 1982, 195–199; Hess 1974, 134–137; McFarland 1980, 346–347; Kuniholm 1980, 317–326; *FR* 1946 7: 378, 413.

46. Hess 1974, 133, 138–141; Kuniholm 1980, 306 n., 313–315, 326–335, 337–342.

47. Clifford's report is printed in Arthur Krock, *Memoirs: Sixty Years on the Firing Line*, 419–482; *FR* 1946 7: 529–532, 523–525.

48. Pfau 1977, 359–372; McFarland 1980, 348–350; see also George V. Allen, "Mission to Iran," unpublished manuscript, George V. Allen Papers, HSTL; and Allen to Jernegan, 21 Jan. 1948, Allen-Angier Family Papers, Perkins Library, Duke University.

49. The quotation is from Joseph Marion Jones, *The Fifteen Weeks*, 11; see also Jones, *The Fifteen Weeks*, 132–142, 156–157; *FR* 1946 7: 1–6, 240–245, 894–897; Paterson 1973, 183–193; Kuniholm 1980, 17–70, 359–382; on Greece, see Wittner 1982, 52–63, 311–312; on Turkey, see Leffler 1985.

50. *FR* 1947 5: 902–904, 891–893, 934–936.

51. Ibid., 893–894, 936–937.

52. Ibid., 934–936, 916, 951–952, 958–960, 966, 968–969.

53. Ibid., 969–972, 981–982; Elwell-Sutton 1955, 118–119; McFarland 1980, 350; Louis 1984, 68–69.

54. *FR* 1945 8: 727–730, 742–743, 784–785; Cohen 1982, 55–59. The Harrison report is printed in *DSB*, 30 Sept. 1945, 456–463. Clark Clifford's account of Truman's policy on Palestine emphasizes the State Department's opposition to the president's policies; see Clifford, "Factors Influencing President Truman's Decision to Support Partition and Recognize the State of Israel."

55. Cohen 1982, 96–115; Truman *Memoirs* 2: 143–153; *FR* 1946 7: 631–633, 652–657.

56. *FR* 1946 7: 708–717; Cohen 1982, 141–170.

57. *FR* 1946 7: 18–27; Miller 1980, 154–157; see also Loftus to Levy and Henderson, 4 Feb. 1946, 890F.6363/2-446; Merriam, "Memo of Meeting with ARAMCO Officials," 15 Feb. 1946, 867N.6363/2-1546; Henderson and Loftus to SecState, 11 Mar. 1946, 867N.6363/3-1146.

58. *FR* 1947 5: 662–665; Miller 1980, 179.

59. Hoffman, MemCon, 29 July 1947, 890F.6363/7-2947; Duce to Villard, 6 Aug. 1947, 890F.6363/8-647; and Miller 1980, 180–181.

60. Mattision to Henderson, 28 Aug. 1947; Hoffman to Loftus, 3 Sept. 1947; and Henderson and Kennedy to Lovett, 4 Sept. 1947; all in 890F.6363/8-2847; George to Garrett, 26 Aug. 1947, 890F.6363/10-2747; Miller 1980, 181; on Forrestal, see Rogow 1963, 189–190, 341.

61. *FR* 1947 5: 1143; Cohen 1982, 260–267.

62. *FR* 1947 5: 665–666, 1153–1158 (part of this document is quoted in Miller, 1980, 186–187); Henderson to Lovett, 24 Sept. 1947, 890F.6363/10-2747. On Henderson's background, see Kuniholm 1980, 237–244; and Podet 1978.

63. Ball to Krug, 17 Sept. 1947, Interior Records, Central Classified File, 1–322, Oil and Gas Division, Administration, part 2; for the announcement, see 890F.6363/10-2747. The decision was reached in a meeting on September 22; see the records of the meeting in 890F.6363/10-2747, and by Max Ball in Interior Records, Minerals and Fuels Division, Box 7, "OGD."

64. For Brown's argument, see Brown to Harriman, 24 Sept. 1947; Wherry to Harriman, 23 Sept. 1947; both in 890F.6363/10-2747; see also the testimony in U.S. Congress, Senate, Special Committee to Study the Problems of American Small Business, *Problems of American Small Business: Part 21: Oil Supply and Distribution Problems:* 5, 2271–2301; Forrestal's executive session testimony is recorded in *Forrestal Diaries*, 323–324; for Lovett's testimony, see Eakens to Thorp, 16 Mar. 1948, 890.6363/3-1648.

65. *FR 1947* 5: 1147–1150, 1177–1178, 1180; see also Mazuzan 1975, 169–170; Truman *Memoirs* 2: 156–157; Donovan 1977, 325–326.

66. *FR 1947* 5: 1153–1158; (part of this document is quoted in Miller 1980, 186–187); "The Problem of Palestine," JCS 1684/5, 10 Oct. 1947, RG 165, Records of the War Department General and Special Staff, ABC 092.3 Palestine (9 May 46), sec. 1; CIA, "The Current Situation in Palestine," ORE-49, 20 Oct. 1947, Truman Papers, PSF: Intelligence, Box 254.

67. For the king's letter, see *FR 1947* 5: 1212–1213; Ibn Saud is quoted in Merriam to Satterthwaite and Henderson, 13 Nov. 1947, 890F.6363/11-1347; CIA, "The Consequences of the Partition of Palestine," ORE-55, 28 Nov. 1947, Truman Papers, PSF: Intelligence, Box 254.

68. Cohen 1982, 292–300. The vote was 33–13 in favor of partition with ten abstentions. *FR 1947* 5: 668, 1289–1290, 1335–1338; Wilson 1979, 124–130; see the memorandums by William A. Eddy, 2 and 8 Dec. 1947 in OSD Records, CD 6-1-43; also Eddy Memo, 10 Dec. 1947 in JCS Records, Admiral Leahy Files, 56 Palestine, "Arabs, Jews"; Tuck to SecState, 17 Dec. 1947, 890.6363/12-1747; *Forrestal Diaries* , 356–357; Miller 1980, 188–189.

69. Forrestal had been head of the investment banking firm of Dillon, Read and Company. Rogow 1963, 178–195; *Forrestal Diaries*, 344–349, 356–357, 359–365, 376–377; W. John Kenney (under secretary of the navy) to Forrestal, 3 Jan. 1948, which encloses a memorandum from Philip Kidd of ARAMCO, 29 Dec. 1947, OSD Records, CD 7-1-8.

70. *FR 1947* 5: 1313–1314; *FR 1948* 5: 545–554, 573–581, 600–603, 619–625; Cohen 1982, 345–350.

71. "The Position of the United States with Respect to Palestine," 17 Feb. 1948, Clark Clifford Papers, Box 12, HSTL (I wish to thank Aaron Miller for drawing my attention to this document); Ganin 1979, 149–154.

72. *FR 1948* 5: 690–696; Ganin 1979, 157–160.

73. Sanger (NEA) MemCon with Kidd, 26 Dec. 1947, 890F.6363/12-2647; Kidd to Kenney, n.d., OSD Records, CD 7-1-8; Henderson to Lovett, 6 Jan. 1948, 890F.6363/1-648; Kelly to Hill, 27 Feb. 1948, NSRB Records, Entry 1, Box 3, "Middle East Oil"; see also the study by the Army-Navy Petroleum Board dated 11 Dec. 1947, NSRB Records, Entry 1, Box 3, "Middle East Oil."

74. Miller 1980, 196; Kelly to Hill, 27 Feb. 1948, cited in note 73; Swanson to Collisson, 9 Jan. 1948, Interior Records, Collisson Files, Box 14, "Office File."

75. "Military and Strategic Value to Our National Security of Middle East Oil Development," JCS 1684/11, JCS Records, CCS 678 (3-6-47), sec. 1; McCone to Forrestal, 24 Mar. 1948, OSD Records, CD 7-1-19; Forrestal Diary, 26 Mar. 1948, vol. 10: 2162; Miller 1980, 197–198.

76. *FR 1948* 5: 705–707; Donovan 1977, 371–379; and Cohen 1982, 354–362, 367, are the best accounts; see also Clifford, "Factors Influencing Truman's

Decision on Israel," 33–36; Jessup MemCon, 21 Apr., 1948, RG 59, Rusk-McClintock Files, 1947–1949, Box 2, "Palestine, Apr. 1–19, 1948;" *Forrestal Diaries*, 410–411; Forrestal to Marshall, 27 Apr. 1948; and JCS 1684/12; both in JCS Records, P & O 091 Palestine TS (27 Apr 48).

77. See Clifford's account of this meeting in "Factors Influencing Truman's Decision on Israel," 38–41. See also Donovan 1977, 379–385; Cohen 1982, 379–386. The Soviets also recognized Israel; on Soviet policy, see Krammer 1973.

78. Tuck to SecState, 18 May 1948, 890.6363/5-1848; Henderson MemCon, 28 May 1948, 890F.6363/5-2848; Childs to SecState, 9 June 1948, 890F.6363/6-948; U.S. Embassy Cairo, "Summary of Oil Developments during May," 890F.6363/6-248; Miller 1980, 200–201.

79. *FR* 1948 5: 1020–1021, 1034–1036, 1042 n., 1060–1061, 1217–1218, 1221–1222.

80. Eakens to Thorp (plus attachment), 9 July 1948, 800.6363/7-848.

81. Lovett to Hill, 20 Aug. 1948, NSRB Records, Entry 1, Box 3, "Middle East Oil."

82. *FR* 1948 5: 23–24, 39–40; *Forrestal Diaries*, 551; Kegan to Blaisdell, 15 Sept. 1948, NSRB Records, Office of the Chairman, Box 3, "Middle Eastern Oil"; Miller 1980, 285; *Oil Lift Hearings*, 1481–1485.

83. *FR* 1948 5: 2–3; "Demolition of Oil Facilities in the Middle East," JCS Records, CCS 600.6 Middle East (1-26-48); Leahy to Forrestal, 8 Apr. 1948, OSD Records, CD 7-1-9; see also the SANACC 398 Series in SWNCC Records, Box 71, "677 Oil, Middle East"; Rosenberg 1976; and Hoskins, *Middle East Oil*, 108–109.

Chapter 6 Containing Economic Nationalism

1. Forrestal to Marshall, 6 Nov. 1947, SWNCC Records, Box 53, SWNCC 289 Series, 463.7 Gasoline and Oil; Joint Logistics Committee, "Petroleum Reserves in South America," 10 Mar. 1947, ANPB Records, Army-Navy Petroleum Board 38, Box 6, Folder 2; "Petroleum Reserves in South America," SANACC 289/6, 6 Oct. 1947, ANPB Records, Army-Navy Petroleum Board 38, Box 11, Folder 7; Circular Telegram, 9 July 1947, 810.6363/7-947; see also "Petroleum Resources in South America," SWNCC 289; and "Petroleum Reserves in the Western Hemisphere," 6 Aug. 1947; both in SWNCC Records, Box 53, SWNCC 289 Series, 463.7 Gasoline and Oil; Eakens to Brown, 9 Apr. 1948, 810.6363/4-948; and *FR* 1948 9: 244–248, 250–255.

2. See the replies to a State Department survey in the 810.6363 file: /7-2147 (Mexico); /7-2547 (Colombia); /8-447 (Peru); /8-447 (Chile); /8-547 (Brazil); /8-2847 (Bolivia); /10-3147 (Venezuela); and *FR* 1948 9: 286–290 (Argentina); on U.S. views see "Principles of Laws Governing Petroleum Development in Latin America," TPC D-3/19, 22 Apr. 1947, SWNCC Records, Box 53, SWNCC 289 Series, 463.7 Gasoline and Oil; *FR* 1947 8: 243–244, 250–255; Hoffman to Schaetzel, 22 Sept. 1947, PED Records, Box 7, "Latin America-Misc."; Sappington, "The Department's Position as It Relates to Latin America," 12 Feb. 1945, 810.6363/2-1245.

3. The quotation is from Coppock, "Rough Draft of a General Policy Statement with Respect to Encouragement of Private Enterprise and Private Property System in Other Countries," 27 June 1947, PED Records, Box 1, "U.S. Foreign Oil Policy

from 1945"; John A. Loftus, "Petroleum in International Relations," *DSB,* 5 Aug. 1945, 174; see also the testimony of Willard Thorp in U.S. Congress, House, Committee on Armed Services, Special Subcommittee on Petroleum, *Petroleum for National Defense,* 314–315; and Paterson 1973, 1–19.

4. For the U.S. military's estimate of the importance of Venezuelan oil, see Ordway to Timberman, 2 Feb. 1948, RG 319, Records of the Army Staff, 463 TS; Pigott to Loftus, "Venezuelan Petroleum Production," 21 Jan. 1947, 831.6363/ 1-2147; "Survey of Current Petroleum Situation in Venezuela," 4 Nov. 1948, OSS/SD Records, Report No. 4796-PV; Statement by Creole Petroleum Company, 14 Dec. 1948, ECEFP Records, "PPC Documents 6–10"; Peters, "The Petroleum Industry in Venezuela," 3 Mar. 1949, 831.6363/3-349; *The International Petroleum Cartel,* 185; Lieuwen 1954, 108.

5. On the coup, see Burggraaff 1972, 52–78, 206–208. See also Rabe 1982, 94-96; Lieuwen 1954, 102-103; and Alexander, 195–218.

6. Tugwell 1975, 44–45; Betancourt, 1979, 127; Salazar-Carrillo 1976, 36, 59, 72, 77; Dawson to SecState, 26 Jan. 1946, 831.6363/1-2546; Maleady to SecState, 24 Apr. 1946, 831.6363/4-2446.

7. *FR* 1945 9: 1405–1417; Rabe 1982, 73–74, 97; Lieuwen 1954, 103.

8. *FR* 1946 11: 1330–1334; Dawson to SecState, 10 Jan. 1946, 831.6363/1-1046; Flack MemCon, 2 and 14 May 1946, 831.6363/5-146 and 831.6363/5-1446; Rabe 1982, 102; Tugwell 1975, 45–46; *New Horizons,* 799. On oil company profits, see Salazar-Carrillo 1976, 205.

9. *FR* 1946 11: 1331; Dawson to SecState, 17 Jan. 1946, 831.6363/1-1746; Dawson to Corrigan, 22 Jan. 1946, Frank P. Corrigan Papers, General Correspondence, "Allen Dawson," FDRL; see also the several reports of assurances from Betancourt and Pérez Alfonso, 831.6363/1-2346, /1-2546, /3-2746, /2-446. and /2-1146; on the Pantepec case see *FR* 1946 11: 1337–1343; and Rabe 1982, 102–103.

10. On the negotiations, see the reports by Maleady, 831.6363/4-2746, /5-646, /5-2046, and /5-2746. See also Wells to Briggs, 28 May 1946, 831.6363/5-2846; Corrigan to SecState, 4 June 1946, 831.6363/5-3146; *FR* 1946 11: 1344–1348; Patterson to SecState, 7 June 1946, JCS Records, CCS 463.7 South America (4-17-46), sec. 2; Strong to Asst. Chief of Staff, 3 June 1946, RG 107, Records of the Office of the Secretary of War, Patterson Papers, Office of the Assistant Secretary of War, Classified, 091 Venezuela. Secondary accounts include Rabe 1982, 103–105; Lieuwen 1954, 104–105; Fanning 1954, 86–93.

11. On the issue of royalty oil see "Survey of Current Petroleum Situation in Venezuela," 4 Nov. 1948, cited in note 4; and the reports by Maleady, 831.6363/ 6-2646, /8-246, /9-1046 and /9-2446; Russell MemCon, 26 Sept. 1946, 831.6363/ 9-2646; Corrigan to Russell and Braden, 1 Oct. 1946, 831.6363/10-146; *FR* 1946 11: 1352–1353; Corrigan to SecState, 4 Feb. 1947, 831.6363/2-447; Lieuwen 1954, 106–107; Tugwell 1975, 46.

12. *FR* 1946 11: 1349–1355; Maleady to SecState, 20 Nov. and 16 Dec. 1947, 831.6363/11-2047 and 831.6363/12-1647; Post to Espy, Mills and Eakens, 3 June 1948, 831.6363/6-348; "Survey of Current Petroleum Situation in Venezuela," 4 Nov. 1948, cited in note 4; *FR* 1948 9: 763–764; *New Horizons,* 729; Alexander 1982, 259–261. Fanning 1954, 94-97, provides an extensive discussion of the new law.

13. *FR* 1948 9: 756–759, 763–764; "Survey of Current Petroleum Situation in Venezuela," 4 Nov. 1948, cited in note 4; McGinnis to Krieg, Mills, and Daniel, "Current Venezuelan Oil Policy," 29 Sept. 1948, 831.6363/9-2048; Krieg to McGinnis, Mills, and Woodward, "Venezuelan Congress Debates Petroleum Policy and Rockefeller VBEC," 9 Nov. 1948, 831.6363/11-148; Betancourt 1979, 137–144; Lieuwen 1954, 108–109; *New Horizons*, 727; Alexander 1982, 261–263.

14. Maleady to SecState, 31 Oct. 1947, 810.6363/10-3147; SecState to US Embassy Caracas, 27 July 1948, 831.6363/7-248; *FR* 1948 9: 761–763; "Survey of Current Petroleum Situation in Venezuela," 4 Nov. 1948, cited in note 4; Mann, "Memo on Venezuelan Oil Policy," 7 Mar. 1949, 812.6363/3-749; Lieuwen 1954, 107–108; Alexander 1982, 263–264.

15. Landau 1985 provides the best discussion of U.S. policy. See also "Current U.S. Policy toward Venezuela," 20 Feb. 1946, Report No. 3579, *O.S.S./State Department Intelligence and Research Reports: Latin America, 1941–1961* (Washington, D.C.: University Publications of America, 1979); Pigott to Loftus, 21 Jan. 1947, 831.6363/1-2147; CIA, "The Venezuelan Elections of 14 December 1947," 5 Jan. 1948, ORE 65, *CIA Research Reports: Latin America, 1946–1976* (Frederick, Md.: University Publications of America, 1982); *FR* 1948 9: 761–763.

16. Landau 1985, 25–27, 48, 52–54, 59–63, 88; CIA, "The Venezuelan Elections of 14 December 1947," 5 Jan. 1948, ORE 65; CIA, "Vulnerability to Sabotage of Petroleum Installations in Venezuela, Aruba, and Curaçao," 14 May 1948, ORE 31-48; both in *CIA Research Reports: Latin America, 1946–1976*; *FR* 1948 9: 759–761, 766. On Betancourt's "flirtation" with Communism, see Alexander 1982, 67–87.

17. *FR* 1948 9: 133–147; Landau 1985, 67–83; Rabe 1982, 112–116; Hellinger 1984, 39–40, 48–50; Kolb 1974, 41–57; Lieuwen 1954, 109–110; see also, "It's Hot in Venezuela," *Fortune* 39 (May 1949): 101–107, 150–164. See also the extensive discussion in Alexander 1982, 265–315; and Burggraaff 1972, 79–111, 209–211.

18. Burggraaff 1972, 116–138; Baloyra 1974, 47–49; Lieuwen 1954, 111–112; Maull 1980, 211; "United States Trade in Crude Oil and Petroleum Products," 12 Aug. 1949, 811.6363/8-1249; Salazar-Carrillo 1976, 205.

19. Krieg MemCon, 30 Dec. 1948, 831.6363/12-3048; Krieg MemCon, 28 Jan. 1949, 831.6363/1-2849; Peters to McGinnis, Krieg, and Mills, 25 Mar. 1949, 831.6363/3-2549; Krieg MemCon, 25 Nov. 1949, 831.6363/11-2549.

20. Sollenberger, "Mexico's Oil Economy," 10 Sept. 1946, 812.6363/9-1246; Eakens to Nitze, "Further Meeting on Mexican Oil Situation," 24 Sept. 1947, 812.6363/9-2447.

21. Conkwright to Bohan, 15 July 1946, 812.6363/10-146; Meyer 1974, 145–146, 149–150; "Mexico May Adopt Progressive Role in Petroleum," *Oil Forum* 1 (June 1947): 149–150, 164. Alemán planned sweeping changes in all aspects of Mexican life; see Cline 1953, 307–312; and Vazquez and Meyer 1982, 189–190.

22. "American Capital in the Mexican Petroleum Industry," 19 Aug. 1946, PED Records, Box 18, "Mexico 1945–47." The companies represented were Jersey, Socony-Vacuum, Gulf, Sinclair, and the Texas Company.

23. *FR* 1946 11: 1008–1016. This and the following paragraph are based on this document. The sections omitted by *FR* are in 812.6363/8-2746.

24. McCollum to Loftus, 8 July 1946, 812.6363/7-846; Geist MemCon, 29 Aug. 1946, 812.6363/8-2946. For a detailed discussion of Mexican petroleum law, see "Constitutional and Legal Provisions Regarding the Operations of Foreign Petroleum Companies in Mexico," 5 Apr. 1949, Report No. 4917-PV, *O.S.S./State Department Intelligence and Research Reports: Latin America, 1941–1961.*

25. Thurston to Braden, 21 Oct. 1946; and Braden to Thurston, 12 Nov. 1946; both in 812.6363/10-2146; Braden to Thurston, 7 Nov. 1946, 812.6363/10-2846.

26. *FR* 1947 8: 788–791.

27. Eakens to Thorp, 9 July 1947, 812.6363/7-947; Thurston to SecState, 21 July 1947, 810.6363/7-2147; *FR* 1947 8: 793–795.

28. Reveley to Wright, 3 Sept. 1947; and Wright to Lovett, 3 Sept. 1947; both in 812.6363/8-2747; *FR* 1947 8: 793–795; Eakens to Thurston, 2 Oct. 1947, 812.6363/10-247. Ambassador Espinosa apparently had made a similar offer to Braden in 1946, but had been turned down; see Braden to Thurston, 22 Oct. 1946, 812.6363/10-146.

29. Forrestal Manuscript Diary, 25 Aug. 1947, vol. 8: 1787, Xerox at the Operational Archives, Naval Historical Division, Washington Navy Yard; Lovett to Wright, 29 Aug. 1947, 812.6363/8-2747; Leva Minutes of Meeting on Mexican Oil, 1 Oct. 1947, OSD Records, CD 9-1-5; Hoffman MemCon, 3 Oct. 1947, 812.6363/10-347; Thurston to SecState, 10 Oct. 1947, 812.6363/10-1047; *FR* 1947 8: 795–798.

30. *FR* 1947 8: 795–798.

31. Ibid., 798–802.

32. Hoffman, "Proposed Contract between PEMEX and Foreign Private Oil Companies for Petroleum Operations in Mexico," TPC D-23, 4 Mar. 1948, OSD Records, CD 7-1-17; Kane 1981, 52.

33. Hoffman, "Proposed Contract," 4 Mar. 1948, OSD Records, CD 7-1-17; Thurston to Reveley, 2 Mar. 1948, 812.6363/3-248; McLean MemCon, 9 Mar. 1948, 812.6363/3-948.

34. In addition to the documents cited in note 33, see "Boiling Oil," *Fortune* 38 (May 1948): 107–109. On the Cities Service deal, see "Opening Wedge to Private U.S. Cooperation with Mexico," *Oil Forum* 2 (May 1948): 195; *FR* 1948 9: 603–604; and Powell 1956, 48–49, 66. Donnelly to SecState, 8 June 1948, 812.6363/6-848, warns of the impact on Venezuela. The Texas Company also tried unsuccessfully to make a deal with Bermúdez, see Ogarrio to Bermúdez, 17 June 1948, 812.6363/6-2848; Andrews Memo for Files, 25 Oct. 1948, OSD Records, CD 7-1-2.

35. Southeastern Oil, Inc., Memo for the Export-Import Bank, Jan. 1948, Senate Records, Select Committee on Small Business, "Southeastern Oil Memo"; Reveley to Daniels, 28 Jan. 1948, 812.6363/1-2848; Ohly to Forrestal, 10 Feb. 1948, OSD Records, CD 7-1-13; Excerpts of TelCon between Leva and Jones, 9 Mar. 1948, NSRB Records, Office of the Chairman, Box 3, "Oil in Mexico"; Eakens to Thorp, 16 Mar. 1948, 890.6363/3-1648. On Southeastern Oil see also McBride MemCon, 21 June 1948, 812.6363/6-2148.

36. Wherry to SecState, 3 June 1948, Senate Records, Select Committee on Small Business, "Mexican Oil Development"; *FR* 1948 9: 604–606.

37. *FR* 1948 9: 606–610; Thurston to Reveley, 2 Sept. 1948; and Turkel

MemCon, 2 Sept. 1948; both in 812.6363/9-248; Brown to Thurston and Bohan, 2 Sept. 1948, 812.6363/9-348; Meyer 1974, 150–151.

38. See the documents cited in note 37.

39. *FR* 1948 9: 616–619 is a good summary of State Department thinking on this matter.

40. Eakens MemCon, 20 Sept. 1948, 812.6363/9-2048; Eakens to Thurston, 11 Oct. 1948, 812.6363/10-1148; Andrews to Forrestal (plus attachment), 25 Oct. 1948, and 12 Nov. 1948, OSD Records, CD 7-1-2; *FR* 1948 9: 612–616.

41. U.S. Congress, House, Committee on Interstate and Foreign Commerce, *Fuel Investigation: Mexican Petroleum*, 12–17 (hereafter *Mexican Petroleum Report*); see also Braun to Thurston, 17 Dec. 1948, 812.6363/12-2048.

42. *Mexican Petroleum Report*, 4–11; see also Arthur Krock, "The Vital Question of Mexican Oil," *NYT*, 28 Dec. 1948.

43. *FR* 1948 9: 612–616; Eakens MemCon, 4 Feb. 1949, 812.6363/2-449; Eakens MemCon, 2 Mar. 1949, 812.6363/3-249; Reveley to Thurston, 16 Mar. 1949, 812.6363/3-1649.

44. Eakens MemCons, 3 and 5 Jan. 1949, 812.6363/1-349; Eakens MemCon, 3 Feb. 1949, 812.6363/2-349; Hoffman MemCon, 14 Jan. 1949, 812.6363/1-1449; Thurston to SecState, 6 Mar. 1949, 812.6363/3-649; Eakens MemCon, 15 Mar. 1949, 812.6363/3-1549; Meyer 1974, 151–152.

45. Eakens MemCon, 15 Mar. 1949, 812.6363/3-1549; Reveley to Thurston, 27 Jan. 1949, 812.6363/1-2749; Eakens MemCon, 25 Feb. 1949, 812.6363/2-2549; Reveley to Thurston, 23 Mar. 1949, 812.6363/3-2349; Thorp to Steelman, 25 Mar. 1949, 812.6363/3-2549; *FR* 1949 2: 671–673; Hoffman, "PEMEX Request for Eximbank Loans," 12 Aug. 1949, 812.51/8-1549. On developments in Canada, see *New Horizons*, 723–726; and Halford L. Hoskins, *Middle East Oil in United States Foreign Policy*, 37–40.

46. Petroleum Policy Committee Minutes, 12 Jan. 1949, ECEFP Records, "PPC Minutes 1–38;" Armed Services Petroleum Board, "The Mexican Oil Situation," 15 Mar. 1949, OSD Records, CD 7-1-2; Eakens, "PEMEX Loan Proposal," PPC D-11, OSD Records, CD 7-1-2; Mann, "Memo on Venezuelan Oil Policy," 7 Mar. 1949, 812.6363/3-749; *FR* 1949 2: 673–674.

47. Petroleum Policy Committee Minutes, 20 May 1949, PPC D-11, OSD Records, CD 7-1-2; *FR* 1949 2: 674–675. On the "era of good feeling," see Cline 1953, 312–317. Senator Truman had visited Mexico City in 1939 as a guest of the Mexican Congress. On U.S. investment, see Newfarmer and Mueller 1975, 47–49.

48. See the undated paper marked "For Reber: Petroleum Loan to Mexico," 812.6363/5-2749; Webb to Johnson, 8 June 1949, OSD Records, CD 7-1-2; "Department of State Position on PEMEX Loan Proposal," 10 June 1949, OSD Records, CD 7-1-2; "National Military Establishment Position on Oil Loan to Mexico," 14 June 1949; and Carpenter Memo of Meeting on Mexican Oil, 17 June 1949; both in OSD Records, CD 7-1-2; Eakens Memo of Meeting on PEMEX Loan Problem, 13 June 1949, 812.6363/6-1349.

49. Thorp to Webb, 15 June 1949, Truman Papers, PSF: Mexico, Box 130; Thorp to Webb, "Loans to PEMEX," 24 June 1949, 812.51/7-2649.

50. Thorp to Webb, "Protection of the Interests of American Individuals and Companies Having Unsettled Petroleum Matters Pending in Mexico," 29 June

1949; and Thorp to Webb, Background Information on the Sabalo Case, 29 June 1949; both in 812.6363/6-2949; Hoffman MemCon, 13 Apr. 1949, 812.6363/4-1349; Thurston to Alemán, 6 June 1949, Truman Papers, OF 212-A (1950), Box 802; Hermida Ruiz 1974, 155, 161–162, 166. Hermida Ruiz reproduces many Sullivan and Cromwell memorandums.

51. *FR* 1949 2: 675–677. Truman had approved the Aide Mémoire, see Acheson MemCon, 30 June 1949, RG 59, Office of the Executive Secretariat, Box 3, "MemCons with the President, 1949" (hereafter ExecSec Records).

52. Reveley to Thurston, 6 July 1949, 812.6363/7-649.

53. *FR* 1949 2: 683–685, 699–690.

54. Martin MemCon, 16 Aug. 1949, 812.51/8-1649; *FR* 1949 2: 686–687; Charles A. Wolverton, "The Case of Mexican Oil Development," *Congressional Record* (26 Aug. 1949). The State Department defended Miller; see Acheson Memo for the President, 8 Aug. 1949, Truman Papers, OF 212-A, Box 802, enclosing a letter from Acheson to Congressman Crosser, dated 8 Aug. 1949. Acheson did not tell the president, or indicate to Crosser, that the letter was drafted by Miller himself; see 812.51/8-449. On Sullivan and Cromwell's dealings with the Mexican government see the memorandums sent to Truman by Senator Wayne Morse, Truman Papers, OF 212-A (1950), Box 802; see also, Sigurd Scholle (vice-president, Southeastern Oil Company), "Mexico at Crossroads of Her Petroleum Affairs," *Oil Forum* 3 (Aug. 1949): 325–328.

55. Acheson to Steelman, 26 Aug. 1949, 812.51/8-2649; Discussion on Mexican Oil, n.d., Truman Papers, PSF: Mexico, Box 130; Truman to Acheson, 29 Aug. 1949, Acheson Papers, MemCons, Box 64, "August-September 1949."

56. Reveley to Miller, 6 Sept. 1949, Truman Papers, PSF: Mexico, Box 130; *FR* 1949 2: 687–690; Barber to SecState, 21 Oct. 1949, 812.6363/10-2549.

57. Draft Memo for the President, 8 Nov. 1949, 812.51/11-849; *FR* 1949 2: 693 n.; Barber to the Acting SecState, 9 Nov. 1949; and Thorp to the Acting SecState, 9 Nov. 1949; both in 812.51/11-949.

58. *FR* 1949 2: 694–696.

59. Wellman MemCon, 17 Oct. 1949, 812.6363/10-2549; *FR* 1949 2: 698–700; Eakens to Brown, 6 Dec. 1949, 812.6363/12-649; Thorp to Webb, 27 Dec. 1949, 812.6363/12-2749; Bermúdez 1963, 43–46, 223–225; "Output vs. Demand Race in Mexico Intensifies," *Oil Forum* 4 (July 1950): 266, 276; Koppes 1980.

60. *FR* 1949 2: 696–698; Humelsine to Webb, 25 Nov. 1949, ExecSec Records, Box 3, "MemCons with the President, 1949"; *FR* 1950 2: 937–939.

61. Kane 1981, 64; *FR* 1950 2: 937 n. According to the editors of *FR*, Truman's handwritten note was attached to a memorandum from Acheson dated 23 Jan. 1950. On Truman's attitude toward business, see Griffith 1981, 303.

62. *FR* 1950 2: 939 n., 950–953; Kane 1981, 64–66; Eakens MemCon, 24 Feb. 1950, ECEFP Records, Box 70, "Misc."; Mann, "PEMEX Loan Negotiations"; "Meeting with Mexican Ambassador," 8 Mar. 1950; and Elsey Memo for Files, 8 Mar. 1950; all three in George M. Elsey Papers, Box 60, "Mexican Affairs," HSTL; Elsey to Murphy, 21 Apr. 1950, Truman Papers, PSF: Mexico, Box 130; Elsey Memo for Files, 27 Apr. 1950, Elsey Papers, Box 60, "Mexican Affairs," HSTL.

63. *FR* 1950 2: 951 n., 954 n., 953–954; Murphy Memo for the President, 14 June 1950, Truman Papers, PSF: Mexico, Box 130.

64. Ray to Mann, 7 Aug. 1950, Truman Papers, PSF: Mexico, Box 130; *FR* 1950 2: 959, 959 n.; Elsey Memo for Files, 9 Aug. 1950, Truman Papers, PSF: Mexico, Box 130; Kane 1981, 68–69.

Chapter 7 Securing the Middle East

Portions of this chapter first appeared in *Business History Review* 58 (Autumn 1984), and are reprinted here with the express permission of the editor of the *Business History Review.*

1. Keohane 1982b, 166–167. Keohane is referring to an earlier period but the point is still valid.

2. See the discussion of the sterling area in *New Horizons*, 695–697, 703, 708; see also Economic Cooperation Administration, *The Sterling Area: An American Analysis*; and Gardner 1956.

3. Block 1977, 76–83; Yergin 1977, 303–309; Alpert 1951, 288–310; Zupnick 1957, 39, 64, 85–86, 118–120; on Europe's economic situation in general, see Sanford 1980, 1–8.

4. *FR* 1947 3: 204–219; see also Jackson 1979; Paterson 1970, 78–96. Block 1977, 82–83.

5. Halford L. Hoskins, *Middle East Oil in United States Foreign Policy*, 41–43; Odell 1970, 108–109; Alpert 1951, 296–308, 387; Jensen 1967, 1–20; on Eastern Europe, see Chester 1983, 91–94; for the view of leading oil experts Wallace Pratt and Joseph Pogue, see Pogue to Robert McConnell, 14 Jan. 1948; and Pratt to McConnell, 23 Jan. 1948; both in NSRB Records, Office File of the Chairman, General Correspondence, "Fuel and Fuel Oil"; Report of the United States Committee on Petroleum, 5 Nov. 1947, PED Records, Box 22, "CEEC Reports;" Collisson to Cooper, 7 Jan. 1948, Interior Records, Cooper Papers, Box 15, "ERP 1948"; "Chapter G-Petroleum," 2 Jan. 1948, Interior Records, Collisson Papers, Box 14, "Office File"; *Forrestal Diaries*, 356–357; U.S. Congress, House, Committee on Interstate and Foreign Commerce, *Petroleum Study*, 400.

6. Mendershausen 1950, 8–11; Zupnick 1957, 119–120; Frank 1966, 29–36; *Oil Lift Hearings*, 104. The price of crude oil f.o.b. Persian Gulf (36° API), though varying slightly from company to company, had risen from $1.05 a barrel in 1945 to $2.22 per barrel by the spring of 1948 when ECA began operations.

7. Mutual Security Agency, "ECA and MSA Relations with International Oil Companies Concerning Petroleum Prices," 15 Aug. 1952, in U.S. Congress, Senate, Select Committee on Small Business, *The Impact of Monopoly and Cartel Practices on Small Business*, 140 (hereafter cited as "ECA Price History"); *Petroleum Study*, 148, 409–411; Walter Levy, "Petroleum under the ECA Program," ECA Records, 53 A 441, Mutual Security Agency, Office of Administration, Box 265, "Petroleum, Procurement Outside the United States"; and *Oil Forum* 2 (Oct. 1948): 397–398. See also the assessment in Katzenstein 1976, 40–41. On Hoffman, see Sanford 1980, 131–133, 231–233.

8. "ECA Price History," 140, 150; *International Petroleum Cartel*, 21–33. The oil shortage quickly turned into a surplus in 1949; see the discussion in chapter 5.

9. "ECA Price History," 140–141; George W. Stocking, "Pricing Middle East Crude Oil, " 10 Sept. 1949, ECA Records, 53 A 405, Box 1, "Oil, Petroleum,

etc." On ECA's problems with the oil companies over the price of oil, see the detailed discussion in Painter 1984, 364–371.

10. "ECA Price History," 141, 147–148; Penrose 1968, 185–188. The profits of producing companies, such as ARAMCO, did fall because of lower prices and decreased sales; see Maull 1980, 211.

11. U.S. Congress, House, Select Committee on Small Business, *Effects of Foreign Oil Imports on Independent Domestic Producers*, 523, 529; Minutes, Petroleum Policy Committee, 9 Feb. 1949, ECEFP Records, Box 70, "PPC M–15"; E. Groen, "The Significance of the Marshall Plan for the Petroleum Industry in Europe—Historical Review of the Period 1947–1950," in U.S. Congress, Senate and House, Select Committees on Small Business, *Report on the Third World Petroleum Congress*, 50–55. "Up from the Ashes," *The Lamp* 33 (Sept. 1951): 2–5, discusses the rehabilitation of European refineries; see also Katzenstein 1976, 40–41.

12. Working Group on Britain (WGB), "United States Petroleum Policy and the United Kingdom," 23 Aug. 1949, PED Records, Box 2, "Near East Oil"; "The Sterling Dollar Oil Problem," 27 Apr. 1949, 800.6363/4-2749; Nitze to SecState, 12 May 1949, 841.6363/5-1249; see also "British and U.S. Battle for Oil Markets," *Business Week* (Nov. 19, 1949): 125; and "Oil Supply—and Demand," *Economist* 157 (Oct. 8, 1949): 797–798.

13. Minutes, Petroleum Policy Committee, 13 Apr. 1949, ECEFP Records, "PPC Minutes, 1948–49;" Nitze to SecState, 27 Apr. 1949; and "The Sterling Dollar Oil Problem," 27 Apr. 1949; both in 800.6363/4-2749; "United States Trade in Crude Oil and Petroleum Products," 12 Aug. 1949, 811.6363/8-1249; Eugene Holman (president of Jersey), "Imported Oil Flood Unsound and Foolhardy," *Oil Forum* 3 (July 1949): 289–292; Ernest O. Thompson (head of the Texas Railroad Commission), "Middle East Oil Presents Problems for U.S. Producers," *Oil Forum* 3 (Mar. 1949): 123–126; and Thompson, "Oil Conservation Progress in Texas during 1949," *Oil Forum* 3 (Nov. 1949): 487–488; FR 1950 5: 33; Adelman 1972, 150–151; Vietor 1984, 96–99. See also the numerous exchanges between the State Department and Congressmen from oil states in the 811.6363 file for 1949; and two articles by the editor of *Oil Forum*, T. Orchard Lisle, "Import-Export Oil Divides American Oil Industry," *Oil Forum* 3 (Mar. 1949): 115–118; and (Oct. 1949): 412–414, 431.

14. On Levy's background, see the introduction to his collected essays, Levy 1982. Levy, "Statement of Issues Relating to Dollar and Sterling Oil As They May Be Raised by American Oil Companies," 14 Mar. 1949, ECA Records, 53 A 405, Box 1, "Oil, Petroleum, etc."; *Effects of Oil Imports*, 529–531; Groen, "Significance of the Marshall Plan," 40–46, 56; "Chapter G-Petroleum," 2 Jan. 1948, Interior Records, Collisson Papers, Box 14, "Office File."

15. On the divisions within ECA, see Henderson to Levy, 17 Mar. 1949; ECA Industry Division, "Potential Impact of the Petroleum Expansion Plans of Participating Countries on United States Security Interests and United States Foreign Commercial Policy," 6 May 1949; and M.E. Locker, "The Application of ECA Policy to the Petroleum Program," 11 May 1949; all in ECA Records, 53 A 441, Office of the Controller, Box 7, "Petroleum"; Ad Hoc Committee on the Problem of Dollar-Sterling Oil, 3 Mar. 1949, ECA Records, 53 A 405, Box 1, "Oil,

Petroleum, etc."; B. B. Biggs (executive officer, ANPB), "Meeting at ECA," 3 Mar. 1949; and Biggs, "Implications of ECA Equipment Policy on Future Position of Dollar versus Sterling Oil," 14 Mar. 1949; both in ANPB Records, ANPB 38, Box 8, F. 4.

16. Funkhouser to SecState, 5 Apr. 1949, 890.6363/4-549; Nitze MemCon, 9 Apr. 1949, 800.6363/4-949; Nitze to SecState, 12 May 1949, 841.6363/5-1249; WGB, "U.S. Petroleum Policy and the U.K.," 23 Aug. 1949, PED Records, Box 2, "Near East Oil."

17. WGB, "U.S. Petroleum Policy and the U.K.," 23 Aug. 1949, cited in note 16; Biggs to Levy, 14 Mar. 1949, ANPB Records, ANPB 38, Box 8, F.4; Biggs to SecDef, 2 May 1949, OSD Records, CD-7-1-9; Biggs statement, 12 Sept. 1949, ECEFP Records, "PPC Documents 11-19/4c."

18. See the table in *Petroleum Study*, 108–109; *Effects of Oil Imports*, 531–532; Mendershausen 1950, 13; Anderson to Hoffman et al., 23 May 1949, ECA Records 53 A 441, Box 87; "Statement on Estimated European Consumption in Fiscal 1953 and on Refining Capacity as Planned by the European Countries," attached to Anderson to Hoffman et al., 23 May 1949; Anderson to Hoffman et al., 13 June 1949, ECA Records, 53 A 405, Box 1. (I wish to thank Ethan Kapstein for drawing my attention to this document.)

19. Ad Hoc Committee on the Problem of Dollar-Sterling Oil, 3 Mar. 1949, ECA Records, 53 A 405, Box 1, "Oil, Petroleum, etc."; *Effects of Oil Imports*, 532; Levy testimony, *Petroleum Study*, 394-441; Hoffman testimony, *Petroleum Study*, 150–151; "World Oil in Turmoil," *Fortune* (Feb. 1950): 111; *NYT*, 21 Oct. 1949.

20. See the comments by Emilio Collado of Jersey in Record of Action, Investment Panel, Advisory Committee on Overseas Territories, 25 Oct. 1949, William L. Clayton Papers, HSTL. (I wish to thank William Burr for drawing my attention to this document.) *New Horizons*, 697, 703; Mendershausen 1950, 7–9; "British and U.S. Battle for Oil Markets," *Business Week* (19 Nov. 1949): 125.

21. "The Sterling Oil Problem," 27 Apr. 1949, 800.6363/4-2749; "Discrimination by the United Kingdom against Oil Operations of United States Companies," 26 Sept. 1949, RG 353, Records of the Interdepartmental and Intradepartmental Committees of the Department of State, Joint Petroleum Discussions, Box 83, "JPD Documents 7–12" (hereafter JPD Records); *New Horizons*, 705–706; Charles Allen Neal, "Cause and Effect of Argentine Oil Deal As Seen through British Eyes," *Oil Forum* 3 (Aug. 1949): 335–336, 347, and (Sept. 1949): 386–387.

22. Holman to Hoffman, 9 Mar. 1949, and 9 Sept. 1949, ECA Records 53 A 405, Box 1; Socony-Vacuum, "United States Oil Company Operations Abroad As Affected by the Sterling Dollar Position and the OEEC Program," 13 May 1949, ECA Records 53 A 405, Box 1; *Oil Forum* 4 (Sept. 1950): 337–338.

23. WGB, "U.S. Petroleum Policy and the U.K.," 23 Aug. 1949; and Funkhouser to Childs, 23 Sept. 1949; both in PED Records, Box 2, "Near East Oil"; Eakens to Beale, 2 Sept. 1949, 841.6363/5-1849.

24. "The Dollar Crisis," *The Round Table* 39 (Sept. 1949): 303–308; "United Kingdom: The Economic Crisis," *The Round Table* 39 (Sept. 1949): 344–346;

Block 1977, 94–95; "Devaluation of Sterling: The Decision and its Consequences," *The Round Table* 40 (Dec. 1949): 8–14; "Report on Sterling: Before and after Devaluation," *The Round Table* 40 (Sept. 1950): 308–319.

25. Block 1977, 94–95; Under Secretary's Meeting, "The United Kingdom Economic Crisis—Factual Background and Possible Remedies," 17 Aug. 1949, RG 59, Office of the Executive Secretariat, Box 9, Minutes, Memos 2/3/49–12/30/49; Mendershausen 1950, 20; Funkhouser to Childs, 23 Sept. 1949, PED Records, Box 2, "Near East Oil"; "The Significance of Petroleum in the British Balance of Payments," 13 Sept. 1949, ECEFP Records, Box 69, "PPC Documents."

26. "The Caltex Plan," 9 Sept. 1949, 841.6363/9-1449; Jersey's plan, dated 12 Aug. 1949, is in JPD Records, "JPD Documents 1–6"; the Socony plan was sent to the State Department on 19 Sept. 1949, 811.6363/9-1949. See also the discussion in Keohane 1982a, 59–60.

27. Eakens to Labouisse, 20 Oct. 1949, 800.6363/10-2049; Acheson to US Embassy London, 2 Nov. 1949, 841.6363/9-2849; U.S. Working Group, U.S.-U.K.-Canada Petroleum Discussions, "ECA Proposal for a U.S. Position on Petroleum," 26 Sept. 1949, PED Records, Box 2, "Near East Oil"; "Position Paper Setting Forth Policy for Talks on Petroleum Matters," 18 Nov. 1949, JPD Records, "JPD Documents 1–6"; Acheson to U.S. Embassy London, 21 Nov. 1949, 841.6363/11-1649. Keohane 1982a, 60, points out that the U.S. position was basically the Caltex plan.

28. George A. Eddy, "Personal Comments on Sterling Oil Competition," 25 Oct. 1949; and Eddy to Labouisse, 16 Dec. 1949; both in 800.6363/12-1649; see also Eakens MemCon, 13 May 1949, 811.6363/5-1349.

29. Webb to U.S. Embassy London, 25 Nov. 1949, 841.6363/11-2549; Moline, "Meeting on Displacement of Dollar Oil in Sterling Area by Surplus British Production," 28 Nov. 1949, 841.6363/11-2849; U.S. Embassy London to SecState, 29 Nov. 1949, 841.6363/11-2849; Moline MemCon, 2 Dec. 1949, 841.6363/12-249; Holmes to SecState, 16 Dec. 1949, 840.50 Recovery/12-1649; FR 1950 5: 9–10; "Saving on Dollar Oil," *Economist* 157 (24 Dec. 1949): 1424–1425; NYT, 25 Dec. 1949 and 4 Jan. 1950.

30. "Memo on Trend of Middle East and Venezuelan Crude Production," 28 Feb. 1950, NSRB Records, Entry 31, J2-14, "Crude Oil"; "Working Paper, Near East Conference, Petroleum," 20 Dec. 1949, PED Records, Box 2,"Near East Oil"; Anwalt MemCon, 27 Dec. 1949, 890F.6363/12-2749.

31. "Dollars, Sterling, and Oil," *The Lamp* 32 (Mar. 1950): 2; *Petroleum Study*, 222.

32. Mendershausen, "Dollar Shortage," 3; FR 1950 5: 9–10; Chester 1983, 97; NYT, 31 Jan., 1 and 2 Feb., and 7 Mar. 1950.

33. FR 1950 5: 34–35; see also Deimel to Labouisse, 3 Oct. 1949; and "Working Paper, Near East Conference, Petroleum," 20 Dec. 1949; both in PED Records, Box 2, "Near East Oil."

34. Mendershausen 1950, 17–19; *Petroleum Study*, 222–224; FR 1950 5: 34–35; *Oil Forum* 4 (Jan. 1950): 12–14, 50; *New Horizons*, 708; NYT 9 Feb., 30 and 31 Mar., and 5 Apr. 1950.

35. Mendershausen 1950, 26–32; *New Horizons*, 710–713; Oil and Gas Division, Department of the Interior, "Paper Dealing with Petroleum Imports and

Exports," 19 July 1950, Interior Records, Minerals and Fuels Division, "Oil and Gas Division"; *NYT*, 28 Apr. 1950. A key element in the settlement was an end to gasoline rationing in Britain; see *Gaitskell Diary*, 184–185.

36. This account of the fifty-fifty decision, though agreeing in detail with the account in Anderson 1981, differs somewhat in emphasis. Rather than focusing on whether the decision to grant ARAMCO a tax credit was "correct," this account views the decision as indicative of the constraints placed on U.S. action by a corporatist foreign oil policy.

37. On Getty and Aminol, see Shwadran 1959, 392–393; *Oil Forum* 1 (Sept. 1947): 228–229, 238; and *Oil Forum* 2 (Aug. 1948): 34.

38. Hill to SecState, 15 Aug. 1949, 890F.6363/8-1549; see also the testimony of Fred Davies, chairman of the board of ARAMCO, in *Oil Lift Hearings*, 1428; and Anderson 1981, 188–189.

39. Sanger MemCon, 20 July 1949; and Mattison MemCon, 22 July 1949; both in 890F.6363/7-2049. A letter was sent to the Treasury Department on July 28, 890F.6363/7-2049. See also the testimony of Treasury official George Eddy in *Oil Lift Hearings*, 1443–1446. Eddy conferred with the U.S. Embassy before discussing the matter with the Saudis.

40. Graham to McGhee, 2 Sept. 1949, 890F.6363/9-249. Anderson 1981, who otherwise provides a full discussion of the fifty-fifty issue, does not mention this memorandum, which contradicts his argument that the decision to grant ARAMCO a tax credit was "routine."

41. Davies's testimony, *Oil Lift Hearings*, 1431; FR 1950 5: 52 n., 52, 56–57, 59–60; Sanger MemCon, 20 July 1949, 890F.6363/7-2049; Stocking 1970, 144–146.

42. FR 1950 5: 62–68, 75–76.

43. Ibid., 76–96. See the comments on this paper in *Multinational Hearings*, part 7: 140–149.

44. FR 1950 5: 76–96.

45. See the summary of the September 11 meeting in *MNC Documents*, 341–345.

46. FR 1950 5: 106 (the editors of FR give November 6 as the date of this document, but other documents indicate that the meeting took place November 3); FR 1950 5: 111; see McGhee's testimony in *Multinational Hearings*, part 4: 83–99 (the quotation is from page 87); and Anderson 1981, 194–196, who disputes the charge of NSC involvement, but incorrectly attributes it to Senator Frank Church, rather than to McGhee.

47. *MNC Documents*, 345–348. The Saudi decree is printed in *MNC Documents*, 374–377.

48. FR 1950 5: 118 n., 118–121; FR 1951 5: 276–278. The decree is printed in *MNC Documents*, 377–378; the December agreement is in *MNC Documents*, 372–374. As a 1957 report by the staff of the Joint Congressional Committee on Internal Revenue noted, ARAMCO was the only oil company operating in the country at the time of the decree; see *MNC Documents*, 350–358.

49. *Multinationals Report*, 85; the IRS ruling is printed in *MNC Documents*, 358.

50. See McGhee's testimony in *Multinational Hearings*, part 4: 83–99.

Chapter 8 The Limits of Corporatist Foreign Policy

1. "The Progressive Threat to the U.S. Position in Foreign Oil," 25 July 1952, Oscar Chapman Papers, Box 102, "Petroleum Cartel File," HSTL. The paper is unsigned, but appears to be an early draft of a paper by Levy entitled "Perils to the Free World's Essential Oil Supplies," which was sent to Chapman in September 1952; see Interior Records, Central Classified File, 1–188, part 33. On the impact on U.S. policy in the Middle East, see Eveland 1980; on the impact on U.S. policy in Latin America, see Immerman 1982.

2. See the various statements on U.S. policy in the Middle East in FR 1951 5: 6–11, 21–27, 50–76, 257–264.

3. On Iran's negotiations with AIOC, see Ferrier 1977, 104–107; Elwell-Sutton 1955, 80–87, 167–186; Ford 1954, 48–50; Iranian Documents, 13–15, 20–25; "Oil Nationalization in Iran and its Potential Impact on Other Oil Concessions in the Near East," 28 June 1951, OSS/SD Records, Report No. 5563; "Basic Factors Leading to the Nationalization of the Oil Industry in Iran," 6 Mar. 1952, OSS/SD Records, Rept. No. 5683; Ramazani 1975, 186–196; and DIA 1951, 471–475. On the composition of the National Front, see Cottam 1964, 264–268. There are several spellings of the Iranian leader's name. I have adopted Mossadegh as the most common.

4. FR 1950 5: 14–15, 97–99, 550–551, 569–570, 593–600, 634–635; FR 1951 5: 70–72; Ferrier 1977, 107–108; C. M. Woodhouse, Something Ventured, 107; George C. McGhee, Envoy to the Middle World, 319–324; DIA 1951, 475; CIA, "The Current Crisis in Iran," SE-3, 16 Mar. 1951, Truman Papers, PSF: National Security Council Meetings, No. 87. The British position is explained in detail in Louis 1984, 642–657.

5. Truman's views are recorded in Arthur Krock, Memoirs: Sixty Years on the Firing Line, 261–262; Acheson, Present at the Creation, 505–507; McLellan 1976, 387; McGhee, Envoy, 331–332; FR 1949 6: 545–551; see also Department of State to President Truman, 8 Nov. 1949, Truman Papers, PSF: Foreign Affairs, "Iran"; FR 1951 5: 268–276, 309–315; NSC 107, 14 Mar. 1951, NSC Documents; U.S. Congress, House, Committee on Foreign Affairs, Selected Executive Session Hearings of the Committee, 1951–56: Vol. 16: The Middle East, Africa, and Inter-American Affairs, 29, 61–62, 68, 77, 79 (hereafter House ExecSess Hearings); CIA, "Current Crisis in Iran," 16 Mar. 1951, Truman Papers, PSF: NSC Mtg. 87; NSC 107/1, 6 June 1951; and NSC 107/2, 27 June 1951; both in NSC Documents.

6. FR 1951 5: 268–276; "Oil Nationalization in Iran and Its Potential Impact on Other Oil Concessions in the Near East," 28 June 1951, OSS/SD Records, Report No. 5563. See also Krock, Memoirs, 261–262.

7. Louis 1984, 655–657; McGhee, Envoy, 325–336; House ExecSess Hearings, 30–31, 63–65; FR 1951 5: 291, 295, 296–297; DSB, 23 Apr. 1951, 657–660; Carrollton Press, Declassified Documents Reference Service, vol. R, Doc. 771 E; CIA, "Current Crisis in Iran," 16 Mar. 1951, Truman Papers, PSF: NSC Mtg. 87; Acheson quoted in McLellan 1976, 387. See also the early report on "Mossadeq As Leader of a Potential Popular Movement in Iran," 9 June 1950, OSS/SD Records, Report No. 5272; DIA 1951, 486–488; I.C.J. Pleadings, Anglo-Iranian Oil Co. Case (United Kingdom v. Iran), 8–19, 45–63, 64–125.

8. Acheson, *Present at the Creation*, 506; NSC 107/1, 6 June 1951, *NSC Documents*; "Chronology of Developments in AIOC Dispute," 28 Nov. 1951, OSS/SD Records, Report No. IR 5675.1; Louis 1984, 657–666.

9. NSC 107/1 Annex, 20 June 1951, *NSC Documents*; *FR 1951* 5: 309–315; *DSB*, 28 May 1951, 851; *DSB* 4 June 1951, 891–892; *DIA 1951*, 490–491; Louis 1984, 668–671.

10. *Iranian Documents*, 28–29; Grady to Mossadegh, 20 June 1951, Henry F. Grady Papers, Box 1, Iran, HSTL; Henry F. Grady, "Adventures in Diplomacy," (unpublished memoirs), 35–37, Grady Papers, Box 5, HSTL. Grady, who was replaced in September, was very critical of U.S. policy in Iran; see his articles, "The Real Story of Iran," *U.S. News and World Report* 30 (19 Oct. 1951): 13–17; and "What Went Wrong in Iran?" *Saturday Evening Post* 24 (5 Jan. 1952): 30, 56–58; "Chronology of AIOC Dispute," 28 Nov. 1951, OSS/SD Records, Report No. IR 5675.1; Acheson, *Present at the Creation*, 506–507; *DIA 1951*, 526–531; *ICJ Pleadings*, 715–716.

11. NSC 107/2, 27 June 1951, *NSC Documents*; *Multinationals Report*, 58 ; Acheson, *Present at the Creation*, 507–508; *DIA 1951*, 498–501; Morrison quoted in McLellan 1976, 388–389.

12. On Mossadegh, see Cottam 1964, 263, 269–270; *NYT* 28 Sept. 1952; and Acheson, *Present at the Creation*, 503–504. On the talks, see the reports from Harriman to Truman and Acheson, 17, 19 (3 telegrams), 24 and 25 July 1951, Truman Papers, PSF: Iran: Harriman (hereafter PSF: Harriman); and the colorful account in Vernon Walters, *Silent Missions*, 241–259; Walter Levy's account is in Middle East Institute, *The Present Situation in Iran: A Survey of Political and Economic Problems Confronting the Country* (Washington: Middle East Institute, 1953), 18–21.

13. The "Harriman formula" is printed in *DIA 1951*, 501. Basically, Harriman got the Iranians to agree that acceptance of the principle of nationalization referred to the general law of March and not the more detailed implementation law of April. McGhee to U.S. Embassy London, 24 July 1951; Harriman to Truman and Acheson (2 telegrams), 25 July 1951; Acheson to Harriman and Grady, 26 July 1951; Harriman to Truman and Acheson (2 telegrams), 27 July 1951; Acheson to U.S. Embassy London, 27 July 1951; Gifford to Acheson, 1 Aug. 1951; all in Truman Papers, PSF: Harriman. According to Louis 1984, 671–676, military intervention remained a possibility until the withdrawal from Abadan in September.

14. *DIA 1951*, 502–504; Harriman to Truman and Acheson, 3 Aug. 1951, Truman Papers, PSF: Harriman; Louis 1984, 678–681. See the assessments of the Stokes proposal by Shwadran 1959, 149; and Elwell-Sutton 1955, 251–252, who notes that "put very succinctly, the British attitude was that, in return for their recognizing the principle of nationalisation, the Persian government should forego its insistence on that principle."

15. See the reports by Harriman to Truman and Acheson, 13, 14, 16, 17, 18, and 19 Aug. 1951, Truman Papers, PSF: Harriman; *DIA 1951*, 504–506. Harriman admitted that the plan could result in "camouflage for the complete return of British control," unless adequate safeguards were included in the final agreement.

16. Harriman to Truman and Acheson, 19 Aug. 1951, Truman Papers, PSF: Harriman. The head of the Iranian delegation made an eloquent reply to Harriman; see *Iranian Documents*, 39–43.

17. Harriman to Truman and Acheson, 23, 24, and 28 Aug. 1951, PSF: Harriman, HSTL; "Chronology of AIOC Dispute," 28 Nov. 1951, OSS/SD Records, Report No. IR 5675.1.

18. Walden 1962, 75–76; "Chronology of AIOC Dispute," 28 Nov. 1951, OSS/SD Records, Report No. IR 5675.1; Acheson, *Present at the Creation*, 509–510. According to Louis 1984, 686–689, only concern over the U.S. reaction prevented the British from militarily intervening. On the British political situation, see *DIA 1951*, 506–519, 521–526, 531–540; House ExecSess Hearings, 86–87, 107–108.

19. *FR 1951* 5: 268–276; House ExecSess Hearings, 30, 74; Assistant Deputy Administrator, Foreign Petroleum Operations, Petroleum Administration for Defense, *Report to the Secretary of the Interior and Petroleum Administrator for Defense* (hereafter *PAD Report*); C. Stribling Snodgrass and Arthur Kuhl, "U.S. Petroleum's Response to the Iranian Shutdown." Snodgrass was the author of the *PAD Report*. See also the secondary account in Lubell 1963, 4–8.

20. *PAD Report*. The Petroleum Administration for Defense was the Korean War counterpart to the World War II Petroleum Administration for War. Like PAW, PAD operated closely with the oil industry; see Goodwin 1981, 107–114; and Bruce K. Brown, *Oil Men in Washington: An Informal Account of the Organization and Activities of the Petroleum Administration for Defense during the Korean War, 1950–1952*.

21. *PAD Report*; Burton Kaufman points out that the oil companies demanded and won a clause in the agreement stating that the agreement would not be construed to require them to take any actions in conflict with prior obligations; see Kaufman 1978, 43; Prewitt to Corwin Edwards, 24 July 1951, RG 122, Records of the Federal Trade Commission, General Records Section Files, 1914–55, Request 13; see also Goodwin 1981, 114–115.

22. "The Middle East after Abadan: Oil Production without Persia," *The Round Table* 44 (Dec. 1953): 32–34; Walden 1962, 99–100; *Present Situation in Iran*, 11–12. The process of helping AIOC replace Iranian oil was facilitated by the close links among the major oil companies.

23. Acheson, MemCon, 10 Oct. 1951, Acheson Papers, MemCons, Box 66, "October 1951," HSTL. The oilmen present were Eugene Holman (Jersey), R. G. Follis (SOCAL), Sidney A. Swensrud (Gulf), Brewster B. Jennings (Socony), and W.S.S. Rodgers (Texas).

24. "The Anglo-Iranian Problem," NSC 117, 18 Oct. 1951, Truman Papers, PSF: NSC Mtg. 125; see also JSC to SecDef, 10 Oct. 1951, JCS Records, CCS 1951–53, 092 Iran (4-23-48), sec. 4.

25. House ExecSess Hearings, 102–104, 110; State Department Briefing Memo, 22 Oct. 1951, Truman Papers, PSF: Iran; U.S. Congress, Senate, Committee on Foreign Relations, *Executive Sessions (Historical Series): Vol. 3: Eighty-second Congress, First Session, 1951*, 456–459 (hereafter Senate ExSess Hearings). For the context of State Department policy, see *DSB*, 25 Oct. 1951, 612–615; and McGhee, *Envoy*, 392–393.

26. *DIA, 1951*, 540–546 prints extracts from Mossadegh's speech before the United Nations. McGhee, *Envoy*, 390–404 details the negotiations with Mossadegh; see also Walters, *Silent Missions*, 259–263; Acheson, MemCon, 24 Oct.

1951, Acheson Papers, Box 66, "October 1951," HSTL; interview with George C. McGhee, 20 Feb. 1980; House ExecSess Hearings, 91–92.

27. Acheson, *Present at the Creation*, 510–511; see Penrose 1968, 25–52, for a discussion of internal pricing.

28. See the discussion in Louis 1984, 682–685.

29. Acheson to McGhee, 10 Nov. 1951; and Acheson to Webb, 14 Nov. 1951; both in Truman Papers, PSF: Iran; "The Iranian Situation," n.d., attached to Acheson to Webb, 14 Nov. 1951; Acheson, *Present at the Creation*, 510–511; Anthony Eden, *Full Circle: The Memoirs of Anthony Eden*, 198–202. Churchill is quoted in Acheson, MemCon, 7 Jan. 1952, Acheson Papers, MemCons, "January 1952," HSTL.

30. Steering Group on U.S.-U.K. Talks, "Iran," 5 Jan. 1952, Truman Papers, PSF: Papers Prepared for Middle East Problems; Acheson, MemCon, 7 Jan. 1952, Acheson Papers, MemCons, "January 1952," HSTL. The U.S. provided Iran $1.6 million in 1951 and $23.4 million in 1952. Most of the aid was Point IV and military assistance; see *DSB*, 15 Dec. 1952, 940–941. In contrast, the British received $300 million in financial assistance in early 1952 to help cover the dollar costs of replacing Iranian oil; see Acheson's testimony in Senate ExSess Hearings, vol. 4, 157–161.

31. Mason and Asher 1973, 598–610 (chapter written by Harold N. Graves, Jr.); Franks to FO, 13 Mar. 1952, FO 371/98686/EP15314/51. On U.S. and British views on the World Bank effort, see the documents in Fo 371/98608. (I wish to thank Nina J. Noring for sharing with me the results of her research in British Foreign Office records.) See also "World Bank and Persian Oil," *Economist* 162 (9 Feb. 1952): 328; *DIA 1952*, 337–338.

32. "Man of the Year," *Time*, 7 Jan. 1952, 18–21; "Mossadeq and the Current Iranian Elections," 10 Jan. 1952, OSS/SD Records, Report No. 5735; "Iran: An Estimate of Possible Political Developments," 16 May 1952, Report No. 5881, OSS/SD Records; JSC to SecDef 25 June 1952, JCS Records, CCS 1951–53 092 Iran (4-23-48); Franks to FO, 26 July 1952, FO 371/98691/EP15314/182; Middleton to FO, 28 July 1952, FO 371/98691/EP15314/189 and /190; Franks to FO, 29 and 31 (two) July 1952, FO 371/98691/EP15314/195, /198, 199; Acheson MemCons, 21 and 31 July 1952, Acheson Papers, MemCons, "July 1952," HSTL; Acheson, *Present at the Creation*, 679–680; House ExecSess Hearings, 119–120; *DIA 1952*, 338–341. See also "Persia Seeks a Master," *Economist* 164 (19 July 1952): 171–172; "Two Cheers for Dr Mossadegh," *Economist* 164 (26 July 1952): 213; Zabih 1982, 56–66; and Abrahamian 1982, 270–272.

33. Elwell-Sutton 1955, 282–283; FO to Tehran, 13 Sept. 1952, FO 371/98697/EP15314/333; *DIA 1952*, 341–352; "Mossadegh: Fanatic or Strategist," *Economist* 164 (6 Sept. 1952): 535–536; *NYT* 26 Sept. 1952. Acheson later admitted that the joint proposal had been a "psychological blunder"; Acheson MemCon, 24 Sept. 1952, Acheson Papers, MemCons, "September 1952," HSTL.

34. Franks to FO, 3 Oct. 1952, FO371/98700/EP15314/404; Acheson, *Present at the Creation*, 681–682. On the October 8 meeting, see *Multinationals Report*, 60–61; see also NSC 136, 6 Nov. 1952, *NSC Documents*.

35. For background on the cartel case, see *FR 1952–1954* 1: 1259–1289; and

Kaufman 1978, 19–37. On industry opposition, see *Multinationals Report*, 57–58; Duce to Chapman, 23 Oct. 1952, Chapman Papers, Oil Cartel File, HSTL; House ExecSess Hearings, 220–222; "Memo Re Oil Cartel Investigation," n.d., Truman Papers, PSF: Oil; "Oil Firms Face Hottest Fight Since 1911 Standard Oil Breakup," *Newsweek*, 1 Sept. 1952, 49–51; see also the material in the Spingarn Papers, "FTC: Attacks on Oil Cartel Report" file, HSTL.

36. Franks to FO, 9 Oct. 1952, FO371/98700/EP15314/429 mentions Acheson's antitrust concerns; *Consortium Documents*, 17–21; Kaufman 1978, 44–45.

37. *Consortium Documents*, 22–23.

38. *Consortium Documents*, 24–26; NSC 136, 6 Nov. 1952; and NSC 136/1, 20 Nov. 1952; *NSC Documents*; Acheson, *Present at the Creation*, 683–684.

39. Byroade MemCon, 20 Nov. 1952, Acheson Papers, MemCons, "November 1952," HSTL; Jebb to FO, 22 Nov. 1952, FO 371/98703/EP15314/693, contains the British version of the meeting; see also Acheson, *Present at the Creation*, 682–683. For the 6 December announcement, see *DSB*, 15 Dec. 1952, 946. Eden had just met with President-elect Eisenhower, who seemed to be more sympathetic to the British position; see the MemCon of the meeting dated 4 Dec. 1952, FO 371/98703/EP15314/503G; see also Steel to FO, 6 Dec. 1952, FO 371/98703/EP15314/504.

40. *Consortium Documents*, 26–28; Acheson, *Present at the Creation*, 684.

41. See the British summaries of Henderson's telegrams in FO 371/104606. See also *Present Situation in Iran*, 22–23; Acheson, *Present at the Creation*, 684–685; Transcript, Loy W. Henderson Oral History Interview, 14 June and 5 July 1973, HSTL; and Eden, *Full Circle*, 209–210.

42. "Persian Oil," 17 Jan. 1953 (summary of Henderson's meeting with Mossadegh 15 Jan.), FO 371/104609/EP1531/70; Elwell-Sutton 1955, 287–288; Shwadran 1959, 140–142; Eden, *Full Circle*, 210–211.

43. "Persian Oil," 17 Jan. 1953 FO 371/104609/EP1531/74; Makins to FO, 27 Jan. 1953, FO 371/104610/EP1531/114; "Persian Oil," 6 Mar. 1953, FO 371/104614/EP1531/196; House ExecSess Hearings, 159; Shwadran 1959, 140–142; Eden, *Full Circle*, 211–213; Transcript, Loy W. Henderson Interview, John Foster Dulles Oral History Project, Princeton University; "No More Talks With Mossadegh," *Economist* 166 (28 Mar. 1953): 856.

44. Kermit Roosevelt, *Countercoup: The Struggle for the Control of Iran*, 107–108, 110, 114–116, 119–135; Powers 1980, 437–440. On earlier British activity, see Woodhouse, *Something Ventured*, 110–124.

45. Dwight D. Eisenhower, *Mandate for Change*, 159–163; Makins to FO, 7 Mar. 1953, FO 371/104614/EP1531/197; Dulles 1963, 221, 234; Harter 1980, 16–18; Woodhouse, *Something Ventured*, 131 shares Henderson's assessment. See also Henderson's oral history interview cited in note 42; Roosevelt, *Countercoup*, 2–3, 11; Powers 1980, 437–440; Roy Melbourne, "America and Iran in Perspective: 1953 and 1980," 14–15. (Melbourne was head of the political section in the U.S. Embassy in Iran during and following the Mossadegh years.) See also Rubin 1980, 55–57. For different perspectives on the situation in Iran, see "Iran's Political and Economic Prospects," 9 Jan. 1953, OSS/SD Records, Report No. 6126; and, in general, Cottam 1964 and 1970.

46. "The *Rose Mary*'s Test Run," *Economist* 163 (21 June 1952): 800; "International Law for the *Rose Mary*," *Economist* 166 (7 Jan. 1953): 134; "Mossadegh's Dangerous Deal," *Economist* 166 (24 Jan. 1953): 191–192; Walden

1962, 84–85, 96–97, 100–102; *Present Situation in Iran*, 12–13; Elwell-Sutton 1955, 296–300. On Jones's visit to Iran in August and September 1952, see *NYT* 26 Aug. and 26 Sept. 1952. On concern over Jones's intentions, see "Extract from Record of Meeting Held in the State Department," 6 Mar. 1953; and Makins to FO, 7 Mar. 1953; FO 371/104614/EP1531/197. The United States was also concerned over the possiblity of independents moving large quantities of Iranian oil; see Makins to FO, 9 Mar. 1953, FO 371/104614/EP1531/200.

47. *DSB*, 20 July 1953, 74–76; Eisenhower, *Mandate for Change*, 161–162; House ExecSess Hearings, 147–148; the request was not unexpected, see "Persian Request for Aid," 7 Apr. 1953, FO 371/104615/EP1531/240; and Makins to FO, 8 May 1953, FO 371/104615/EP1531/253.

48. Cottam 1964, 224. See the U.S. memorandum dated 21 July 1953, FO 371/104569/EP1015/194; and the FO Minutes, 27 July 1953, FO 371/104569/EP1015/197; Zabih 1982, 97–115; "Mr. Mossadegh Sees It Through," *Economist* 168 (25 July 1953): 262–263; Ambrose 1981, 200–207. On Zahedi, see Zabih 1982, 21, 29, 77.

49. For contemporary accounts, see the U.S. memorandum dated 19 Aug. 1953, FO 371/104570/EP1015/213; FO Minutes, 20 Aug. 1953, FO 371/104570/EP1015/215; and FO Brief for the Cabinet, 25 Aug. 1953, FO 371/104570/EP1015/224, which carefully avoids mention of outside assistance to the shah's supporters. Kwitny 1984, 164–177, prints extensive extracts from an eyewitness account by *New York Times* correspondent Kennett Love. See also Eisenhower, *Mandate for Change*, 163–164; Roosevelt, *Countercoup*, 169–197; Melbourne, "American and Iran," 14–16; and *Present Situation in Iran*, 3–7 and 23–26. Secondary accounts include Elwell-Sutton 1955, 307–313; Rubin 1980, 83–86; Zabih 1982, 116–126; Cottam 1970, 13–15; and Ambrose 1981, 207–214.

50. In his memoirs, President Eisenhower refers to "our representatives on the spot who were working actively with the Shah's supporters"; *Mandate for Change*, 164. Roosevelt, *Countercoup*, 199; Woodhouse, *Something Ventured*, 130; for Love's assessment, see Kwitny 1984, 164. In contrast, Cottam 1964, 229, argues that Mossadegh "could not have been overthrown if significant elements of the population had not lost faith in his leadership." See also Zabih 1982, 139–143, who details British involvement.

51. NSC 175, 21 Dec. 1953; NSC 175 Annex, 21 Dec. 1952; both in *NSC Documents*; *DIA 1953*, 354–355.

52. Eisenhower, *Mandate for Change*, 166; NSC 175 Annex, 21 Dec. 1953, *NSC Documents*; *Present Situation in Iran*, 15–17; Howard C. Gary, "Iran and the Future of Oil," 1–2; "Middle East After Abadan," 34–35.

53. NSC 175 Annex, 21 Dec. 1953, *NSC Documents*; *Consortium Documents*, 49–50, 54–55; *Multinationals Report*, 66. Walter Levy had recommended the creation of a multinational consortium in the summer of 1951; see Louis 1984, 677 n.

54. On Truman's decision, see *FR 1952–1954* I: 1338–1345; *Consortium Documents*, 33; Engler 1961, 213–216; see also the accounts in Kaufman 1978, 45–47, 145–147; and Goodwin 1981, 126–127.

55. Kaufman 1978, 55; *Consortium Documents*, 33–46; *FR 1952–1954* I: 1346–1351.

56. *Consortium Documents*, 54–55; NSC 175 Annex, 21 Dec. 1953, *NSC Documents*.

57. L.A.C. Fry, "Persia," 31 Oct. 1953, FO 371/104585/EP1051/40; *Consortium Documents*, 56–61, 77–79; NSC 175 Annex, 21 Dec. 1953, *NSC Documents*.

58. *Consortium Documents*, 60–76.

59. Ibid., 73–76; *Multinationals Report*, 72.

60. *Consortium Documents*, 76–77, 82–83; *DSB*, 27 Dec. 1954, 985–986; Dulles Telephone Conversation (TelCon) with Secretary Wilson, 17 Mar. 1954; Dulles TelCon with Secretary Humphrey, 23 Mar. 1954; John Foster Dulles Telephone Transcripts, Dwight D. Eisenhower Library/Princeton University (hereafter Dulles Telephone Transcripts); Eden, *Full Circle*, 218.

61. Dulles TelCon with Secretary Wilson, 17 Mar. 1954; Dulles TelCon with Secretary Humphrey, 22 Mar. 1954, 24 Mar. 1954; Dulles Telephone Transcripts; Shwadran 1959, 183; Eden, *Full Circle*, 218; "The Persian Settlement: Its Significance for the Middle East," *The Round Table* 44 (Sept. 1954): 331.

62. The formula used to determine production levels insured that production would be kept at levels compatible with the majors' interests. The presence of all seven majors in one company facilitated the majors' control of overall supply conditions, and hence the price of oil in world markets; see Blair 1976, 103–108. See the colorful account of the negotiations in Mosley 1973, 223–228. See also Shwadran 1959, 183–186; Walden 1962, 111–112; *DIA 1954*, 95–116.

63. Elwell-Sutton 1955, 315–327; Ramazani 1975, 269–272; Saikal 1980, 48–51, 97–100.

64. *Consortium Documents*, 89–94; Kaufman 1978, 59–60; on the fate of the suit, see 64–114.

65. These were only the costs to the United States. The costs to Iran were, and continue to be, high. Cottam 1970, 3, argues that "American intervention altered Iranian historical trends so fundamentally that the impact may well be felt for generations."

Conclusion: The Political Economy of Foreign Oil

1. Michael J. Hogan has argued that this "substantial similarity" betweeen the arrangements arrived at in the 1920s and those of the 1940s–1950s points to the high degree of continuity in U.S. economic diplomacy in the twentieth century; letter to the author, 15 Mar. 1985; see also Hogan 1984.

2. On the major's control of the world oil economy in the early 1950s, see *International Petroleum Cartel*, 21–33.

3. On the changing structure of the world oil economy, see Jacoby 1974; regarding the oil import program, see the recent accounts in Barber 1981 and Vietor 1984, 91–145.

4. Commoner 1976 contains information on both consumption and reserves; on consumption, see Darmstadter 1972; and Darmstadter and Landsberg, 1976; on the shift in investment, see Gols 1962. Snell 1974 argues that General Motors's control of automobile, bus, truck, and locomotive production was a major factor in the displacement of rail and bus transportation by cars and trucks. This displacement was an important factor in the rise of oil consumption. On the "Snell thesis," see St. Clair 1981; and Yago 1984.

5. On the origins and development of OPEC, see Seymour 1980; Danielsen 1982; Schneider 1983; and U.S. Department of State, Bureau of Public Affairs,

Office of the Historian, "OPEC and the Changing Structure of the World Oil Economy, 1960–1983." Anderson and Boyd 1983, is an interesting, but flawed, attempt to understand the decisions and events that undermined U.S. control of the world oil economy.

6. On the ideological orientation of U.S. business, see the stimulating remarks by Vogel 1978; and the detailed study by Sutton et al. 1956. On the importance of understanding the economic context of business-government relations, see Becker 1982, viii; and Vietor 1984, 9. The divisions in the oil industry were paralleled by similar divisions in the minerals industry; see Eckes 1979 and Randall 1980.

7. For a different view, see Gaddis 1983, 174; and Vietor 1984, 11.

8. For a similar argument in the context of business divisions over the Employment Act of 1946, see Griffith 1984.

9. On the divisions within the U.S. government, see the provocative interpretation in Schurmann 1974, part 1.

10. Brody 1975 stresses the conservative nature of the U.S. mobilization effort in World War II.

11. Koppes 1982, 263.

12. I wish to thank Michael Hogan for helping me clarify this point. On the Justice Department's position, see *Consortium Documents*, 29–33.

13. Kaufman 1978, 3–18; and Kaufman 1981.

14. See the discussion in Andrews 1975 and Katzenstein 1976.

15. The term is from Lindblom 1977, 187–205. However, this observation is based upon careful examination of a wide range of studies on the political economy of the United States in the twentieth century.

16. Bachrach and Baratz 1963 and 1969; see also Lukes 1974.

17. The quotation is from Temkin 1983, 454; see also Hacker 1969 and Connolly 1969.

18. Graham 1974 and 1976 discuss the obstacles to public planning; see also Lindberg 1977.

19. Maier 1977; Williams 1962 argues that such decisions to look abroad for solutions to the imbalance between consumption and reserves, rather than to confront the problem at home, have been characteristic of U.S. public policy since the turn of the century.

20. Lindberg 1977, 326–328.

Appendix 1: Closing the Door

1. *MNC Documents*, 176–178; Anderson 1981, 179–180.

2. *Multinationals Report*, 78; *MNC Documents*, 167–168.

3. *MNC Documents*, 195–205. Anderson 1981, 180–181, considers "coat-tail riding" the only important issue.

4. *MNC Documents*, 251–253.

5. Ibid., 95–96, 191–192, 251–253. See the explanation in Adelman 1972, 83. Anderson 1981, 183, incorrectly attributes the charge of curtailment of competition to the Church Committee report rather than to Caltex.

6. *MNC Documents*, 191–192, 215–216.

7. Ibid.

8. Ibid., 95–96, 178–179, 195–205; *Multinationals Report*, 77–78.

9. *MNC Documents*, 210–211, 221–227, 235, 237–239, 241; *Multinationals Report*, 79.

10. *MNC Documents*, 244, 247–249, 271, 274–276, 278; *Multinationals Report*, 80. In January 1948, the ARAMCO board had raised the price to $1.30 a barrel; *MNC Documents*, 246; 36° API Midcontinent crude oil rose from $1.87 a barrel in March 1947 to $2.57 a barrel in December 1947; see *Oil Lift Hearings*, 104.

11. *MNC Documents*, 223–227, 265–266, 273–277; 278–280. See chapter 5 for a discussion of the Palestine issue and chapter 7 for a discussion of oil and the Marshall Plan.

Glossary of Terms Relating to the Petroleum Industry

AFFILIATE. A company whose relation to another company is basically similar to, but looser than, the relationship of parent and subsidiary. One company is affiliated with another if there is a bond of common ownership between them, directly or through or by another company, and if the operations of the two are subject to some common management or planning.

ALLOWABLE. The amount of oil or gas a well is allowed to produce under state conservation regulations.

API GRAVITY. The standardized specific gravity of crude oil and refined products established by the American Petroleum Institute (API), used worldwide. Specific gravity is relative density, a comparison of the density of a volume of oil to the density of the same volume of water. The gravity of water on the API scale is 10°. Very heavy crude oil is 11° or 12°, very light crude is 40°; gasoline is about 60°. The higher the API gravity, the lighter the crude. High gravity crudes are generally considered more valuable.

BARREL. The standard measure in the U.S. petroleum industry; equal to 42 U.S. gallons, 35 Imperial (British) gallons, or between 280 and 380 pounds, depending upon API gravity.

BOOK VALUE. The value of capital stock as indicated by the excess of assets over liabilities; the value of equipment without regard to the actual or potential value of the product made by the equipment.

C.I.F. The initials refer to cost, insurance, and freight. The price includes in a lump sum the cost of the goods and the insurance and freight to the named destination.

CRUDE. Liquid petroleum in its natural state, before refining or processing. Crude oils range from very light (high in gasoline) to very heavy (high in residual

Sources: U.S. Congress, Senate, Committee on Foreign Relations, Subcommittee on Multinational Corporations, *Glossary of Terms Relating to the Petroleum Industry*, Committee Print (Washington, D.C.: U.S. Government Printing Office, 1974); Robert O. Anderson, *Fundamentals of the Petroleum Industry* (Norman: University of Oklahoma Press, 1984), 338–349.

oils). Sour crude is high in sulfur content. Sweet crude is low in sulfur and therefore usually more valuable.

DOWNSTREAM. Oil is said to flow downstream from well head to gas pump. All operations taking place after crude oil is produced are considered downstream activities.

F.O.B. "Free on Board" indicates that the price includes loading aboard ship but does not include the cost of insurance or freight to its intended destination.

HOT OIL. Oil produced in excess of the amount permitted by state regulatory agencies.

INDEPENDENT. In international oil, an independent is any company other than one of the seven or eight international majors. In the United States, an independent is usually one of the smaller nonintegrated companies involved in only one phase of the oil industry. In economic terms, an independent is a firm whose activities are not large enough to affect the market.

INTEGRATED COMPANY. An oil company that is involved in more than one of the several steps involved in producing and processing oil from its crude state to ultimate use: exploration, production, transportation, refining, distribution, and retail sales.

MAJOR. In international oil, the majors are the seven largest companies that control most of the non-Communist world's oil—Standard Oil Company (New Jersey), Anglo-Iranian Oil Company, the Shell group, the Texas Company, Gulf, Socony-Vacuum Oil Company, and Standard Oil of California. In the United States, the term *majors* usually applies to the twenty largest integrated companies. In economic terms, a major is a firm that must consider the impact of its activities on the market.

MAXIMUM EFFICIENT RATE (MER). The production rate consistent with maximum recovery of reserves without damage to the reservoir.

MIDCONTINENT CRUDE. Oil produced mainly in Kansas, Oklahoma, and North Texas.

OIL-IN-PLACE. Deposits of oil whose existence has been discovered through exploration. (Compare Proved reserves)

PRIMARY RECOVERY. The extraction of oil from a well using only the natural gas or water pressure in the reservoir to force the petroleum to the surface, without pumping or other assistance.

PRODUCT. One of the products made from crude oil by the refining process, such as kerosene, diesel fuel, or gasoline.

PRORATIONING. The allocation of production shares to individual producers by state regulatory agencies.

PROVED RESERVES. As defined by the American Petroleum Institute, proved reserves are the estimated quantities of crude oil that geological and engineering data demonstrate with reasonable certainty to be recoverable from known reservoirs under existing economic and operating conditions. Proved reserves are the small part of oil-in-place which has been readied for production by the drilling and equipping of wells. They represent only the amount of oil that can be produced at existing operating costs; they in no way represent the magnitude of oil that is ultimately recoverable.

RESERVE/PRODUCTION RATIO. The ratio of proved reserves to annual production expressed in years supply. The size of proved reserves and the volume of production are determined by investment and production policies as well as by geological and engineering considerations.

RESIDUAL OIL. The fraction of crude oil remaining after gasoline and other low-boiling constituents have been distilled off during the refining process. Known as #6 fuel oil and Bunker C oil, residual oil is widely used under boilers to produce steam and generate electricity.

ROYALTY. Payment to the landowner or mineral rights owner on each unit of resource produced.

SECONDARY RECOVERY. The introduction of water or gas into a well to supplement the natural reservoir drive and force additional oil to the surface.

SYNFUEL. Fuel produced through chemical conversion of natural hydrocarbon substances such as coal and oil shale.

TAX PAID COST. The cost of production plus royalties and taxes.

TON. The common measure of quantity of petroleum outside the United States is the metric ton (2,204.62 lbs.). Because the common U.S. measure, the barrel, is a measure of volume rather than weight, no single conversion factor applies to all crude oils. The normal range is 6.5 to 8.5 barrels per metric ton.

TRANSFER PRICE. The price assigned to raw materials and intermediate products that the various affiliates of vertically integrated firms sell to each other.

UPSTREAM. Activities concerned with finding and producing petroleum.

WILDCAT. A well drilled in an area that has not previously produced oil or gas.

Selected Bibliography

Collections of Personal Papers

Dean Acheson Papers. Harry S. Truman Library. Independence, Mo.

George V. Allen Papers. Harry S. Truman Library. Independence, Mo.

Allen-Angier Family Papers. Perkins Library, Duke University. Durham, N.C.

Max W. Ball Papers. Harry S. Truman Library. Independence, Mo.

Bernard Baruch Papers. Mudd Library. Princeton University. Princeton, N.J.

Adolph A. Berle Papers. Franklin D. Roosevelt Library. Hyde Park, N.Y.

Oscar L. Chapman Papers. Harry S. Truman Library. Independence, Mo.

William L. Clayton Papers. Harry S. Truman Library. Independence, Mo.

Clark M. Clifford Papers. Harry S. Truman Library. Independence, Mo.

Thomas Connally Papers. Manuscript Division, Library of Congress. Washington, D.C.

Frank P. Corrigan Papers. Franklin D. Roosevelt Library. Hyde Park, N.Y.

Ralph K. Davies Papers. Harry S. Truman Library. Independence, Mo.

John Foster Dulles Telephone Transcripts (Xeroxes from the Dwight D. Eisenhower Library, Abilene, Kans.). Firestone Library. Princeton University. Princeton, N.J.

George M. Elsey Papers. Harry S. Truman Library. Independence, Mo.

James V. Forrestal Diary (Xerox from Princeton University). Operational Archives, Naval Historical Division, Navy Yard. Washington, D.C.

James V. Forrestal Papers. Mudd Library. Princeton University. Princeton, N.J.

Warner W. Gardner Papers. Harry S. Truman Library. Independence, Mo.

Henry F. Grady Papers. Harry S. Truman Library. Independence, Mo.

Harry L. Hopkins Papers. Franklin D. Roosevelt Library. Hyde Park, N.Y.

Harold L. Ickes Papers and Manuscript Diary. Manuscript Division, Library of Congress. Washington, D.C.

Joseph M. Jones Papers. Harry S. Truman Library. Independence, Mo.

W. John Kenney Papers. Harry S. Truman Library. Independence, Mo.

Arthur Krock Papers. Mudd Library. Princeton University. Princeton, N.J.

James M. Landis Papers. Manuscript Division, Library of Congress. Washington, D.C.

Breckinridge Long Papers. Manuscript Division, Library of Congress. Washington, D.C.

George S. Messersmith Papers. University of Delaware. Newark, Del.
Charles S. Murphy Papers. Harry S. Truman Library. Independence, Mo.
Sumner T. Pike Papers. Harry S. Truman Library. Independence, Mo.
Franklin D. Roosevelt Papers. Franklin D. Roosevelt Library. Hyde Park, N.Y.
Stephen J. Spingarn Papers. Harry S. Truman Library. Independence, Mo.
Harry S. Truman Papers. Harry S. Truman Library. Independence, Mo.

Records at the United States National Archives, Washington, D.C.

National Security Council Numbered Papers
Record Group 46. Records of the U.S. Senate
Record Group 48. Records of the Office of the Secretary of the Interior
Record Group 59. Records of the Department of State
    Decimal Files
    Harley A. Notter Files
    Office of the Executive Secretariat
    Records of the Petroleum Division
    Records of the Research and Analysis Branch (OSS) and the Bureau of
    Intelligence and Research (State Department)
    Records of the State-War-Navy Coordinating Committee
    Rusk-McClintock Files
Record Group 80. Records of the Department of the Navy
Record Group 107. Records of the Office of the Secretary of War
Record Group 122. Records of the Federal Trade Commission
Record Group 165. Records of the War Department General and Special Staffs
Record Group 218. Records of the Joint Chiefs of Staff
Record Group 226. Records of the Office of Strategic Services, Research and
    Analysis Branch
Record Group 234. Records of the Reconstruction Finance Corporation
Record Group 250. Records of the Office of War Mobilization and Reconversion
Record Group 253. Records of the Petroleum Administration for War
Record Group 304. Records of the National Security Resources Board
Record Group 319. Records of the Army Staff
Record Group 330. Secretary of Defense Office Files
Record Group 334. Records of Interservice Agencies
Record Group 353. Records of the Interdepartmental and Intradepartmental
    Committees of the Department of State

Miscellaneous Archival Collections

British Foreign Office Political Correspondence. Class FO 371. Public Records
    Office. London, England.
Office Files of the Assistant Secretary for Economic Affairs and the Under Secretary
    for Economic Affairs, 1944–1948 (Clayton-Thorp). Harry S. Truman Library.
    Independence, Mo.
Record Group 286. Records of the Agency for International Development. Federal
    Records Center. Suitland, Md.
Records of the Secretary of the Navy/Chief of Naval Operations. Operational
    Archives, Naval Historical Division, Navy Yard. Washington, D.C.

Published Collections of Documents

Alexander, Yonah, and Nanes, Allan, eds. *The United States and Iran: A Documentary History.* Frederick, Md.: University Publications of America, 1980.

CIA Research Reports: Latin America, 1946–1976. Frederick, Md.: University Publications of America, 1980.

Declassified Documents Reference Service, Vol. R. Washington, D.C.: Carrollton Press, 1976.

Documents of the National Security Council, 1947–1977. Washington, D.C.: University Publications of America, 1980.

Folliot, Denise, ed. *Documents on International Affairs, 1951.* London: Royal Institute of International Affairs, 1954.

———. *Documents on International Affairs, 1952.* London: Royal Institute of International Affairs, 1955.

———. *Documents on International Affairs, 1953.* London: Royal Institute of International Affairs, 1956.

———. *Documents on International Affairs, 1954.* London: Royal Institute of International Affairs, 1957.

Iranian Embassy. *Some Documents on the Nationalization of the Oil Industry in Iran.* Washington, D.C.: Iranian Embassy, 1951.

O.S.S./State Department Intelligence and Research Reports: Latin America, 1941–1961. Washington, D.C.: University Publications of America, 1979.

U.S. Congress. Senate. Committee on Foreign Relations. Subcommittee on Multinational Corporations. *A Documentary History of the Petroleum Reserves Corporation, 1943–1944.* Washington, D.C.: U.S. Government Printing Office, 1974.

———. *The International Petroleum Cartel, the Iranian Consortium, and U.S. National Security.* Washington, D.C.: U.S. Government Printing Office, 1975.

———. *Multinational Corporations and United States Foreign Policy.* Part 8. 93rd Cong., 2d sess., 1975.

U.S. Department of State. *Foreign Relations of the United States.* Washington: U.S. Government Printing Office, 1959–1984.

Oral History Interviews

Mervin L. Bohan (Truman Library)

Emilio Collado (Truman Library)

Loy W. Henderson (Truman Library)

Loy W. Henderson (John Foster Dulles Oral History Project, Princeton University)

Loy W. Henderson (John J. Harter. "Mr. Foreign Service on Mossadegh and Wristonization: An Interview with Loy W. Henderson." *Foreign Service Journal* 57 [November 1980]: 16–20).

Harry N. Howard (Truman Library)

Marx Leva (Truman Library)

George C. McGhee (Truman Library)

George C. McGhee (Interview with author)

Stephen J. Spingarn (Truman Library)

Memoirs and Diaries

Acheson, Dean Gooderham. *Present at the Creation: My Years in the State Department*. New York: W. W. Norton & Co., 1969.

Alemán Valdés, Miguel. *La Verdad de Petróleo in México*. México, D.F.: Biografica Gandesa, 1977.

Blum, John Morton, ed. *The Price of Vision: The Diary of Henry A. Wallace, 1942–1946*. Boston: Houghton Mifflin Co., 1973.

Bernays, Edward T. *Biography of an Idea: Memoirs of Public Relations Counsel Edward T. Bernays*. New York: Simon & Schuster, 1965.

Brown, Bruce K. *Oil Men in Washington: An Informal Account of the Organization and Activities of the Petroleum Administration for Defense During the Korean War, 1950–1952*. N.p.: Evanil Press, 1965.

Clifford, Clark M. "Factors Influencing President Truman's Decision to Support Partition and Recognize the State of Israel." In *The Palestine Question in American History*, pp. 24–45. Edited by Clark M. Clifford, Eugene V. Rostow, and Barbara W. Tuchman. New York: Arno Press, 1978.

Dulles, Allen Welsh. *The Craft of Intelligence*. New York: Harper & Row, 1963.

Eddy, William A. *F.D.R. Meets Ibn Saud*. New York: American Friends of the Middle East, 1954.

Eden, Anthony. *Full Circle: The Memoirs of Anthony Eden*. Boston: Houghton Mifflin Co., 1962.

Eisenhower, Dwight D. *The White House Years: Mandate for Change, 1953–1956*. Garden City, N.Y.: Doubleday & Co., 1963.

Feis, Herbert. *Seen from E.A.: Three International Episodes*. New York: Alfred A. Knopf, 1947.

Hull, Cordell. *Memoirs*. 2 vols. New York: Macmillan Co., 1948.

Ickes, Harold L. *Fightin' Oil*. New York: Alfred A. Knopf, 1943.

Israel, Fred, ed. *The War Diaries of Breckinridge Long*. Lincoln: University of Nebraska Press, 1966.

Jones, Jesse H. *Fifty Billion Dollars: My Thirteen Years with the RFC (1932–1945)*. New York: Macmillan Co. 1951.

Jones, Joseph Marion. *The Fifteen Weeks*. New York: Viking Press, 1955.

Krock, Arthur. *Memoirs: Sixty Years on the Firing Line*. New York: Funk & Wagnalls, 1968.

McGhee, George C. *Envoy to the Middle World: Adventures in Diplomacy*. New York: Harper & Row, 1983.

Melbourne, Roy. "America and Iran in Perspective: 1953 and 1980." *Foreign Service Journal* 57 (April 1980): 10–17.

Millis, Walter, ed. *The Forrestal Diaries*. New York: Viking Press, 1951.

Millspaugh, Arthur C. *Americans in Persia*. Washington: Brookings Institution, 1946.

Roosevelt, Kermit. *Countercoup: The Struggle for the Control of Iran*. New York: McGraw-Hill, 1979.

Truman, Harry S. *Memoirs: Year of Decision*. Garden City, N.Y.: Doubleday & Co., 1955.

———. *Memoirs: Years of Trial and Hope, 1946-1952*. Garden City, N.Y.: Doubleday & Co., 1956.

Walters, Vernon A. *Silent Missions*. Garden City, N.Y.: Doubleday & Co., 1978.
Williams, Philip M., ed. *The Diary of Hugh Gaitskell, 1945–1956*. London: Jonathan Cape, 1983.
Woodhouse, C. M. *Something Ventured*. London: Granada, 1982.

Congressional Hearings and Reports

U.S. Congress. House. Committee on Armed Services. Special Subcommittee on Petroleum. *Petroleum for National Defense*. 80th Cong., 2d sess., 1948.
———. *Report of Investigation of Petroleum in Relation to National Defense*. Washington, D.C.: U.S. Government Printing Office, 1948.
U.S. Congress. House. Committee on Foreign Affairs. *Extension of the European Recovery Program*. 81st Cong., 1st sess., 1949.
———. *Selected Executive Session Hearings of the Committee, 1951–56: Vol. 16: The Middle East, Africa, and Inter-American Affairs*. Washington, D.C.: U.S. Government Printing Office, 1980.
U.S. Congress. House. Committee on Interstate and Foreign Commerce. *Effects of Foreign Oil Imports on Independent Domestic Producers*. H. Rept. 2344. 81st Cong., 2d sess., 1950.
———. *Fuel Investigation: Mexican Petroleum*. H. Rept. 2470. 80th Cong., 2d sess., 1949.
———. *Fuel Investigation: Petroleum and the European Recovery Program*. H. Rept. 1438. 80th Cong., 2d sess., 1948.
———. *Petroleum Investigation (Gasoline and Rubber)*. 78th Cong., 2d sess., 1942.
———. *Petroleum Study*. 81st Cong., 2d sess., 1950.
———. *Petroleum Study: Petroleum Imports*. H. Rept. 2055. 81st Cong., 2d sess., 1950.
U.S. Congress. House. Select Committee on Foreign Aid. *Petroleum Requirements and Availabilities*. H. Rept. 1149. 80th Cong., 1st sess., 1947.
U.S. Congress. House. Select Committee on Small Business. *Effects of Foreign Oil Imports on Independent Domestic Producers*. 81st Cong. 1st sess., 1949, and 81st Cong., 2d sess., 1950.
U.S. Congress. Senate and House. Select Committees on Small Business. *The Third World Petroleum Congress*. Washington, D.C.: U.S. Government Printing Office, 1952.
U.S. Congress. Senate. Committee on Foreign Relations. *Executive Sessions (Historical Series): Vol. 1: Eightieth Congress, First and Second Sessions, 1947–48*. Washington, D.C.: U.S. Government Printing Office, 1976.
———. *Executive Sessions (Historical Series): Vol. 3: Eighty-second Congress, First Session, 1951*. Washington, D.C.: U.S. Government Printing Office, 1976.
———. *Executive Sessions (Historical Series): Vol. 4: Eighty-second Congress, Second Session, 1952*. Washington, D.C.: U.S. Government Printing Office, 1976.
———. *Petroleum Agreement with Great Britain and Northern Ireland*. 80th Cong., 1st sess., 1947.
U.S. Congress. Senate. Committee on Foreign Relations. Subcommittee on Multinational Corporations. *Glossary of Terms Relating to the Petroleum In-*

dustry. Committee Print. Washington, D.C.: U.S. Government Printing Office, 1974.

——. *Multinational Corporations and United States Foreign Policy*. 93rd Cong., 2d sess., 1975.

——. *Multinational Oil Corporations and U.S. Foreign Policy*. Washington, D.C.: U.S. Government Printing Office, 1975.

U.S. Congress. Senate. Committee on Naval Affairs. *Nomination of Edwin W. Pauley for Appointment as Under Secretary of the Navy*. 79th Cong., 2d sess., 1946.

U.S. Congress. Senate. Committee on the Judiciary and Committee on Interior and Insular Affairs. *Emergency Oil Lift Program and Related Oil Problems*. Joint Hearings. 85th Cong., 2d sess., 1957.

U.S. Congress. Senate. Select Committee on Small Business. *The Impact of Monopoly and Cartel Practices on Small Businesses*. 82d Cong., 2d sess., 1952.

——. *The International Petroleum Cartel: Staff Report to the Federal Trade Commission*. Washington, D.C.: U.S. Government Printing Office, 1952.

U.S. Congress. Senate. Special Committee Investigating the National Defense Program. *Investigation of the National Defense Program: Part 41: Petroleum Arrangements with Saudi Arabia*. 80th Cong., 1st sess., 1948.

——. *Navy Purchases of Middle East Oil*. Report no. 440, pt. 5, 80th Cong., 2d sess., 1948.

——. *Report of Subcommittee Concerning Investigation Overseas: Section 1: Petroleum Matters*. S. Rept. 10, pt. 15. 78th Cong., 2d sess., 1944.

U.S. Congress. Senate. Special Committee Investigating Petroleum Resources. *American Petroleum Interests in Foreign Countries*. 79th Cong., 1st sess., 1946.

——. *Investigation of Petroleum Resources in Relation to the National Welfare*. S. Rept. 179. 79th Cong., 1st sess., 1945.

——. *Investigation of Petroleum Resources in Relation to the National Welfare*. S. Rept. 9. 80th Cong., 1st sess., 1947.

——. *Investigation of Petroleum Resources: New Sources of Petroleum in the United States*. 79th Cong., 1st sess., 1946.

——. *Petroleum Requirements—Postwar*. 79th Cong., 1st sess., 1946.

——. *Wartime Petroleum Administration under the Petroleum Administration for War*. 79th Cong., 1st sess., 1946.

U.S. Congress. Senate. Special Committee to Study the Problems of American Small Business. *Problems of American Small Business: Part 21: Oil Supply and Distribution Problems: 5*. 80th Cong., 1st sess., 1948.

Miscellaneous United States Government Publications

Economic Cooperation Administration. *The Sterling Area: An American Analysis*. London: Economic Cooperation Administration, 1951.

Energy Resources and National Policy. Report of the Energy Resources Committee to the National Resources Committee. Washington, D.C.: U.S. Government Printing Office, 1939.

U.S. Department of the Interior. *National Resources and Foreign Aid*. Washington, D.C.: U.S. Government Printing Office, 1947.

U.S. Department of State. Department of State *Bulletin*.
———. Bureau of Public Affairs. Office of the Historian. "OPEC and the Changing Structure of the World Oil Economy, 1960–1983." *Historical Issues* No. 10. April 1983.
———. Office of Public Affairs. "Venezuela: Oil Transforms a Nation." February 1953.
United States v. Standard Oil Company of California, the Texas Company, Bahrein Petroleum Company, Ltd., California-Texas Company Ltd., Caltex Oceanic, Ltd., and Mideast Crude Sales Company. 155 F. Supp. 121.

Miscellaneous Contemporary Materials

Brodie, Bernard. "American Security and Foreign Oil." *Foreign Policy Reports* 23 (1 March 1948): 298–311.
Feis, Herbert. *Petroleum and American Foreign Policy*. Stanford: Food Research Institute, Stanford University, 1944.
Gary, Howard C. "Iran and the Future of Oil." *Foreign Policy Bulletin* 33 (1 October 1953): 1–2.
Groen, E. "The Significance of the Marshall Plan for the Petroleum Industry in Europe—Historical Review of the Period 1947–1950." In *The Third World Petroleum Congress*, pp. 37–73. U.S. Congress. Senate and House. Select Committees on Small Business. Washington, D.C.: U.S. Government Printing Office, 1952.
Hoskins, Halford L. *Middle East Oil in United States Foreign Policy*. Public Affairs Bulletin No. 89. Washington, D.C.: Library of Congress Legislative Reference Service, 1950. Reprint ed. Westport, Conn.: Hyperion Press, 1976.
Independent Petroleum Association of America. *Petroleum in the Western Hemisphere: Report of the Western Hemisphere Oil Committee of the Independent Petroleum Association of America*. Washington, D.C.: Independent Petroleum Association of America, 1952.
International Court of Justice. The Hague. *I.C.J. Pleadings, Anglo-Iranian Oil Co. Case (United Kingdom v. Iran)*.
Levy, Walter J. "The Past, Present, and Future Price Structure of the International Oil Trade." In *The Third World Petroleum Congress*, pp. 21–37. U.S. Congress. Senate and House. Select Committees on Small Business. Washington, D.C.: U.S. Government Printing Office, 1952.
MacMahon, Arthur W., and Dittman, W. R. "The Mexican Oil Industry Since Expropriation I. *Political Science Quarterly* 57 (March 1942): 28–50.
———. "The Mexican Oil Industry Since Expropriation II. *Political Science Quarterly* 57 (June 1942): 161–188.
Mendershausen, Horst. "Dollar Shortage and Surplus Oil in 1949–1950." *Essays in International Finance* 11 (November 1950).
Middle East Institute. *The Present Situation in Iran: A Survey of Political and Economic Problems Confronting the Country*. Washington, D.C.: Middle East Institute, 1953.
Pogue, Joseph. *Oil in Venezuela*. New York: Chase National Bank, 1949.
Sinclair, Angus. "Iranian Oil." *Middle East Affairs* 2 (June-July 1951): 213–224.
Snodgrass, C. Stribling, and Kuhl, Arthur. "U.S. Petroleum's Response to the Iranian Shutdown." *Middle East Journal* 5 (Autumn 1951): 501–504.

Standard Oil Company (New Jersey). *Some Facts on the Search for Oil in the United States of America.* New York: Standard Oil Company (New Jersey), 1944.

———. *Standard Oil Company (New Jersey) and Middle Eastern Oil Production: A Background Memorandum on Company Policies and Actions.* New York: Standard Oil Company (New Jersey), 1947.

Voskuil, Walter H., and Meyers, Hope M. "Can United States Oil Reserves Meet the Postwar Demand?" In *American Interests in the War and the Peace.* New York: Council on Foreign Relations, 1945.

Contemporary Newspapers and Periodicals

Atlantic Monthly
American Magazine
Business Week
Chicago Sun
Christian Science Monitor
Economist
Fortune
The Lamp
Nation
New York Times
Newsweek
Oil and Gas Journal
Oil Forum (known as *International Oilman* before 1947)
The Round Table
U.S. News and World Report

Books, Articles, and Dissertations

Abrahamian, Ervand. *Iran between Two Revolutions.* Princeton: Princeton University Press, 1982.

Adelman, M. A. *The World Petroleum Market.* Baltimore: Johns Hopkins University Press, 1972.

Adler, Selig. "American Policy vis-à-vis Palestine in the Second World War." In *World War II: An Account of Its Documents,* pp. 43–58. Edited by James E. O'Neill and Robert W. Krauskopf. Washington, D.C.: Howard University Press, 1976.

Alexander, Robert J. *Rómulo Betancourt and the Transformation of Venezuela.* New Brunswick, N.J.: Transaction Books, 1982.

Alford, Robert R. "Paradigms of Relations between State and Society." In *Stress and Contradiction in Modern Capitalism: Public Policy and the Theory of the State,* pp. 145–160. Edited by Leon N. Lindberg, Robert Alford, Colin Crouch, and Claus Offe. Lexington, Mass.: Lexington Books, 1975.

Alpert, Paul. *Twentieth-Century Economic History of Europe.* New York: Henry Schuman, 1951.

Ambrose, Stephen E. *Eisenhower: The President.* New York: Simon & Schuster, 1984.

———. *Ike's Spies: Eisenhower and the Espionage Establishment.* Garden City, N.Y.: Doubleday & Co., 1981.

Anderson, Irvine H. *Aramco, the United States, and Saudi Arabia: A Study in the Dynamics of Foreign Oil Policy, 1933–1950.* Princeton: Princeton University Press, 1981.
————. "Lend-Lease for Saudi Arabia: A Comment on Alternative Conceptualizations." *Diplomatic History* 3 (Fall 1979): 413–423.
————. *The Standard-Vacuum Oil Company and United States East Asian Policy, 1933–1941.* Princeton: Princeton University Press, 1975.
Anderson, Jack, and Boyd, James. *Fiasco.* New York: Times Books, 1983.
Anderson, Robert O. *Fundamentals of the Petroleum Industry.* Norman: University of Oklahoma Press, 1984.
Andrews, Bruce. "Social Rules and the State as a Social Actor." *World Politics* 27 (July 1975): 521–540.
Bachrach, Peter, and Baratz, Morton. "Decisions and Nondecisions: An Analytical Framework." *American Political Science Review* 57 (September 1963): 634–642.
————. "Two Faces of Power." In *The Bias of Pluralism*, pp. 51–64. Edited by William E. Connolly. New York: Atherton Press, 1969.
Bain, Kenneth Ray. *The March to Zion: United States Policy and the Founding of Israel.* College Station: Texas A & M University Press, 1979.
Baloyra, Enrique. "Oil Policies and Budgets in Venezuela, 1938–1968." *Latin American Research Review* 9 (Summer 1974): 28–72.
Barber, William J. "The Eisenhower Energy Policy: Reluctant Interventionism." In *Energy Policy in Perspective: Today's Problems, Yesterday's Solutions*, pp. 221–261. Edited by Craufurd D. Goodwin. Washington, D.C.: Brookings Institution, 1981.
Baram, Philip J. *The Department of State in the Middle East, 1919–1945.* Philadelphia: University of Pennsylvania Press, 1978.
Barnet, Richard J. *Intervention and Revolution: America's Confrontation with Insurgent Movements around the World.* New York: World Publishing Co., 1968.
————. *The Roots of War.* Baltimore: Penguin Books, 1973.
Beck, Peter J. "The Anglo-Persian Oil Dispute of 1932–33." *Journal of Contemporary History* 9 (October 1974): 123–151.
Becker, William H. *The Dynamics of Business-Government Relations: Industry and Exports, 1893–1921.* Chicago: University of Chicago Press, 1982.
Bermúdez, Antonio J. *The Mexican Petroleum Industry: A Case Study in Nationalization.* Stanford: Institute of Hispanic American and Luso-Brazilian Studies, 1963.
Betancourt, Rómulo. *Venezuela: Oil and Politics.* Boston: Houghton Mifflin Co., 1979.
Blair, John. *The Control of Oil.* New York: Pantheon Books, 1976.
Block, Fred L. *The Origins of International Economic Disorder: A Study of United States International Monetary Policy from World War II to the Present.* Berkeley and Los Angeles: University of California Press, 1977.
Bohi, Douglas R., and Russell, Milton. *Limiting Oil Imports: An Economic History and Analysis.* Baltimore: Johns Hopkins University Press, 1978.
Brand, Donald R. "Corporatism, the NRA, and the Oil Industry." *Political Science Quarterly* 98 (Spring 1983): 99–118.

Brody, David. "The New Deal and World War II." In *The New Deal: The National Level*, pp. 267–309. Edited by John Braeman, Robert H. Bremner, and David Brody. Columbus: Ohio State University Press, 1975.

Buhite, Russell D. *Patrick J. Hurley and American Foreign Policy*. Ithaca: Cornell University Press, 1973.

Burggraaff, Winfield J. *The Venezuelan Armed Forces in Politics, 1935–1959*. Columbia: University of Missouri Press, 1972.

Carnoy, Martin. *The State and Political Theory*. Princeton: Princeton University Press, 1984.

Chandler, Alfred D., Jr. *Strategy and Structure: Chapters in the History of the American Industrial Enterprise*. Cambridge: MIT Press, 1962.

———. *The Visible Hand: The Managerial Revolution in American Business*. Cambridge: Harvard University Press, 1977.

———, and Galambos, Louis. "The Development of Large-Scale Economic Organizations in Modern America." *Journal of Economic History* 30 (March 1970): 201–217. (Reprinted in Perkins 1977.)

Chester, Edward W. *United States Oil Policy and Diplomacy: A Twentieth-Century Overview*. Westport, Conn.: Greenwood Press, 1983.

Cline, Howard F. *The United States and Mexico*. Cambridge: Harvard University Press, 1953.

Cohen, Michael J. *Palestine and the Great Powers, 1945–1948*. Princeton: Princeton University Press, 1982.

Collins, Robert M. "American Corporatism: The Committee for Economic Development, 1942–1964." *The Historian* 44 (February 1982): 151–173.

———. *The Business Response to Keynes, 1929–1964*. New York: Columbia University Press, 1981.

———. "Positive Business Responses to the New Deal: The Roots of the Committee for Economic Development, 1933–1942." *Business History Review* 52 (Autumn 1978): 369–391.

Commoner, Barry. *The Poverty of Power: Energy and the Economic Crisis*. New York: Alfred A. Knopf, 1976.

Connolly, William E. "The Challenge to Pluralist Theory." In *The Bias of Pluralism*, pp. 3–34. Edited by William E. Connolly. New York: Atherton Press, 1969.

Cook, Roy C. *Control of the Petroleum Industry by Major Oil Companies*. U.S. Congress. Temporary National Economic Committee. *Investigation of Concentration of Economic Power*. Monograph No. 39. Washington, D.C.: U.S. Government Printing Office, 1941.

Cottam, Richard. *Nationalism in Iran*. Pittsburgh: University of Pittsburgh Press, 1964.

———. "The United States, Iran, and the Cold War." *Iranian Studies* 3 (Winter 1970): 2–22.

Cowhey, Peter F. *The Problems of Plenty: Energy Policy and International Politics*. Berkeley and Los Angeles: University of California Press, 1985.

Cronon, E. David. *Josephus Daniels in Mexico*. Madison: University of Wisconsin Press, 1960.

Cuff, Robert D. *The War Industries Board: Business-Government Relations during World War I*. Baltimore: Johns Hopkins University Press, 1973.

Danielsen, Albert L. *The Evolution of OPEC.* New York: Harcourt Brace Jovanovich, 1982.

Darmstadter, Joel. "Energy Consumption: Trends and Patterns." In *Energy, Economic Growth, and the Environment,* pp. 155–223. Edited by Sam H. Schurr. Baltimore: Johns Hopkins University Press, 1972.

———, and Landsberg, Hans H. "The Economic Background." In *The Oil Crisis,* pp. 15–37. Edited by Raymond Vernon. New York: W. W. Norton & Co., 1976.

———, Teitelbaum, P. D. and Polach, J. G. *Energy in the World Economy: A Statistical Review of Trends in Output, Trade, and Consumption Since 1925.* Baltimore: Johns Hopkins Press, 1971.

DeGolyer & McNaughton. *Twentieth-Century Petroleum Statistics.* Dallas: DeGolyer & McNaughton, 1980.

DeNovo, John A. *American Interests and Policies in the Middle East, 1900–1939.* Minneapolis: University of Minnesota Press, 1963.

———. "The British Factor in Harold L. Ickes's Frustrated Design for a United States Oil Policy, 1943–1944." In *Oil, the Middle East, North Africa, and the Industrial States: Developmental and International Dimensions,* pp. 79–98. Edited by Klaus Jürgen Gantzel and Helmut Mejcher. Paderborn: Ferdinand Schöningh, 1984.

———. "The Culbertson Economic Mission and Anglo-American Tensions in the Middle East, 1944–1945." *Journal of American History* 63 (March 1977): 913–936.

———. "The Federal Government as Manager of Petroleum Resources, 1940–1942." Paper presented at the Annual Meeting of the American Historical Association, Dallas, Texas, 1977.

———. "The Movement for an Aggressive American Oil Policy Abroad, 1918–1920." *American Historical Review* 61 (July 1956): 854–876.

———. "Petroleum and the United States Navy before World War I." *Mississippi Valley Historical Review* 41 (March 1955): 641–656.

———. "Researching American Relations with the Middle East: The State of the Art, 1970." In *The National Archives and Foreign Relations Research,* pp. 243–264. Edited by Milton O. Gustafson. Athens: Ohio University Press, 1974.

Dewhurst, J. Frederick, and Associates. *America's Needs and Resources: A New Survey.* New York: Twentieth Century Fund, 1955.

Dirlam, Joel B. "The Petroleum Industry." In *The Structure of American Industry,* rev. ed., pp. 236–273. Edited by Walter Adams. New York: Macmillan Co., 1954.

Donovan, Robert J. *Conflict and Crisis: The Presidency of Harry S. Truman, 1945–1948.* New York: W. W. Norton & Co., 1977.

Eckes, Alfred E. *The United States and the Global Struggle for Minerals.* Austin: University of Texas Press, 1979.

Eden, Lynn. "Capitalist Conflict and the State: The Making of United States Military Policy in 1948." In *Statemaking and Social Movements: Essays in History and Theory,* pp. 233–261. Edited by Charles Bright and Susan Harding. Ann Arbor: University of Michigan Press, 1984.

Elwell-Sutton, Lawrence Paul. *Persian Oil: A Study in Power Politics.* London: Lawrence & Wishart, 1955.

Engler, Robert, ed. *America's Energy: Reports from the Nation on 100 Years of Struggle for Democratic Control of Our Resources.* New York: Pantheon Books, 1980.

———. *The Politics of Oil: Private Power and Democratic Directions.* Chicago: University of Chicago Press, 1961.

Eveland, Wilbur Crane. *Ropes of Sand: America's Failure in the Middle East.* New York: W. W. Norton & Co., 1980.

Fagen, Richard R. "Introduction." In *Capitalism and the State in U.S.–Latin American Relations*, pp. 1–22. Edited by Richard R. Fagen. Stanford: Stanford University Press, 1979.

Falk, Richard. "Beyond Controversy: A Liberal Balancing Act of Cold War Accounts." *Reviews in American History* 6 (June 1978): 139–145.

Fanning, Leonard M. *Foreign Oil and the Free World.* New York: McGraw-Hill, 1954.

Fatemi, Nasrollah Saifpour. *Oil Diplomacy: Powderkeg in Iran.* New York: Whittier Books, 1954.

Ferrier, Ronald, "The Development of the Iranian Oil Industry." In *Twentieth-Century Iran*, pp. 93–128. Edited by Hossein Amirsadeghi. London: Heinemann, 1977.

Ferguson, Thomas. "From Normalcy to New Deal: Industrial Structure, Party Competition, and American Public Policy in the Great Depression" *International Organization* 38 (Winter 1984): 41–94.

Ford, Alan W. *The Anglo-Iranian Oil Dispute of 1951–1952: A Study of the Role of Law in the Relations of States.* Berkeley and Los Angeles: University of California Press, 1954.

Frank, Helmut J. *Crude Oil Prices in the Middle East: A Study in Oligopolistic Price Behavior.* New York: Frederick A. Praeger, 1966.

Frankel, Paul H. *Essentials of Petroleum: A Key to Oil Economics.* London: Frank Cass & Co., 1946.

Frey, John W., and Ide, H. Chandler. *A History of the Petroleum Administration for War 1941–1945.* Washington, D.C.: U.S. Government Printing Office, 1946.

Fusfeld, Daniel R. "The Rise of the Corporate State in America." In *The Economy as a System of Power: Papers from the Journal of Economic Issues*, vol. 1, *Corporate Systems*, pp. 139–160. Edited by Warren J. Samuels. New Brunswick, N.J.: Transaction Books, 1979. (Originally published 1972.)

Gaddis, John Lewis. *The United States and the Origins of the Cold War, 1941–1947.* New York: Columbia University Press, 1972.

———. "The Emerging Post-Revisionist Synthesis on the Origins of the Cold War." *Diplomatic History* 7 (Summer 1983): 171–190.

Galambos, Louis. "The Emerging Organizational Synthesis in American History." *Business History Review* 44 (Autumn 1970): 279–290. (Reprinted in Perkins 1977.)

———. "Technology, Political Economy, and Professionalization: Central Themes of the Organizational Synthesis." *Business History Review* 57 (Winter 1983): 471–493.

Gardner, Lloyd C. *Architects of Illusion: Men and Ideas in American Foreign Policy, 1941–1949.* Chicago: Quadrangle Books, 1970.

————. *Economic Aspects of New Deal Diplomacy.* Madison: University of Wisconsin Press, 1964.

Gardner, Richard N. *Sterling-Dollar Diplomacy.* London: Oxford University Press, 1956.

Garson, G. David. *Power and Politics in the United States: A Political Economy Approach.* Lexington, Mass.: D. C. Heath, 1977.

Gibb, George Sweet, and Knowlton, Evelyn H. *The Resurgent Years, 1911–1927.* Vol. 2 of *History of Standard Oil Company (New Jersey).* New York: Harper & Brothers, 1956.

Giebelhaus, August W. *Business and Government in the Oil Industry: A Case Study of Sun Oil, 1876–1945.* Greenwich, Conn.: JAI Press, 1980.

Gillam, Richard, ed. *Power in Postwar America: Interdisciplinary Perspectives on a Historical Problem.* Boston: Little, Brown & Co., 1971.

Gols, A. George. "Postwar U.S. Foreign Petroleum Investment." In *U.S. Private and Government Investment Abroad,* pp. 409–458. Edited by Raymond F. Mikesell. Eugene: University of Oregon Books, 1962.

Goodwin, Craufurd D., "Truman Administration Policies toward Particular Energy Sources." In *Energy Policy in Perspective: Today's Problems, Yesterday's Solutions,* pp. 62–203. Edited by Craufurd D. Goodwin. Washington, D.C.: Brookings Institution, 1981.

Gormly, James L. "Keeping the Door Open in Saudi Arabia: The United States and the Dhahran Airfield, 1945–1946." *Diplomatic History* 4 (Spring 1980): 189–205.

Gourevitch, Peter Alexis. "The Second Image Reversed: The International Sources of Domestic Politics." *International Organization* 32 (Autumn 1978): 881–912.

Graham, Otis L., Jr. "The Broker State." *Wilson Quarterly* 8 (Winter 1984): 86–97.

————. "The Planning Ideal and American Reality: The 1930s." In *The Hofstadter Aegis: A Memorial,* pp. 257–299. Edited by Stanley Elkins and Eric McKitrick. New York: Alfred A. Knopf, 1974.

————. *Toward a Planned Society: From Roosevelt to Nixon.* New York: Oxford University Press, 1976.

Green, David. *The Containment of Latin America: A History of the Myths and Realities of the Good Neighbor Policy.* Chicago: Quadrangle Books, 1971.

Griffith, Robert. "Dwight D. Eisenhower and the Corporate Commonwealth." *American Historical Review* 87 (February 1982): 87–122.

————. "Forging America's Postwar Order: Politics and Political Economy in the Age of Truman." Unpublished paper, August 1984.

————. "Harry S Truman and the Burden of Modernity." *Reviews in American History* 9 (September 1981): 295–306.

————. "Truman and the Historians: The Reconstruction of Postwar American History." *Wisconsin Magazine of History* 59 (Autumn 1975): 20–50.

Hacker, Andrew, "Power to Do What?" In *The Bias of Pluralism,* pp. 67–80. Edited by William E. Connolly. New York: Atherton Press, 1969.

Harbutt, Fraser. "American Challenge, Soviet Response: The Beginning of the Cold War, February-May 1946." *Political Science Quarterly* 96 (Winter 1981–1982): 623–639.

Harris, Nigel. *Competition and the Corporate Society: British Conservatives, the State and Industry, 1945–1964.* London: Methuen & Co., 1972.

Hartshorn, J. E. *Politics and World Oil Economics: An Account of the International Oil Industry in Its Political Environment.* New York: Frederick A. Praeger, 1962.

Hawley, Ellis W. "The Discovery and Study of a 'Corporate Liberalism.'" *Business History Review* 52 (Autumn 1978): 309–320.

――――. "Herbert Hoover and American Corporatism, 1929–1933." In *The Hoover Presidency: A Reappraisal*, pp. 101–119. Edited by Martin L. Fausold and George T. Mazuzan. Albany: State University of New York Press, 1974a.

――――. "Herbert Hoover, the Commerce Secretariat, and the Vision of an 'Associative State,' 1921–1928." *Journal of American History* 61 (June 1974b): 116–140.

――――. "The New Deal and Business." In *The New Deal: The National Level*, pp. 50–82. Edited by John Braeman, Robert H. Bremner, and David Brody. Columbus: Ohio State University Press, 1975.

――――. "New Deal und 'Organisierte Kapitalismus' in internationaler Sicht." In *Die Grosse Krise in Amerika: Vergleichende Studien zur politischen Sozialgeschichte, 1929–1939*, pp. 9–39. Edited by Heinrich August Winkler. Göttingen: Vandenhoeck & Ruprecht, 1973.

――――. *The New Deal and the Problem of Monopoly: A Study in Economic Ambivalence.* Princeton: Princeton University Press, 1966.

Heilbroner, Robert. *The Limits of American Capitalism.* New York: Harper & Row, 1965.

Hellinger, Daniel. "Populism and Nationalism in Venezuela: New Perspectives on Acción Democrática." *Latin American Perspectives* 43 (Fall 1984): 33–59.

Hermida Ruiz, Angel J. *Bermúdez y la Batalla por el Petroleo.* México, D.F.: Costa Amic, 1974.

Hess, Gary R. "The Iranian Crisis of 1945–46 and the Cold War." *Political Science Quarterly* 89 (March 1974): 117–146.

Hogan, Michael J. "American Marshall Planners and the Search for a European Neocapitalism." *American Historical Review* 90 (February 1985): 44–72.

――――. *Informal Entente: The Private Structure of Cooperation in Anglo-American Economic Diplomacy, 1918–1928.* Columbia: University of Missouri Press, 1977.

――――. "Informal Entente: Public Policy and Private Management in Anglo-American Petroleum Affairs." *Business History Review* 48 (Summer 1974): 187–205.

――――. "Revival and Reform: America's Twentieth Century Search for a New Economic Order Abroad." *Diplomatic History* 8 (Fall 1984): 287–310.

Immerman, Richard H. *The CIA in Guatemala: The Foreign Policy of Intervention.* Austin: University of Texas Press, 1982.

Jackson, Scott. "Prologue to the Marshall Plan: The Origins of the American Commitment for a European Recovery Program." *Journal of American History* 65 (March 1979): 1043–1068.

Jacoby, Neil H. *Multinational Oil.* New York: Macmillan Co., 1974.

Jensen, Walter G. *Energy in Europe, 1945–1980.* London: G. T. Fowler & Co., 1967.

Johnson, Arthur M. "Continuity and Change in Government-Business Relations." In *Change and Continuity in Twentieth-Century America*, pp. 191–219. Edited by John Braeman, Robert H. Bremner, and Everett Walters. Columbus: Ohio State University Press, 1964.

Jones, G. Gareth. "The British Government and the Oil Companies, 1912–1924: The Search for an Oil Policy." *Historical Journal* 20 (September 1977): 647–672.

Kane, N. Stephen. "Corporate Power and Foreign Policy: Efforts of American Oil Companies to Influence United States Relations with Mexico, 1921–1928." *Diplomatic History* 1 (Spring 1977): 170–198.

———. "The United States and the Development of the Mexican Petroleum Industry, 1945–1950." *Inter-American Economic Affairs* 35 (Summer 1981): 45–72.

Kaplan, Lawrence S. "Ethnic Politics, the Palestine Question, and the Cold War." In *Culture and Diplomacy: The American Experience*, pp. 242–266. Edited by Morrell Heald and Lawrence S. Kaplan. Westport, Conn.: Greenwood Press, 1977.

Kapstein, Ethan B. "Alliance Energy Security, 1945–1983." *Fletcher Forum* 8 (Winter 1984): 91–116.

Karl, Barry. *The Uneasy State: The United States from 1915 to 1945.* Chicago: University of Chicago Press, 1983.

Katzenstein, Peter J. "Conclusion: Domestic Structures and Strategies of Foreign Economic Policy." *International Organization* 31 (Autumn 1977): 879–920.

———. "International and Domestic Structures: Foreign Economic Policies of Advanced Industrial States." *International Organization* 30 (Winter 1976): 1–45.

———. "Introduction: Domestic and International Forces and Strategies of Foreign Economic Policy." *International Organization* 31 (Autumn 1977): 587–606.

Kaufman, Burton I. "Mideast Multinational Oil, U.S. Foreign Policy, and Antitrust: The 1950s." *Journal of American History* 63 (March 1977a): 937–959.

———. "Oil and Antitrust: The Oil Cartel Case and the Cold War." *Business History Review* 51 (Spring 1977b): 35–56.

———. *The Oil Cartel Case: A Documentary Study of Antitrust Activity in the Cold War Era.* Westport, Conn.: Greenwood Press, 1978.

———. "Oil, Security, and America's Involvement in the Middle East in the 1940s." *Reviews in American History* 9 (March 1981): 131–137.

Keddie, Nikki R. *Roots of Revolution: An Interpretive History of Modern Iran.* New Haven: Yale University Press, 1981.

Kent, Marian. "Developments in British Oil Policy in the Inter-War Period." In *Oil, the Middle East, North Africa, and the Industrial States: Developmental and International Dimensions*, pp. 61–77. Edited by Klaus Jürgen Gantzel and Helmut Mejcher. Paderborn: Ferdinand Schöningh, 1984.

Keohane, Robert O. "Hegemonic Leadership and U.S. Economic Policy in the 'Long Decade' of the 1950s." In *America in a Changing World Political Economy*, pp. 49–76. Edited by William P. Avery and David P. Rapkin. New York: Longman, 1982a.

———. "State Power and Industry Influence: American Foreign Oil Policy in the 1940s." *International Organization* 36 (Winter 1982b): 165–183.

Klein, Herbert S. "American Oil Companies in Latin America: The Bolivian Experience." *Inter-American Economic Affairs* 18 (Autumn 1964): 47–72.

Koistinen, Paul A. C. "The 'Industrial-Military Complex' in Historical Perspective: World War I." *Business History Review* 41 (Winter 1967): 378–403.

Kolb, Glenn L. *Democracy and Dictatorship in Venezuela, 1945–1958.* Hamden, Conn.: Archon Books, 1974.

Kolko, Gabriel. *The Politics of War: The World and United States Foreign Policy, 1943–1945.* New York: Random House, 1968.

Kolko, Joyce, and Kolko, Gabriel. *The Limits of Power: The World and United States Foreign Policy, 1945–1954.* New York: Harper & Row, 1972.

Koppes, Clayton R. "Crude Diplomacy: Private Power and Public Purposes in Middle Eastern Oil." *Reviews in American History* 10 (June 1982): 259–264.

———. "The Good Neighbor Policy and the Nationalization of Mexican Oil." *Journal of American History* 69 (June 1982): 62–81.

———. "Mexico, Venezuela, and the Anglo-American Petroleum Order: From Open Door to OPEC." Paper presented at the Annual Meeting of the American Historical Association, San Francisco, Calif., 1983.

———. "Reversing Nationalization: The United States, Mexico, and Oil, 1938-1950." Paper presented at the Annual Meeting of the American Historical Association, Washington, D.C., 1980.

Krammer, Arnold. "Soviet Motives in the Partition of Palestine." *Journal of Palestine Studies* 2 (Winter 1973): 102–119.

Krasner, Stephen D. *Defending the National Interest: Raw Materials Investments and U.S. Foreign Policy.* Princeton: Princeton University Press, 1978.

Kuniholm, Bruce R. *The Origins of the Cold War in the Near East: Great Power Conflict and Diplomacy in Iran, Turkey, and Greece.* Princeton: Princeton University Press, 1980.

Kwitny, Jonathan. *Endless Enemies: The Making of an Unfriendly World.* New York: Congdon & Weed, 1984.

Landau, Christopher Thomas. "The Rise and Fall of Petro-liberalism: United States Relations with Socialist Venezuela, 1945–1948." Senior honors thesis, Harvard University, 1985.

Larson, Henrietta M.; Knowlton, Evelyn H.; and Popple, Charles S. *New Horizons, 1927–1950.* Vol. 3 of *History of Standard Oil Company (New Jersey).* New York: Harper & Row, 1971.

Lear, Linda J. *Harold L. Ickes: The Aggressive Progressive.* New York: Garland, 1981.

Leatherdale, Clive. *Britain and Saudi Arabia, 1925–1939: The Imperial Oasis.* London: Frank Cass, 1983.

Leeman, Wayne A. *The Price of Middle East Oil: An Essay in Political Economy.* Ithaca: Cornell University Press, 1962.

Leffler, Melvyn P. "The American Conception of National Security and the Beginnings of the Cold War, 1945–1948." *American Historical Review* 89 (April 1984): 346–381.

———. *The Elusive Quest: America's Pursuit of European Stability and French Security.* Chapel Hill: University of North Carolina Press, 1979.

———. "From Cold War to Cold War in the Near East." *Reviews in American History* 9 (March 1981): 124–130.

———. "From the Truman Doctrine to the Carter Doctrine: Lessons and Dilemmas of the Cold War." *Diplomatic History* 7 (Fall 1983): 245–266.

———. "Strategy, Diplomacy, and the Cold War: The United States, Turkey, and N.A.T.O., 1945–1952." *Journal of American History* 71 (March 1985): 807–825.

Levy, Walter J. *Oil Strategy and Politics, 1941–1981.* Boulder, Col.: Westview Press, 1982.

Lieuwen, Edwin. *Petroleum in Venezuela: A History.* Berkeley and Los Angeles: University of California Press, 1954.

Lindberg, Leon N. "Comparing Energy Policies: Political Constraints and the Energy Syndrome." In *The Energy Syndrome: Comparing National Responses to the Energy Crisis,* pp. 325–356. Edited by Leon N. Lindberg. Lexington, Mass.: Lexington Books, 1977.

Lindblom, Charles E. *Politics and Markets: The World's Political-Economic Systems.* New York: Basic Books, 1977.

Louis, William Roger. *The British Empire in the Middle East, 1945–1951.* Oxford: Clarendon Press, 1984.

Lowi, Theodore J. "American Business, Public Policy, Case Studies, and Political Theory." *World Politics* 16 (July 1964): 677–715.

Lubell, Harold. *Middle East Oil Crises and Europe's Energy Supplies.* Baltimore: Johns Hopkins Press, 1963.

Lukes, Stephen. *Power: A Radical View.* London: Macmillan, 1974.

Lytle, Mark Hamilton. "American-Iranian Relations, 1941–1947, and the Redefinition of National Security." Ph.D. dissertation, Yale University, 1973.

McBeth, B. S. *Juan Vicente Gómez and the Oil Companies in Venezuela, 1908–1935.* Cambridge: Cambridge University Press, 1983.

McCormick, Thomas J. "Drift or Mastery? A Corporatist Synthesis for American Diplomatic History." *Reviews in American History* 10 (December 1982): 318–330.

———. "The State of American Diplomatic History." In *The State of American History,* pp. 119–141. Edited by Herbert J. Bass. Chicago: Quadrangle Books, 1970.

McFarland, Stephen L. "A Peripheral View of the Origins of the Cold War: The Crises in Iran, 1941–1947." *Diplomatic History* 4 (Fall 1980): 333–351.

McLellan, David S. *Dean Acheson: The State Department Years.* New York: Dodd, Mead & Co., 1976.

McQuaid, Kim. "Corporate Liberalism in the American Business Community, 1920–1940." *Business History Review* 52 (Autumn 1978): 342–368.

Maier, Charles S. "Fictitious Bonds . . . of Wealth and Law: On the Theory and Practice of Interest Representation." In *Organizing Interests in Western Europe: Pluralism, Corporatism, and the Transformation of Politics,* pp. 27–61. Edited by Suzanne Berger. Cambridge: Cambridge University Press, 1981.

———. "The Politics of Productivity: Foundations of American International Economic Policy after World War II." *International Organization* 31 (Autumn 1977): 607–633.

Mark, Eduard M. "Allied Relations in Iran, 1941–1947: The Origins of a Cold War Crisis." *Wisconsin Magazine of History* 59 (Autumn 1975): 51–63.

Mason, Edward S. "The Political Economy of Resource Use." In *Perspectives on Conservation: Essays on America's Natural Resources*, pp. 157–186. Edited by Henry Jarrett. Baltimore: Johns Hopkins Press, 1958.

———, and Asher, Robert E. *The World Bank Since Bretton Woods*. Washington, D.C.: Brookings Institution, 1973.

Maull, Hanns. *Europe and World Energy*. Boston: Butterworths, 1980.

May, Ernest R. "The 'Bureaucratic Politics' Approach: U.S.–Argentine Relations, 1942–1947." In *Latin America and the United States: The Changing Political Realities*, pp. 129–163. Edited by Julio Cotler and Richard R. Fagen. Stanford: Stanford University Press, 1974.

Mazuzan, George T. "United States Policy toward Palestine at the United Nations, 1947–1948: An Essay." *Prologue* 7 (Fall 1975): 163–176.

Mejcher, Helmut. "The International Petroleum Cartel (1928), Arab and Turkish Oil Aspirations and German Policy towards the Middle East on the Eve of the Second World War." In *Oil, the Middle East, North Africa, and the Industrial States: Developmental and International Dimensions*, pp. 27–59. Edited by Klaus Jürgen Gantzel and Helmut Mejcher. Paderborn: Ferdinand Schöningh, 1984.

Messer, Robert. *The End of an Alliance: James F. Byrnes, Roosevelt, Truman, and the Origins of the Cold War*. Chapel Hill: University of North Carolina Press, 1982.

Meyer, Lorenzo. *Mexico and the United States in the Oil Controversy, 1917–1942*. Austin: University of Texas Press, 1977.

———. "La Resistencia al Capital Privado Extranjero; El Caso del Petróleo, 1938–1950." In *Las Empresas Transnacionales en México*, pp. 105–156. Edited by Bernardo Sepúlveda Amor, Olga Pellicer de Brody, and Lorenzo Meyer. México, D.F.: El Colegio de México, 1974.

Mikesell, Raymond F., and Chenery, Hollis B. *Arabian Oil: America's Stake in the Middle East*. Chapel Hill: University of North Carolina Press, 1949.

Miller, Aaron David. *Search for Security: Saudi Arabian Oil and American Foreign Policy, 1939–1949*. Chapel Hill: University of North Carolina Press, 1980.

Mills, C. Wright. *The Power Elite*. New York: Oxford University Press, 1956.

———. *The Sociological Imagination*. New York: Oxford University Press, 1959.

Moore, John Robert. "The Conservative Coalition in the United States Senate, 1942–1945." *Journal of Southern History* 33 (August 1967): 368–376.

Moran, Theodore H. "Foreign Expansion as an Institutional Necessity for United States Corporate Capitalism: The Search for a Radical Model." *World Politics* 25 (April 1973): 369–386.

Mosley, Leonard. *Power Play: Oil in the Middle East*. New York: Random House, 1973.

Motter, T. H. Vail. *The Persian Corridor and Aid to Russia*. Vol. 7 of *The United States Army in World War II*. Washington, D.C.: U.S. Government Printing Office, 1952.

Nash, Gerald D. *United States Oil Policy, 1890–1964: Business and Government in Twentieth-Century America*. Pittsburgh: University of Pittsburgh Press, 1968.

Newfarmer, Richard S., and Mueller, Willard F. *Multinational Corporations in Brazil and Mexico: Structural Sources of Economic and Noneconomic Power.* Report to the Subcommittee on Multinational Corporations of the Committee on Foreign Relations. United States Senate. Washington, D.C.: U.S. Government Printing Office, 1975.

Nordhauser, Norman. "Origins of Federal Oil Regulation in the 1920s." *Business History Review* 47 (Spring 1973): 53–71.

———. *The Quest for Stability: Domestic Oil Regulation, 1917–1935.* New York: Garland, 1979.

Odell, Peter R. *Oil and World Power.* 5th ed. Baltimore: Penguin Books, 1979.

Offe, Claus. "The Theory of the Capitalist State and the Problem of Policy Formation." In *Stress and Contradiction in Modern Capitalism: Public Policy and the Theory of the State*, pp. 125–144. Edited by Leon Lindberg. Lexington, Mass.: Lexington Books, 1975.

Painter, David S. "Oil and the Marshall Plan." *Business History Review* 58 (Autumn 1984): 359–383.

Panitch, Leo. "Recent Theorizations of Corporatism: Reflections on a Growth Industry." *British Journal of Sociology* 31 (June 1980): 159–187.

Paterson, Thomas G. "The Quest for Peace and Prosperity: International Trade, Communism, and the Marshall Plan." In *Politics and Policies of the Truman Administration*, pp. 78–112. Edited by Barton J. Bernstein. Chicago: Quadrangle Books, 1970.

———. *Soviet-American Confrontation: Postwar Reconstruction and the Origins of the Cold War.* Baltimore: Johns Hopkins University Press, 1973.

Penrose, Edith T. *The Growth of Firms, Middle East Oil, and Other Essays.* London: Frank Cass, 1971.

———. *The Large International Firm in Developing Countries: The International Petroleum Industry.* London: George Allen & Unwin, 1968.

Perkins, Bradford. "Reluctant Midwife: America and the Birth of Israel." *Reviews in American History* 9 (March 1981): 138–143.

———. "Sand in the Diplomats' Eyes." *Reviews in American History* 4 (December 1976): 601–606.

Perkins, Edwin J. *Men and Organizations: The American Economy in the Twentieth Century.* New York: G. P. Putnam's Sons, 1977.

Pfau, Richard A. "Avoiding the Cold War: The United States and the Iranian Oil Crisis, 1944." *Essays in History* 18 (1974): 104–114.

———. "Containment in Iran, 1946: The Shift to an Active Policy." *Diplomatic History* 1 (Fall 1977): 359–372.

Philip, George. *Oil and Politics in Latin America: Nationalist Movements and State Companies.* Cambridge: Cambridge University Press, 1982.

Podet, Allen H. "Anti-Zionism in a Key United States Diplomat: Loy Henderson at the End of World War II." *American Jewish Archives* 30 (November 1978): 155–187.

Polenberg, Richard. *War and Society: The United States, 1941–1945.* Philadelphia: J. B. Lippincott Co., 1972.

Pollard, Robert A. and Wells, Samuel F., Jr. "The Era of American Economic Hegemony, 1945–1960." In *Economics and World Power: An Assessment of American Diplomacy Since 1789*, pp. 333–390. Edited by William H. Becker and Samuel F. Wells, Jr. New York: Columbia University Press, 1984.

Powell, J. Richard. *The Mexican Petroleum Industry, 1938–1950.* Berkeley and Los Angeles: University of California Press, 1956.

Powers, Thomas. "A Book Held Hostage." *Nation* 230 (12 April 1980): 437–440.

Pratt, Joseph A. "Creating Coordination in the Modern Petroleum Industry: The American Petroleum Institute and the Emergence of Secondary Organizations in Oil." In *Research in Economic History* 8, pp. 179–215. Edited by Paul Uselding. Greenwich, Conn.: JAI Press, 1983.

———. "The Petroleum Industry in Transition: Antitrust and the Decline of Monopoly Control in Oil." *Journal of Economic History* 40 (December 1980): 815–837.

Prindle, David F. *Petroleum Politics and the Texas Railroad Commission.* Austin: University of Texas Press, 1981.

Rabe, Stephen G. "Anglo-American Rivalry for Venezuelan Oil, 1919–1929." *Mid-America* 58 (April-July 1976): 97–109.

———. "Energy for War: United States Oil Diplomacy in Latin America during World War II." In *Proceedings of the Conference on War and Diplomacy, 1976*, pp. 125–132. Edited by David H. White. Charleston: Citadel, 1976.

———. *The Road to OPEC: United States Relations with Venezuela, 1919–1976.* Austin: University of Texas Press, 1982.

Ramazani, Rouhollah K. *Iran's Foreign Policy, 1941–1973: A Study of Foreign Policy in Modernizing Nations.* Charlottesville: University Press of Virginia, 1975.

Randall, Stephen J. "Harold Ickes and United States Foreign Petroleum Policy Planning, 1939–1945." *Business History Review* 57 (Autumn 1983): 367–387.

———. "Raw Materials and United States Foreign Policy." *Reviews in American History* 8 (September 1980): 413–418.

Ray, Dennis M. "Corporations and American Foreign Relations." *Annals of the American Academy of Political and Social Science* 403 (September 1972): 80–92.

Reagan, Michael D. *The Managed Economy.* New York: Oxford University Press, 1963.

Reed, Peter Mellish. "Standard Oil in Indonesia, 1898–1928." *Business History Review* 32 (Autumn 1958): 311–337.

Ring, Jeremiah J. "American Diplomacy and the Mexican Oil Controversy, 1938–1942." Ph.D. dissertation, University of New Mexico, 1974.

Rogow, Arnold A. *James Forrestal: A Study of Personality, Politics, and Policy.* New York: Macmillan Co., 1963.

Rosenberg, David A. "The United States Navy and the Problem of Oil in a Future War: The Outlines of a Strategic Dilemma, 1945–1950." *Naval War College Review* 29 (Summer 1976): 53–64.

Rosenberg, Emily S. *Spreading the American Dream: American Economic and Cultural Expansion, 1890–1945.* New York: Hill & Wang, 1982.

Rositzke, Harry. *The CIA's Secret Operations: Espionage, Counterespionage, and Covert Action.* New York: Reader's Digest Press, 1977.

Rubin, Barry. *Paved with Good Intentions: The American Experience and Iran.* New York: Oxford University Press, 1980.

Saikal, Amin. *The Rise and Fall of the Shah.* Princeton: Princeton University Press, 1980.

St. Clair, David J. "The Motorization and Decline of Urban Public Transit, 1935–1950." *Journal of Economic History* 41 (September 1981): 579–600.

Salamon, Lester M., and Siegfried, John J. "Economic Power and Political Influence: The Impact of Industry Structure on Public Policies." *American Political Science Review* 71 (September 1977): 1026–1043.

Salazar-Carrillo, Jorge. *Oil in the Economic Development of Venezuela.* New York: Praeger Publishers, 1976.

Sale, Richard T. "America in Iran." *SAIS Review* 3 (Winter 1981–82): 27–39.

Sampson, Anthony. *The Seven Sisters: The Great Oil Companies and the World They Shaped.* New York: Viking Press, 1975.

Sanford, William F., Jr. "The American Business Community and the European Recovery Program." Ph.D. dissertation, University of Texas, 1980.

Schatz, Ronald W. *The Electrical Workers: A History of Labor at General Electric and Westinghouse, 1923–1960.* Urbana and Chicago: University of Illinois Press, 1983.

Schmitt, Karl M. *Mexico and the United States, 1821–1973: Conflict and Coexistence.* New York: John Wiley & Sons, 1974.

Schmitter, Philippe C. "Still the Century of Corporatism?" In *The New Corporatism: Social-Political Structures in the Iberian World,* pp. 85–131. Edited by Frederick B. Pike and Thomas Strich. Notre Dame: University of Notre Dame Press, 1974.

Schneider, Steven A. *The Oil Price Revolution.* Baltimore: Johns Hopkins University Press, 1983.

Schurmann, Franz. *The Logic of World Power: An Inquiry into the Origins, Currents, and Contradictions of World Politics.* New York: Pantheon Books, 1974.

Schurr, Sam H.; Netschert, Bruce C.; Eliasberg, Vera F.; Lerner, Joseph; and Landsberg, Hans H. *Energy in the American Economy, 1950–1975: An Economic Study of Its History and Prospects.* Baltimore: Johns Hopkins Press, 1960.

Seymour, Ian. *OPEC: Instrument of Change.* London: Macmillan Co., 1980.

Shaffer, Edward H. *The Oil Import Program of the United States.* New York: Praeger Publishers, 1968.

Shonfield, Andrew. *Modern Capitalism: The Changing Balance of Public and Private Power.* New York: Oxford University Press, 1965.

Shwadran, Benjamin. *The Middle East, Oil, and the Great Powers, 1959.* New York: Council for Middle Eastern Affairs Press, 1959.

———. *The Middle East, Oil, and the Great Powers.* 3rd ed. New York: John Wiley & Sons, 1973.

Skocpol, Theda. "Political Response to Capitalist Crisis: Neo-Marxist Theories of the State and the Case of the New Deal." *Politics and Society* 10 (1980/2): 155–201.

Smerk, George M. *Urban Transportation: The Federal Role.* Bloomington: Indiana University Press, 1965.

Smith, Robert Freeman. "Businessmen, Bureaucrats, Historians, and the Shaping of United States Foreign Policy." *Reviews in American History* 2 (December 1974): 575–581.

——. *The United States and Revolutionary Nationalism in Mexico, 1916–1932.* Chicago: University of Chicago Press, 1972.

——. "Who's Afraid of SONJ?: Energy and Nationalism in International Relations." *Reviews in American History* 6 (September 1978): 394–399.

Snell, Bradford C. *American Ground Transport: A Proposal for Restructuring the Automobile, Truck, Bus, and Rail Industries.* Washington, D.C.: U.S. Government Printing Office, 1974.

Snetsinger, John G. *Truman, the Jewish Vote, and the Creation of Israel.* Stanford: Hoover Institution Press, 1974.

Stivers, William. "International Politics and Iraqi Oil, 1918–1928: A Study in Anglo-American Diplomacy." *Business History Review* 55 (Winter 1981): 517–540.

——. "A Note on the Red Line Agreement." *Diplomatic History* 7 (Winter 1983): 23–34.

Stobaugh, Robert B. "The Evolution of Iranian Oil Policy, 1925–1975." In *Iran under the Pahlavis*, pp. 201–252. Edited by George Lenczowski. Stanford: Hoover Institution Press, 1978.

Stocking, George W. *Middle East Oil: A Study in Political and Economic Controversy.* Nashville: Vanderbilt University Press, 1970.

Stoff, Michael B. *Oil, War, and American Security: The Search for a National Policy on Foreign Oil, 1941–1947.* New Haven: Yale University Press, 1980.

Sutton, Francis X.; Harris, Seymour; Kaysen, Carl; and Tobin, James. *The American Business Creed.* Cambridge: Harvard University Press, 1956.

Tanzer, Michael. *The Political Economy of International Oil and the Underdeveloped Countries.* Boston: Beacon Press, 1969.

Taylor, Graham D. "Debate in the United States over the Control of International Cartels, 1942–1950." *International History Review* 3 (July 1981): 385–398.

Temkin, Benny. "State, Ecology, and Independence: Policy Responses to the Energy Crisis in the United States." *British Journal of Political Science* 13 (October 1983): 441–462.

Tugwell, Franklin. *The Politics of Oil in Venezuela.* Stanford: Stanford University Press, 1975.

Tulchin, Joseph S. *The Aftermath of War: World War I and U.S. Policy toward Latin America.* New York: New York University Press, 1971.

Vazquez, Josefina Z., and Meyer, Lorenzo. *México Frente a Estados Unidos: Un Ensayo Histórico, 1776–1980.* México, D.F.: El Colegio de México, 1982.

Vernon, Raymond. *Sovereignty at Bay: The Multinational Spread of U.S. Enterprises.* New York: Basic Books, 1971.

Vietor, Richard H. K. *Energy Policy in America Since 1945: A Study of Business-Government Relations.* Cambridge: Cambridge University Press, 1984.

——. "The Synthetic Liquid Fuels Program: Energy Politics in the Truman Era." *Business History Review* 54 (Spring 1980): 1–34.

Vogel, David. "Why Businessmen Distrust Their State: The Political Consciousness of American Corporate Executives." *British Journal of Political Science* 8 (January 1978): 45–78.

Walden, Jerrold L. "The International Petroleum Cartel in Iran—Private Power and the Public Interest." *Journal of Public Law* 11 (Spring 1962): 64–121.

Wall, Bennett H., and Gibb, George S. *Teagle of Jersey Standard.* New Orleans: Tulane University Press, 1974.

Weil, Martin. *A Pretty Good Club: The Founding Fathers of the U.S. Foreign Service.* New York: W. W. Norton & Co., 1978.

Wilkins, Mira. *The Maturing of Multinational Enterprise: American Business Abroad 1914–1970.* Cambridge: Harvard University Press, 1974.

———. "The Oil Companies in Perspective." In *The Oil Crisis,* 159–178. Edited by Raymond Vernon. New York: W. W. Norton & Co., 1975.

Williams, William Appleman. *History as a Way of Learning.* New York: New Viewpoints, 1973.

———. *The Tragedy of American Diplomacy.* Rev. ed. New York: Dell Publishing Co., 1962.

Williamson, Harold F., and Andreano, Ralph Louis. "Integration and Competition in the Oil Industry: A Review Article." *Journal of Political Economy* 69 (April 1961): 381–385.

———; Daum, Arnold R.; and Klose, Gilbert C. *The American Petroleum Industry: The Age of Energy 1899–1959.* Evanston, Ill.: Northwestern University Press, 1959.

Wilson, Evan M. *Decision on Palestine: How the U.S. Came to Recognize Israel.* Stanford: Hoover Institution Press, 1979.

Wilson, Joan Hoff. *Ideology and Economics: U.S. Relations with the Soviet Union, 1918–1933.* Columbia: University of Missouri Press, 1974.

Winkler, Heinrich August, ed. *Organisierte Kapitalismus: Voraussetzungen und Anfänge.* Göttingen: Vandenhoeck & Ruprecht, 1974.

Wittner, Lawrence S. *American Intervention in Greece, 1943–1949.* New York: Columbia University Press, 1982.

Wolfe, Alan. *The Limits of Legitimacy: Political Contradictions of Contemporary Capitalism.* New York: Free Press, 1977.

Wolin, Sheldon S. *Politics and Vision: Continuity and Innovation in Western Political Thought.* Boston: Little, Brown & Co., 1960.

Wood, Bryce. *The Making of the Good Neighbor Policy.* New York: Columbia University Press, 1961.

Woodward, Sir Ernest Llewellyn. *British Foreign Policy in the Second World War.* London: H. M. Stationery Office, 1962.

———. *British Foreign Policy in the Second World War.* Vol. 2. London: H. M. Stationery Office, 1971.

———. *British Foreign Policy in the Second World War.* Vol. 4. London: H. M. Stationery Office, 1975.

Wright, Denis. *The Persians Amongst the English: Episodes in Anglo-Persian History.* London: I. B. Tauris & Co., 1985.

Wyman, Donald L. "Dependency and Conflict: U.S. Relations with Mexico, 1920–1975." In *Diplomatic Dispute: U.S. Conflict with Iran, Japan, and Mexico,* pp. 83–141. Edited by Robert L. Paarlberg. Cambridge: Center for International Affairs, Harvard University, 1978.

Yago, Glenn. *The Decline of Transit: Urban Transportation in German and U.S. Cities, 1900–1970.* Cambridge: Cambridge University Press, 1984.

Yergin, Daniel. *Shattered Peace: The Origins of the Cold War and the National Security State*. Boston: Houghton Mifflin Co., 1977.

Zabih, Sepehr. *The Mossadegh Era: Roots of the Iranian Revolution*. Chicago: Lake View Press, 1982.

Zupnick, Elliot. *Britain's Postwar Dollar Problem*. New York: Columbia University Press, 1957.

Index

Abd al-Aziz ibn Abd al Rahman al Faisal
   al Saud. *See* Ibn Saud
Acción Democrática, 22, 129–130, 133–
   135
Acheson, Dean, 17, 30, 66, 90, 114, 148;
   policy toward Iran, 173–177, 181–
   190
Alemán Valdés, Miguel, 136, 138, 140,
   145–146, 148, 152
Allen, George V., 113–115
American Independent Oil Company
   (AMINOL), 165, 196
American Petroleum Institute (API), 3, 7,
   73
Andrews, M. E., 142–143
Anglo-American Oil Agreement, 202–203;
   background, 47–51, 54–56;
   negotiations, 59–64; opposition to,
   62–67; revision and rejection, 67–74;
   and Soviet Union, 239
Anglo-Iranian Oil Company (AIOC), 14,
   15, 47, 115, 186, 199; dispute with
   Iran, 171, 173–197, 205–206; and
   great oil deals, 102, 105–108, 203.
   See also Anglo-Persian Oil Company
Anglo-Persian Oil Company (APOC), 4–5,
   7–8. *See also* Anglo-Iranian Oil
   Company
Antitrust, 3, 12, 15; and Anglo-American
   Oil Agreement, 47, 65–69; and great
   oil deals, 106; and Iran, 180, 186–
   188, 193–197, 206
Arabian American Oil Company
   (ARAMCO): and aid to Saudi
   Arabia, 90–95; and fifty-fifty
   agreement, 165–171, 205; and great
   oil deals, 102–110, 203–204; internal
   price dispute, 211–214; and Palestine,
   117–118, 121, 125. *See also*

California Arabian Standard Oil
   Company; Standard Oil Company of
   California; Standard Oil Company
   (New Jersey); Standard Oil Company
   of New York; Texas Company
Armed Services Petroleum Board, 159
Army-Navy Petroleum Board (ANPB), 36,
   38–39, 52
As-Is Agreement (Achnacarry Agreement),
   5, 8, 47, 224
Atlantic Charter, 16–17, 41, 51, 61
Atlantic Refining Company, 4, 103, 196
Austin, Warren R., 125
Ávila Camacho, Manuel, 28, 31, 83
Azerbaijan, 112–114

Bahrain, 8, 14
Ball, Max W., 120
Beaverbrook, Lord Privy Seal, 55, 62, 64
Berle, Adolph A., Jr., 40
Bermúdez, Antonio J. (PEMEX), 136,
   141–142, 144–145, 148
Betancourt, Rómulo, 130–131, 132, 134
Beteta, Ramón, 138–139
Biddle, Francis, 13, 68
Braden, Spruille, 138
Bradley, Gen. Omar, 193
Brewster, Owen, 57, 58, 62
Brodie, Bernard, 101
Brown, John A. (Socony), 22, 46, 232
Brown, Russell (IPAA), 120
Brownell, Herbert, 195, 197
Bullard, Sir Reader, 79
Bullitt, William C., 36–39
Bureau of Mines, 6, 7
Byrnes, James F., 43, 54, 113

Cadman, Sir John (APOC), 5

Independent oil companies (*continued*)
193, 196; and Mexico, 82, 84–85,
137–138, 140–141, 144, 200; and oil
imports, 100, 158; and Petroleum
Reserves Corporation, 56–57
Independent Petroleum Association of
America (IPPA), 56, 70–73, 120
Industry structure, 3; political impact of,
73–74, 95, 201–203, 206–208
International Court of Justice (ICJ), 175–
176, 185
Interstate Oil Compact Commission, 7
Iran: 1953 coup, 189–192, 197, 206;
nationalization of oil industry, 172–
189, 205–206; and Truman Doctrine,
111–116; during World War II, 75–
81
Iranian consortium, 193–197, 206, 264;
table, 196
Iraq Petroleum Company (IPC), 4–5, 8,
47; and great oil deals, 102–110;
table, 4
Israel, 116, 125–126. *See also* Palestine

Jackson, Basil (AIOC), 47–48, 175
Jennings, Brewster (Socony), 169, 260
Joint Chiefs of Staff (JCS), 92; on
importance of Middle East oil, 99,
113, 117, 124, 127; and Petroleum
Reserves Corporation, 39–40, 54, 59;
views on Iran, 113, 125, 181; views
on Palestine issue, 117, 125
Joint ventures: and cartel case, 186, 193–
194, 197, 206; in Iran, 194–197,
206, 264; in Iraq, 4–5; in Kuwait, 7–
8; in Saudi Arabia, 8, 102–110, 203–
204, 211–214; in Venezuela, 17
Jones, Jesse, 27, 33–34
Justice Department, 13, 208; and Anglo-
American Oil Agreement, 68–69,
236; cartel case, 186, 193–197, 206;
and great oil deals, 106–107; and
Iran, 187, 193–197, 206

Kavtaradze, Sergei, 79–80
Kennan, George F.: and Iran, 80–81; and
"long telegram," 112–113; and
Palestine, 122–123
Kirk, Alexander, 33, 35
Klein, Harry T. (Texas Company), 106
Knox, Frank, 33, 41, 58
Korean War, 152, 164, 167, 180
Kuwait, 7–8, 14, 165

Kuwait Oil Company, 7–8, 14, 44, 105–
106

Leahy, William D., 40
Leavell, John H., 78
Lend-lease: to Iran, 76; to Saudi Arabia,
33–35, 37, 86, 88; to United
Kingdom, 16
Leoni, Raúl, 132
Levy, Walter J.: and Economic
Cooperation Administration, 158–
159; and Iran, 172, 176–177
Lloyd, Geoffrey, 49
Loftus, John A., 66, 71, 73
Long, Breckinridge, 87
Lovett, Robert A., 118, 120, 125

McCloy, John J., 132
McGhee, George C., 104; and Iran, 174–
175, 181; and Saudi Arabia, 165–
166, 168–169
McGuire, Paul F., 91
Major oil companies, 3, 206, 208, 223–
224, 264; and Anglo-American Oil
Agreement, 65–66, 73, 202; and Iran,
186–188, 193; and Mexico, 82, 85;
and oil imports, 158; relations with
Economic Cooperation Administration,
153, 156–157; and sterling-dollar oil
controversy, 161; and synthetic fuels,
100; as vehicles of national interest,
1, 16, 25, 95, 116, 136, 153, 172,
198, 199–200, 203–209
Manrique Pacanins, Gustavo, 18–20
Marshall, George C., 88, 125, 128, 154
Marshall Plan, 118, 122, 124, 126, 153–
160. *See also* European Recovery
Program
Medina Angarita, Isaías, 18–19, 121, 131
Mene Grande Oil Company, 14, 129,
131–132. *See also* Gulf Oil
Corporation
Messersmith, George S., 24, 28, 30, 81–
85
Mexico: drilling and participation
contracts, 138–140, 142–152; high
octane refinery, 26–31; management
contract proposals, 24, 29, 81–82,
137; nationalization of oil industry,
8–9, 22–24; oil industry map, 150;
oil loan issue, 81–85, 139, 142–152;
oil policy, 38–39, 147–148; oil

The Johns Hopkins University Press

Oil and the American Century

This book was set in Sabon type by Monotype Composition Co., Inc., from a design by Ann Walston. It was printed on 50-lb. Glatfelter paper and bound by Thomson-Shore, Inc.